MW01600571

ADME Processes in Pharmaceutical Sciences

Alan Talevi • Pablo A. M. Quiroga
Editors

ADME Processes in Pharmaceutical Sciences

Dosage, Design,
and Pharmacotherapy Success

 Springer

Editors
Alan Talevi
Laboratory of Bioactive Research and
Development (LIDeB), Department of
Biological Sciences, Faculty of Exact
Sciences
University of La Plata (UNLP)
La Plata, Buenos Aires, Argentina

Consejo Nacional de Investigaciones
Científicas y Técnicas (CONICET)
La Plata, Buenos Aires, Argentina

Pablo A. M. Quiroga
Quality Control of Pharmaceutical Products,
Pharmaceutical Toxicology, Department of
Biological Sciences, Faculty of Exact Sciences
Universidad Nacional de La Plata
La Plata, Buenos Aires, Argentina

Department of Pharmacological Research
Laboratorios Bagó S.A.
La Plata, Buenos Aires, Argentina

ISBN 978-3-319-99592-2   ISBN 978-3-319-99593-9   (eBook)
https://doi.org/10.1007/978-3-319-99593-9

Library of Congress Control Number: 2018965237

This Springer imprint is published by the registered company Springer Nature Switzerland AG
The registered company address is: Gewerbestrasse 11, 6330 Cham, Switzerland

*A. Talevi dedicates this volume to his parents,
his grandmother, and Ana Paula.*

*P. Quiroga would like to dedicate this book
to his parents, Juan Manuel, Marcelo,
María Guillermina and Marta.*

Preface

For quite a long time, pharmacokinetics was taught as one or two units within pharmacology courses. Today, due to their specific importance, biopharmaceutics and pharmacokinetics are taught as an independent course in every school of pharmacy worldwide, with applications in a diversity of areas and disciplines, from clinical practice to drug discovery, from molecular biology to material sciences. Owing to their impressive and continuous growth, some topics from the field of pharmacokinetics (such as drug metabolism or drug transporters) currently deserve entire specialized volumes.

This book has been conceived as an introduction to pharmacokinetics for pharmacy and medicine students, but also for professionals from other fields circumstantially working in relation to pharmaceutical sciences, including chemists, physicists, bioengineers, and molecular biologists.

This volume is structured in two parts. The first part is dedicated to a classic presentation of absorption, distribution, metabolism, and excretion processes and routes of administration, with an emphasis on qualitative and biological aspects of such topics, case reports discussion, and examples. The second part is dedicated to burgeoning topics associated to pharmacokinetics, including pharmaceutical nanocarriers, pharmacokinetics of biological medical products, pharmacogenomics, bioequivalence studies, drug-drug interactions of pharmacokinetic origin, in vitro and in silico ADME predictions, and drug transporters. The authors come from a diversity of professional areas and include experienced academicians, qualified clinical practitioners, industry professionals, and experts from regulatory agencies. We are proud to count, within the volume contributors, experts from Duquesne University, Indiana University, University of Utah, University of Maryland, and University of Bath, among others.

The editors are thankful to all the people who have made this volume possible. First, we thank all the invited authors who have contributed with their expertise to create authoritative and stimulating chapters for the second part of this volume. We would also like to express our thanks to the editorial team, especially to Carolyn Spence and Jayashree Dhakshnamoorthy, who have always been available to make our job as editors as easy and pleasant as possible, providing valuable advice and assistance.

We hope that this book will serve as a useful resource for newcomers in the field of pharmacokinetics and pharmacology, but also to practicing scientists currently working in the exciting field of drug discovery and development. We also hope we have succeeded in making this volume a pleasant, didactic, and clear introduction to the wonderful, vast world of pharmacokinetics.

La Plata, Buenos Aires, Argentina   Alan Talevi
                                                                                  Pablo A. M. Quiroga

Contents

Part I

The Basics of ADME Processes

Introduction. Biopharmaceutics and Pharmacokinetics

Alan Talevi and Pablo A. M. Quiroga

1.1 Introduction

The terms biopharmaceutics and pharmacokinetics describe two relatively young pharmaceutical sciences. Both have attracted more and more interest in the last decades, as the medical and pharmaceutical communities recognized their effective and potential contributions to design rational dosing recommendations and exploit our therapeutic arsenal in the best possible manner (Hochhaus et al. 2000). The significance of these areas of knowledge has been further enhanced by the attention on the relationship between drug levels in different body compartments and the correspondent pharmacological effects and by the opportunity of predicting in vivo performance of a drug product (typically reflected by plasma drug concentrations or rate and amount of drug absorbed) from in vitro performance (Emami 2006). What is more, last generation pharmaceutical carriers (e.g., pharmaceutical nanocarriers) have recently been introduced which, for the first time in history, could directly participate in and impact on the fate of a drug within the body, providing specific

A. Talevi (✉)
Laboratory of Bioactive Research and Development (LIDeB), Department of Biological Sciences, Faculty of Exact Sciences, University of La Plata (UNLP), La Plata, Buenos Aires, Argentina

Consejo Nacional de Investigaciones Científicas y Técnicas (CONICET), La Plata, Buenos Aires, Argentina
e-mail: atalevi@biol.unlp.edu.ar

P. A. M. Quiroga
Quality Control of Pharmaceutical Products, Pharmaceutical Toxicology, Department of Biological Sciences, Faculty of Exact Sciences, Universidad Nacional de La Plata, La Plata, Buenos Aires, Argentina

Department of Pharmacological Research Laboratorios Bagó S.A., La Plata, Buenos Aires, Argentina
e-mail: pq@biol.unlp.edu.ar

© Springer Nature Switzerland AG 2018
A. Talevi, P. A. M. Quiroga (eds.), *ADME Processes in Pharmaceutical Sciences*,
https://doi.org/10.1007/978-3-319-99593-9_1

(or targeted) distribution and modifying elimination kinetics (Talevi and Castro 2017). Such innovative drug products defy some well-established biopharmaceutical and pharmacokinetic concepts, which should be adapted or expanded to encompass the advances in pharmaceutical technologies. The propagation of complex drug delivery systems explains that today, more than ever, it is necessary that a wider audience gains at least basic biopharmaceutics and pharmacokinetics knowledge, in order to understand the behavior of such systems, assist in their design, and optimize their use.

First of all, what do we mean by biopharmaceutics and pharmacokinetics?

▶ **Definition**

Biopharmaceutics: Barbour and Lipper (2008) have noted that, by introducing the prefix *bio* to the word *pharmaceutics*, it follows that biopharmaceutics study the interdependence of biological aspects of the living organism (i.e., the patient) and the physical/chemical principles that govern the preparation and behavior of the medicinal agent and/or the drug product (Barbour and Lipper 2008). The discipline examines the interrelationship of the physical/chemical properties of the drug, the dosage form in which the drug is delivered, and the route of administration, on the rate and extent of systemic drug absorption (Shargel et al. 2012).

Pharmacokinetics: Pharmacokinetics is a branch of pharmacology that studies, both mathematically and descriptively, how the body affects a drug after administration, through the processes of absorption, distribution, metabolism, and excretion. It has been observed, in a colloquial manner, that pharmacokinetics deals with "what a biosystem does to a compound," that is, with everything that happens to drug molecules within the body aside from the pharmacodynamics events (Testa and Krämer 2006). In his seminal works, Torsten Teorell, who is often regarded as the father of modern pharmacokinetics, observed that the clinicians were often interested in some aspects of drug action (e.g., their specific mechanism of action), while little attention was paid to kinetics, that is, the time relation of drug action. Accordingly, he decided to derive "general mathematical relations from which it is possible, at least for practical purposes, to describe the kinetics of distribution of substances in the body" (Paalzow 1995).

Some authors assimilate both biopharmaceutics and pharmacokinetics into a single discipline. While for conventional drug delivery systems it could make sense to consider both disciplines separately (with the focus of biopharmaceutics in drug release from dosage forms and the focus of pharmacokinetics on those processes happening from absorption onward), such distinction may get blurry when considering last generation drug delivery systems, for which drug release could occur during or after drug distribution.

Conceived either as separate disciplines or as one, all in all biopharmaceutics and pharmacokinetics study what we succinctly call LADME processes, an acronym that refers to the processes of liberation (i.e., drug release from the dosage form), absorption, distribution, metabolism, and excretion. Briefly, *absorption* refers to the drug movement from the absorption site to systemic circulation, *distribution* represents the reversible transfer of drug molecules to the extravascular compartment, *metabolism* represents drug elimination due to biotransformation of drug molecules through enzyme-catalyzed chemical reactions, and *excretion* denotes the physical removal of the drug from the body. ADME processes will be studied in separate chapters of this volume. Metabolism and excretion can be jointly regarded as elimination, whereas distribution, metabolism, and excretion are together referred sometimes as drug disposition.

It is worth differentiating the epistemic object of biopharmaceutics from *biopharmaceuticals*. Biopharmaceuticals (also known as *biological medical products*, *biodrugs,* or, simply, *biologicals*) are pharmaceutical drug products manufactured in, extracted from, or semi-synthesized from biological sources, including vaccines, blood, blood components, gene therapies, tissues, recombinant therapeutic proteins, and living cells used in cell therapy. They can be composed of sugars, proteins, nucleic acids, or complex combinations of these substances or may be living cells or tissues. Biologicals, as any other drug product, may be studied by biopharmaceutics and pharmacokinetics (in fact, a whole chapter of this volume will focus on the kinetics of this type of medications). The procedures to obtain and characterize biologicals fall within the realm of biotechnology and molecular biology.

1.2   Some Practical Definitions: The Notion of Bioavailability

For the best understanding of this and subsequent chapters by those readers who are not familiarized with pharmacology and pharmaceutical glossary, we will next provide some general definitions of terms which would be frequently used throughout this volume.

▶ **Definition** Active pharmaceutical ingredient: The biologically active component of a drug product, that is, the component directly responsible for the pharmacological response. It is also usually referred as *drug*. Some drug products include more than one active pharmaceutical ingredient. Such substances can be used in the diagnosis, cure, mitigation, treatment, or prevention of disease.

Drug product: Through appropriate manufacturing processes, active pharmaceutical ingredients are combined with inactive pharmaceutical ingredients called excipients, which compose the pharmaceutical vehicle, pharmaceutical carrier, or drug delivery system. Closely related to the dosage form concept (physical form in which a drug product is produced and dispensed, e.g., tablet, capsule, syrup, etc.), excipients may contribute to different aspects of the drug products, such as enhancing physical/ chemical or microbial stability, improving organoleptic properties, bulking up solid formulations that contain potent active ingredients in small amounts, enhancing drug dissolution or absorption, etc. The pharmaceutical carrier does not possess intrinsic pharmacological activity, though its composition and manufacturing process impact on drug release and absorption and, thus, in drug pharmacokinetics. Some last generation pharmaceutical carriers may also influence drug distribution and elimination.

Biophase: The effect site of a drug. Physical region (environment) in which the drug target is located.

Systemic treatment or systemic medication: A pharmacological treatment in which the therapeutic agent (the drug) reaches its site of action through the bloodstream. We will consider that a drug molecule has reached systemic circulation once it has left the left ventricle of the heart at least once, that is, once it has reached the aorta.

Topical medication: A medication that is locally applied to the particular place on or in the body where it is intended to elicit its action. Many topical medications are applied directly to the skin. Topical medications may also be inhalational or applied to the surface of tissues other than the skin, such as eye drops applied to the conjunctiva or ear drops placed in the ear.

Therapeutic window: Also known as therapeutic range, it refers to the range of drug concentrations in a bodily fluid (usually, plasma) that provides safe and effective therapy. The lower bound of the therapeutic range is called minimum effective concentration (MEC), whereas the upper bound is called minimum toxic concentration (MTC). Below the lowest concentration of the window, it is likely that the drug will fail to work. If the drug concentration climbs above the therapeutic window, a detrimental intensification of the drug's intended (*on-target*) and unintended (*off-target*) actions will occur. Tabulated limits of the therapeutic window often come from population studies. However, to some extent, each individual has, at a given time point, a unique therapeutic window for each drug, since there are interindividual (and also intraindividual!) variability in drug sensitivity. Drugs with narrow therapeutic windows present small differences between therapeutic and toxic doses. For those drugs, therapeutic drug monitoring (the clinical practice of measuring specific drugs in body fluids at designated intervals) is often performed, to guarantee that the drug achieves the desired concentrations in the patient. When initiating drug therapy with such therapeutic agent, the physician may find it useful to measure the plasma drug concentration and tailor the dosage to the individual, "personalizing" the therapeutic intervention.

Drug target: A biomolecule with which the drug specifically interacts to elicit its therapeutic response. Most drug targets are proteins, though some correspond to other types of biomolecules (e.g., DNA, RNA).

Bioavailability: Bioavailability refers to the extent and rate at which the drug reaches its site of action. As the determination of drug levels in the site of action is not feasible in some cases (for instance, the reader may imagine how invasive would be to measure drug levels in the central nervous system), bioavailability assessment in the site of action is often replaced by measuring systemic bioavailability, i.e., *the extent and rate at which the drug reaches systemic circulation*. This is more convenient, since drug levels are more frequently quantified in serum or plasma. Plasma drug levels have a direct relationship with those in the site of action. This does not mean that drug levels across different organs will be uniform and identical

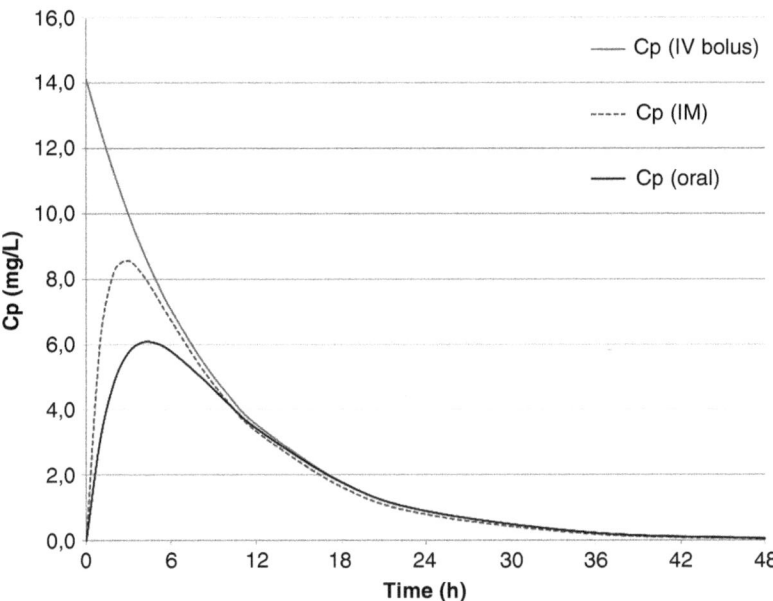

Fig. 1.1 Plasma concentration-time profiles for different routes of administration. Assume that the area under the curve is the same for the three profiles, thus indicating that the total amount of drug that has been absorbed is the same in all cases. Obviously, the kinetic aspect of bioavailability differs in every case. The IV administration implies immediate systemic bioavailability of the entire administered dose; in the other extreme, oral administration implies the slowest absorption, and thus the lowest bioavailability

to those in plasma; it means that, the higher the drug levels in plasma, the higher the drug levels in extravascular tissues. It is important to underline that, although we will frequently focus on drug bioavailability, measuring bioavailability of other types of compounds (e.g., toxins, drug metabolites) could occasionally be of interest. Bioavailability assessment is typically performed from drug plasma concentration-time profiles.

If we take the previous definition of bioavailability into consideration, we will observe that bioavailability has quantitative (extent) and kinetic (rate) components. Figure 1.1 shows a hypothetical situation in which the same dose of a given drug has been administered through different routes of administration (intravenous, intramuscular, oral). Even if, as in the example, the quantitative aspect of bioavailability was the same (as reflected in identical areas under the curve for the three profiles), the kinetic aspect of bioavailability will considerably vary for different administration routes (as reflected by the time at which the drug reaches its maximum plasma concentration). The moment at which the drug levels surpass the MEC and the time period during which the drug levels remain within the therapeutic window will be different in each occasion.

1.3   The Relationship Between Biopharmaceutics and Pharmacokinetics and the Pharmacological Response

Generally speaking, a pharmacologic response will take place when drug molecules interact in a specific manner with drug target molecules. Whereas it is common to speak of the drug target as *receptor*, strictly speaking not all the drug targets correspond to the receptor category. The usually specific recognition event between the drug and the drug target has been classically explained through the key and lock analogy (although now we know that (a) neither the drug nor the target are rigid entities; and (b) considering a systems biology perspective, drug responses can hardly be always explained by a single and punctual interaction event, but by the diversity of events triggered by the interaction of the drug with bodily elements). Speaking in very general terms (more detailed and comprehensive information could be found in a pharmacology textbook), drugs often work by blocking the interaction of a drug target with an endogenous ligand or by inducing a conformational change in the drug target that results in a pharmacologic response.

From the previous comments, it follows that the magnitude of the pharmacological effect will essentially depend on two factors: on the one hand, the number of drug molecules that, at a given moment, are interacting (binding) with the correspondent target copies and, on the other hand, how favorable is (thermodynamically speaking) such interaction. The larger the affinity of the drug for its target, the higher the intrinsic potency of the drug. Note that a maximal response is expected if, at a given moment, all the available copies of the target are occupied by drug molecules (saturated system). Also note that the living system may resort to different strategies to terminate or compensate the action of the drug (e.g., inactivating the drug target, upregulating or downregulating the drug target).

It is interesting to highlight that *the encounter between a drug molecule and a target molecule is a probabilistic event depending on the collision probability between both partners of the drug-target complex. It thus depends on the number of drug molecules neighboring the molecular target and also on the number of copies of the drug target neighboring the drug molecules.*

It is important to meditate on the statement in italics. If we think about it, it explains the relevance of biopharmaceutics and pharmacokinetics. The intensity of the pharmacological response not only depends on the intrinsic potency of the drug but also on how many drug molecules occupy, at a given time point, drug target molecules. This, in turn, directly depends on how many drug molecules are available to the target molecules in the site of action. Even if a drug has a high intrinsic potency, if it does not have access to the biophase in sufficient amount to occupy a pharmacologically relevant proportion of the drug target copies, there will not be pharmacological response.

1.4 The Free Drug Hypothesis

The free drug hypothesis (also called "free drug theory" or "free drug principle" by many authors) provides a conceptual framework to formalize the previous discussion and understand, in subsequent chapters, many issues related to drug distribution and drug elimination mechanisms. It is widely applied in drug discovery and development to establish pharmacokinetic-pharmacodynamic relationships, to predict the therapeutically relevant dose and to monitor drug levels in clinical studies. The hypothesis can be summarized in two prepositions (Smith et al. 2010):

(a) The free drug concentrations at both sides of a biological barrier would be the same if the distribution pseudo-equilibrium (steady state) has been achieved.
(b) The free drug is the species that exerts pharmacological activity (only the collision of free drug with the target is likely to contribute to binding).

The drug concentration in the biophase determines the magnitude of the pharmacological response (at least until maximal response has been achieved). Numerous studies have demonstrated that the average free drug concentration that is present in vivo at the mean efficacious dose is in good agreement with the in vitro potency (see, for instance, Troke et al. (1990) and Yamada et al. (2007)).

But what does free drug mean? In physiologic media, some drug molecules will reversibly bind to components in plasma (primarily, proteins) and in tissues, whereas others will be free (i.e., unbound, interacting with no other things than solvent molecules). Free drug molecules will diffuse across biological barriers much more rapidly than their bound counterparts.

There are many exceptions to the free drug hypothesis (Trainor 2007; Smith et al. 2010), for instance, a drug that has low passive permeability, so that the diffusion into a cell or deep compartment is slow relative to changes in plasma concentration. A major category of exceptions to the hypothesis involves drugs which are substrates for drug transporters. Exceptions to the second part of the hypothesis include drugs whose action involves multiple mechanisms and the activation of target-mediated events and drugs that bind to their targets with a very slow rate of dissociation from the complex, which will sustain receptor occupancy and exert pharmacological effect long after free drug levels have dropped off.

References

Barbour NP, Lipper RA (2008) Introduction to biopharmaceutics and its role in drug development. In: Krishna R, Yu L (eds) Biopharmaceutics applications in drug development. Springer, Boston
Emami J (2006) *In vitro - in vivo* correlation: from theory to applications. J Pharm Pharm Sci 9:169–189
Hochhaus G, Barrett JS, Derendorf H (2000) Evolution of pharmacokinetics and pharmacokinetic/dynamic correlations during the 20th century. J Clin Pharmacol 40:908–917
Paalzow LK (1995) Torsten Teorell, the father of pharmacokinetics. Ups J Med Sci 100:41–46

Shargel L, Wu-Pong S, Yu ABC (2012) Applied Biopharmaceutics & Pharmacokinetics. MacGraw-Hill, New York

Smith DA, Di L, Kerns EH (2010) The effect of plasma protein binding on in vivo efficacy: misconceptions in drug discovery. Nat Rev Drug Discov 9:929–939

Talevi A, Castro GR (2017) Targeted therapies. Mini Rev Med Chem 17:186–187

Testa B, Krämer SD (2006) The biochemistry of drug metabolism--an introduction: part 1. Principles and overview. Chem Biodivers 3:1053–1101

Trainor GL (2007) The importance of plasma protein binding in drug discovery. Expert Opin Drug Discov 2:51–64

Troke PF, Andrews EJ, Pye GW, Richardson K (1990) Fluconazole and other azoles: translation of in vitro activity to *in vivo* and clinical efficacy. Rev Infect Dis 12:S276–S280

Yamada S, Kato Y, Okura T et al (2007) Prediction of alpha 1-adrenoceptor occupancy in the human prostate from plasma concentrations of silodosin, tamsulosin and terazosin to treat urinary obstruction in benign prostatic hyperplasia. Biol Pharm Bull 30:1237–1241

Further Reading

This introductory chapter has discussed some pharmacological concepts in a very brief and general manner. The reader is directed to pharmacology textbooks for a deeper insight into pharmacology basis. Chapter 4 of the volume *Pharmacology: Principles and Practice* (by Hacker et al, Elsevier, 2009) would be an excellent place to find more on ligand binding and tissue response

Drug Absorption

2

Alan Talevi and Carolina Leticia Bellera

2.1 Introduction

▶ **Definition** Absorption is a mass transfer process that involves the movement of unchanged drug molecules from the site of absorption (which often, but not always, coincides with the site of administration) to the bloodstream. We will focus on *systemic absorption*, i.e., the movement of the drug molecules to general circulation.

The extent and rate of drug absorption depend on several factors including physicochemical features of the drug, the pharmaceutical dosage form, and anatomical and physiological factors. In many cases, crossing epithelia or endothelia will become the rate-limiting step of the whole absorption process. For that purpose, drug molecules might exploit the paracellular way or, more commonly, the transcellular way. In the latter, the critical step of drug absorption will usually involve crossing cell membranes. Most drug molecules will cross the cell membrane by simple diffusion, which can be appropriately modeled using Fick's first law. Molecules resembling physiological compounds might also be absorbed by carrier-mediated transport which, to some extent, can be modeled by Michaelis-Menten equation.

A. Talevi (✉)
Laboratory of Bioactive Research and Development (LIDeB), Department of Biological Sciences, Faculty of Exact Sciences, University of La Plata (UNLP), La Plata, Buenos Aires, Argentina

Consejo Nacional de Investigaciones Científicas y Técnicas (CONICET), La Plata, Buenos Aires, Argentina
e-mail: atalevi@biol.unlp.edu.ar

C. L. Bellera
Medicinal Chemistry/Laboratory of Bioactive Research and Development (LIDeB), Faculty of Exact Sciences, Universidad Nacional de La Plata (UNLP), Buenos Aires, Argentina

Consejo Nacional de Investigaciones Científicas y Técnicas (CONICET), Buenos Aires, Argentina

© Springer Nature Switzerland AG 2018
A. Talevi, P. A. M. Quiroga (eds.), *ADME Processes in Pharmaceutical Sciences*,
https://doi.org/10.1007/978-3-319-99593-9_2

Absorption involves overcoming different biological barriers but also surviving pre-systemic chemical or biochemical modifications (e.g., hydrolysis in the gastric medium and/or first-pass metabolism). By considering that a drug molecule has been absorbed as soon as it has left the heart left ventricle with no chemical modifications (i.e., no covalent bond formation or breakage), any pre-systemic phenomenon undergone by drug molecules can be regarded as constitutive of the absorption process.

Establishing the difference between *systemic* and *topical* administrations of a drug is probably the first step to explain the importance of drug absorption. Systemic administration of a medication uses the circulatory system to distribute the drug molecules. In contrast, topical medications are applied to a particular place in the body to treat an ailment in a local manner (typically, the skin or mucous membranes, though topical medications also involve inhalational medications, eye drops, or ear drops, among others). Note that, when systemic administration is used, the whole body (and not only the site of action) is potentially exposed to drug molecules. Conversely, topical administration is intended to circumscribe drug exposure to the affected site of the body and surroundings. Frequently, adverse reactions to medications are due to the interaction of the drug with elements of the body different from the drug molecular target (i.e., off-target interactions). Adverse reactions to topical medications are often local effects, such as irritation or localized allergic reactions. *Systemic* administration is more likely to involve more severe adverse reactions than *topical* administration. Nevertheless, *topical* administration is not always viable when the molecular target is situated in difficult-to-reach organs. Special mention must be made to targeted therapies, where drug molecules are preferably delivered to the site of action through targeted delivery systems (e.g., targeted nanosystems, viral vectors). This later concept will be separately discussed in Chap. 7, which focuses on nanocarriers and drug delivery. Accordingly, this chapter will address conventional medications only.

It follows from the previous discussion that drug absorption is a crucial process whenever resorting to systemic administration, while it is not required (and often, unwanted) when using topical administration: for drugs applied at their target, such as local anesthetics, absorption often terminates the therapeutic effect (Sultatos 2011).

2.2   Factors Affecting Drug Absorption

It should be remembered that, generally speaking, drug molecules are not administered in isolation but incorporated in a pharmaceutical dosage form. However, as discussed in the previous chapter, to overcome biological barriers, the drug must be in its free form; that is, it must have been released from the pharmaceutical carrier and in solution (Sultatos 2011). Only then drug molecules will be able to initiate their movement toward systemic circulation. The previous applies for conventional dosage forms but might not be necessarily true for last-generation pharmaceutical carriers as nanocarriers. Also, note that some dosage forms may provide the drug already in solution (e.g., pharmaceutical solutions).

▶   **Important** When dealing with conventional dosage forms, drug release and dissolution are prerequisites for absorption to occur.

Based on the previous concept and the definition of absorption that has been provided, absorption will depend on:

- *Physicochemical properties* of the drug, fundamentally: Its aqueous solubility, its ionization constant(s), its permeability through biological barriers that it shall encounter before arriving at general circulation (which, in turns, depends mostly on molecular size and lipophilicity), and its affinity to biological systems such as enzymes or transporters that it may come across during absorption.
- The *pharmaceutical dosage form*, which directly influences drug release. Occasionally, the pharmaceutical carrier might include components capable of modulating the drug uptake and/or metabolism. For instance, it has been reported that some common excipients can modify the expression and/or activity of drug transporters and cytochrome p450 enzymes (Tompkins et al. 2010; Hodaei et al. 2014; Al-Mohizea et al. 2014).
- The *anatomical and physiological characteristics* of the site of absorption, including expression levels of drug transporters, pH, anatomical adaptations that may favor absorption, and the characteristics of the tight junctions between adjacent cells of the absorptive tissue. Note that considerable physiological changes that impact on drug absorption can be observed during pathological processes and also with age.
- The *way the medication has been administered*. For instance, the extent and rate of absorption for a drug given through the oral route might vary if the medication has been administered before, along with or immediately after a meal in comparison with administration on an empty stomach. Transdermal absorption could be enhanced by means of different physical stimuli, such as local heat or mechanical friction.

The previous factors should be jointly considered when designing dosage forms or when assessing a medication from a biopharmaceutical viewpoint. The pharmacokinetic processes undergone by a drug after administration are often described through the LADME scheme, liberation, absorption, distribution, metabolism, and excretion, which are considered in separate chapters in the first part of this book.

▶   **Definition** The acronym LADME refers to liberation, absorption, distribution, metabolism, and excretion. These are the pharmacokinetically relevant processes that a drug undergoes after entering the body. Whereas these processes generally follow such sequence, *they should not be regarded as discrete events*. In other words, one process is still occurring, while the next one begins, and in certain occasions (e.g., sustained release delivery systems), all five processes may be occurring simultaneously (in parallel) to different drug molecules.

Conventional drug delivery systems can only impact, in a direct manner, on the processes of Liberation and Absorption, and only indirectly in the remaining ones, by influencing the rate and extent of drug release and absorption. Last-generation delivery systems, in contrast, could possibly modify in a direct manner all the LADME processes.

2.3   The Fluid Mosaic Model and Beyond

Crossing the selective barrier posed by cell membranes is in general the more relevant mass transfer process during absorption. To appreciate the role of biological membranes and tissue barriers in drug absorption, it is convenient to begin with an understanding of cell membranes. While we assume that the reader is familiarized with the generalities of the cell membrane architecture, we will overview them in order to provide a context for understanding drug absorption mechanisms. *Note that the principles governing absorption are similar to those governing distribution.* Accordingly, this and the subsequent sections will be of much value for the comprehension of both Chaps. 2 and 3.

The fluid mosaic model of the cell membrane, devised by Singer and Nicolson (1972) based on thermodynamic considerations and experimental evidence, proposes that phospholipids are the main constituent of cell membranes. This type of lipid (Fig. 2.1) generally consists of two hydrophobic fatty acid "tails" and a hydrophilic "head" consisting of a phosphate group. The fatty acids and the phosphate group are esterified with a glycerol molecule. The phosphate groups can be modified with simple organic molecules such as choline.

The phospholipids provide a mosaic structure predominantly arranged as a fluid bilayer (the matrix of the mosaic), with their phosphate groups in contact with the aqueous phase (either the extracellular medium or the cytosol), while the nonpolar fatty acid chains are sequestered away from contact with water and oriented to the inner space of the bilayer. Many integral proteins (or occasionally, glyco- or lipoproteins) can be found embedded in the membrane (more or less buried in its hydrophobic interior). Integral proteins are referred to as transmembrane if they

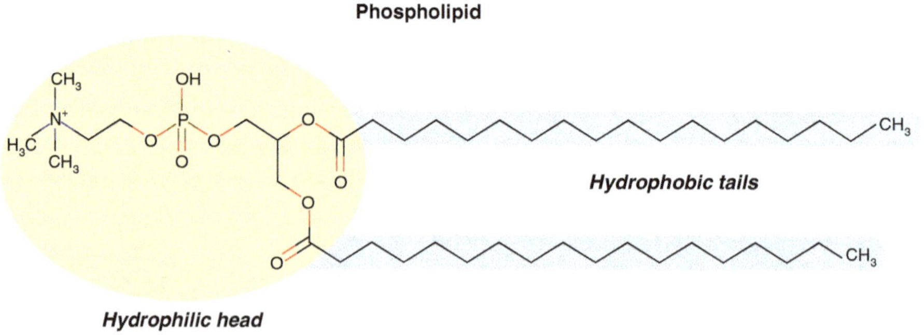

Phospholipid

Fig. 2.1 General structure of a phospholipid, essential constituent of a cell membrane

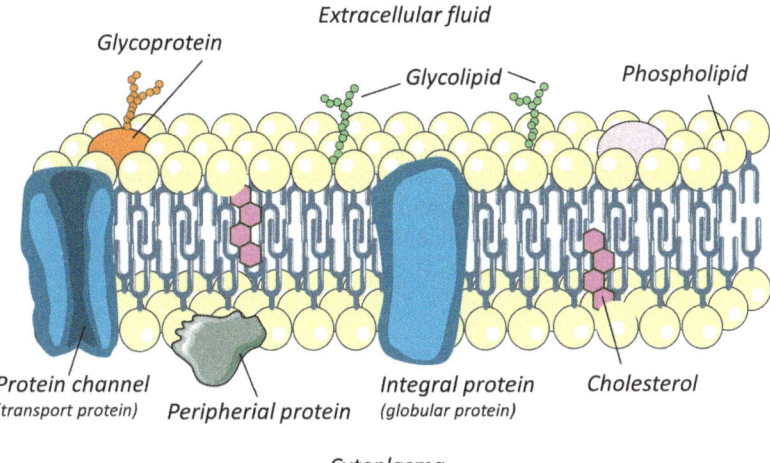

Fig. 2.2 Fluid mosaic model

extend from one side of the membrane to the other. Integral proteins are amphipathic, with highly polar groups protruding from the membrane into the aqueous phase and nonpolar groups embedded within the hydrophobic core of the bilayer. Additionally, peripheral proteins bound to the membrane by weak interactions and not strongly associated with membrane lipids can also be found. Membrane proteins present comparatively high amount of α-helical domains, in contrast with soluble globular proteins that in general exhibit a smaller fraction of α-helix in their native structure.

The arrangement of cell membranes (Fig. 2.2) acts as a bidimensional fluid that, in principle, allows free lateral diffusion of the membrane components. Cholesterol plays a much relevant role regulating membrane fluidity and conferring structural stability.

The model by Singer and Nicolson has been expanded to include important elements of the membrane architecture; some of which had already been envisaged by Singer and Nicolson themselves (Alberts et al. 2002). Among those elements, we may mention membrane asymmetry, the existence of lipid rafts (specialized areas in membranes where proteins and some lipids, primarily sphingolipids and cholesterol, are concentrated), and flip-flop movements.

The cell membrane is permeable to small nonpolar molecules, whereas molecules with high polarity (among them, charged ones), high molecular weight, and/or high conformational freedom will have difficulties crossing it. Cells have developed specialized transport mechanisms by using membrane proteins that include channels and carriers. Embedded in biological membranes, there are carriers that facilitate the entry of their substrates to the cell (*uptake transporters*) and carriers that export their substrates (*efflux transporters*). Channels and some types of carriers carry substrates down a concentration gradient, without direct or indirect ATP consumption. Other carriers require energy to transport their substrates. In some very specific cases, molecules interacting with cell surface receptors trigger the formation of endocytic vesicles, which carry the receptor-bound molecules into the cell (*endocytosis*) or through the cell (transcytosis). This form of transport is more frequent for large molecules (e.g., some therapeutic proteins) or advanced drug delivery systems (e.g., nanoparticles). The combined contribution of all the aforementioned elements (lipid bilayer, channels, uptake and efflux carriers, endocytosis) determines the selective permeability of a given cell to a given molecule. Selective permeability is crucial to regulate the cell inner microenvironment.

2.4   How Can Drugs Be Absorbed?

In principle, any substrate being absorbed could potentially use the paracellular and/or the transcellular transport (Fig. 2.3). The paracellular transport refers to the transfer of substances across the epithelium or endothelium by passing through the intercellular space between adjacent cells. Paracellular transport is exclusively passive; which substances can use this route of transport depends on the tight junction characteristics for the considered epithelium or endothelium. For instance, particularly tight junctions at the blood-brain barrier preclude the paracellular route.

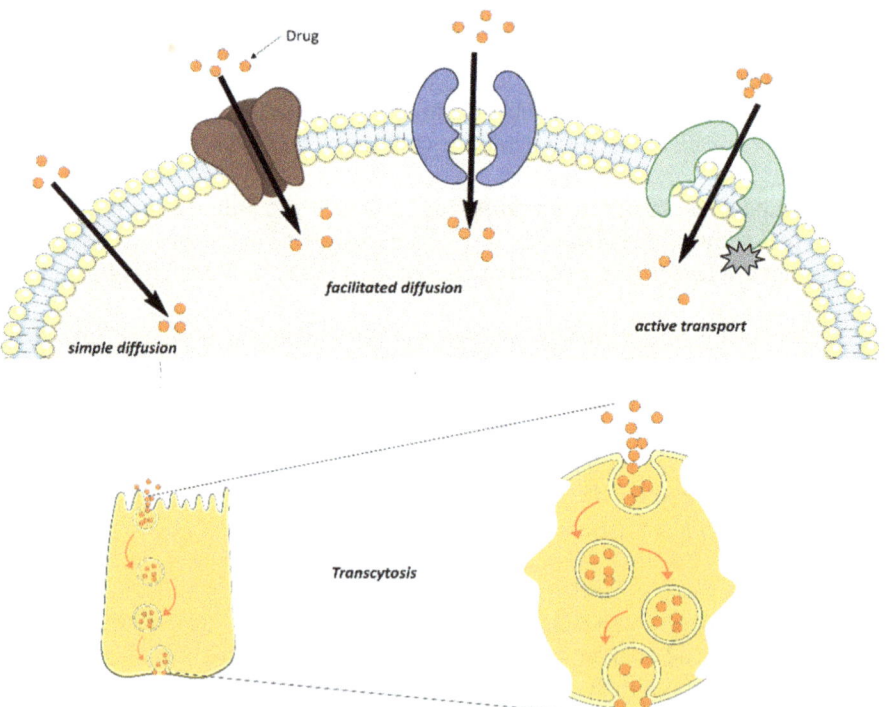

Fig. 2.3 Ways that a drug might employ, depending on its physicochemical and biochemical properties, to cross epithelia or endothelia

In contrast, other tissues present "leaky" tight junctions, thus favoring the paracellular route. When the paracellular route is available, small polar molecules are preferably transported through it.

▶ **Definition** The tight junctions are protein structures which form a selective barrier in the paracellular space between epithelial and endothelial cells, limiting movement of ions, solutes, and water. Tight junction strands represent an impermeable barrier, as a "wall," but a series of permeable aqueous "pores" perforate such wall.

Whereas the structure of the tight junctions is complex and dynamic and varies among different epithelia and capillary beds (González-Mariscal et al. 2003; Bazzoni 2006), evidence points to tight junction-associated claudins as the basis for the selective size, charge, and conductance properties of the paracellular pathway (Van Itallie and Anderson 2004).

Tight junction permeability can be significantly affected by pathological stimuli, but it is also actively investigated as a possible way to enhance drug delivery of hydrophilic drugs (González-Mariscal et al. 2017).

Conversely, the transcellular route offers a wide spectrum of possibilities that expand the diversity of the substrates that may be uptaken by the cells. These possibilities include *simple diffusion* (passive and without intervention of any membrane protein), *facilitated diffusion* (down the concentration gradient either through protein channels or carrier proteins), active transport (up the concentration gradient with energy consumption), and transcytosis.

▶   **Important** By far, the most frequent transfer mechanism involved in drug transport through cell membranes is passive diffusion. To diffuse, drugs should be sufficiently hydrophobic to partition into the lipid bilayer of the plasma membrane. Since most drugs are either weak acids or bases, the ability of a given drug to partition into the membrane will be highly modified by the pH conditions.

2.5   Simple (or Free) Diffusion

In simple diffusion, drug molecules will use the transcellular route spontaneously, without involvement of any membrane protein and down the concentration gradient. Since crossing the cell membrane is usually the slowest step of the whole absorption process (thus governing its kinetics), we will focus on the concentration gradient through the cell membrane. The kinetics of simple diffusion can be adequately modeled using Fick's first law. For the sake of mathematical simplicity, let us imagine the absorption site as a well-defined compartment (which it is usually not) and that a certain amount Q of the drug is homogeneously dispersed (in solution) at the absorption site. In this way, the only diffusion process that we will pay attention to is the movement of the drug molecules from the site of absorption to the inside of the cells. In other words, we will study the drug flux in the orthogonal direction to the epithelium (or endothelium) surface (a direction that we will arbitrarily name x direction). The *instantaneous* rate of absorption corresponding to the change in Q (dQ) during infinitesimal time period dt will be thus given by the following expression:

$$\frac{dQ}{dt} = D \times P \times S \times \frac{dC}{dx} \tag{2.1}$$

D alludes to the coefficient of diffusion through the membrane, and it is expressed in units of surface over units of time (e.g., cm^2/s). It depends on molecular features such as molecular size and shape. For instance, large molecules will diffuse more slowly than small ones, being thus associated to a smaller D. P represents the partition coefficient of the drug between the cell membrane and the surrounding aqueous environment. S refers to the total absorption surface, and dC/dx is the concentration gradient (concentration/distance units). Equation (2.1) shows that, at any moment, the (instantaneous) rate of absorption will be directly proportional to the concentration gradient, which is the driving force of the process.

If we call the thickness of the membrane δ (Fig. 2.4) and we assume that the concentration falls linearly through such thickness, by integrating expression (2.1), we get to:

$$\frac{dQ}{dt} = D \times P \times S \times \frac{\Delta C}{\delta} \tag{2.2}$$

If we now call A the concentration of the drug in the left compartment (representing the site of absorption or donor compartment) and we call B the drug concentration in the right (acceptor) compartment, Eq. (2.2) now takes the form:

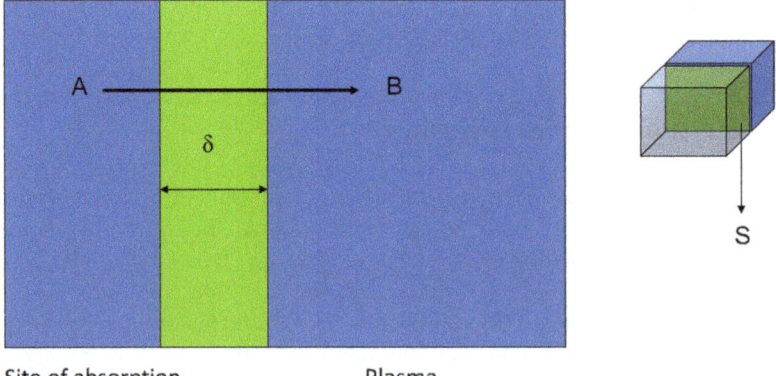

Fig. 2.4 A simple model to describe drug absorption. The site of absorption corresponds to the left (donor) compartment, whereas the right compartment represents the acceptor compartment (that contains absorbed drug)

$$\frac{dQ}{dt} = D \times P \times S \times \frac{(B - A)}{\delta} \tag{2.3}$$

Dividing both sides of the equation by V, the volume in the site of absorption:

$$\frac{dQ}{Vxdt} = \frac{D \times P \times S}{V \times \delta} \times (B - A) \tag{2.4}$$

Note that the factor $(B - A)$ determines if expression (2.4) acquires a negative or positive sign (the remaining factors on the right-hand side of the equation are positive constants at a given pressure and temperature conditions).

Questions

Is the $(B - A)$ factor in Eq. (2.4) smaller or greater than zero?

Answers

Provided that, according to the example in Fig. 2.4, the diffusion is proceeding from the left compartment (the one that has been assimilated to the site of absorption or donor compartment in the example) to the right compartment, A must necessarily be greater than B. Accordingly, the alluded factor is smaller than zero.

Questions

When will the diffusion end?

Answers

In principle, if the considered system were a close system, diffusion would end when A equals B. Take into consideration that, at that point of dynamic equilibrium (identical concentrations at both sides of the diffusion barrier), the mass units of the drug might well be very different in both compartments, since they depend on the compartments' volumes. If for a moment we consider the intravascular fluid as the acceptor compartment, it can be accepted that the volume of the acceptor compartment is much larger than the one of the donor compartment, thus accepting, in a hypothetic equilibrium condition, much more mass units of the drug than the donor compartment.

Moreover, neither the "acceptor compartment" nor any biologic system can be thought of as close systems. As we will discuss in brief, this condition will contribute to an efficient absorption of the administered dose.

Let us express $(B - A)$ as $-(A - B)$, and let us focus on the variation in concentration at the absorption site, instead of the variation in mass units:

$$\frac{dQ}{Vxdt} = \frac{dA}{dt} = -\frac{D \times P \times S}{V \times \delta} \times (A - B) \qquad (2.5)$$

dA/dt has, undoubtedly, negative sign (D, P, S, V, δ are positive quantities, and we have already established that $A > B$). This is reasonable (for the diffusion to occur in the hypothesized sense, the concentration at the site of absorption will be diminishing as time goes by). Also, observe that the instantaneous rate of absorption is directly proportional to the absorptive surface area. This explains the importance of some anatomical adaptations that favor absorption in certain organs by increasing the surface area (i.e., Kerckring valves, villi, and microvilli at the small intestine). The concept is also useful for the design of transdermal dosage forms (see the example below).

Example

Nicotine replacement therapy is one of the most frequent treatments for smoking cessation. It is available in a diversity of dosage forms, including gum, sublingual tablets, nasal spray, inhaler, and transdermal patches.

Table 2.1 shows the impact of nicotine transdermal patches surface area on nicotine bioavailability (the table includes mean data from six healthy male smokers $\pm$ the standard deviation). The bioavailability is measured through the area under the curve in a plot of drug concentration in blood plasma versus time (AUC) and the peak plasma concentration (C_{max}). AUC (from zero to infinity) is an indicator of the total amount of drug that has reached general circulation (total drug exposure over time); C_{max} reflects both the quantitative and kinetic aspects of bioavailability. T_{max} represents the time at which C_{max} is observed, and it speaks of the kinetic component of bioavailability (the smaller T_{max}, the higher the absorption kinetics). The data shown in the table has been extracted from Sobue et al. (2006). The reader may visit Chap. 10 for an extensive discussion on bioavailability.

Note that the bioavailability for a surface area of 40 cm^2 roughly doubles that for a surface area of 20 cm^2

Table 2.1 Comparison of nicotine bioavailability between 20 and 40 cm^2 nicotine transdermal patches

	AUC (ng h/mL)	C_{max} (ng/mL)	T_{max} (h)
Transdermal patch 20 cm^2	155.13 $\pm$ 29.51	9.15 $\pm$ 1.79	13.3 $\pm$ 3.4
Transdermal patch 40 cm^2	280.96 $\pm$ 31.08	17.25 $\pm$ −1.60	8.2 $\pm$ 5.2

Mean $\pm$ SD

Let us go back to Eq. (2.5). As previously stated, P represents the partition coefficient between the majorly hydrophobic cell membrane and the aqueous phase. A lipophilic drug will display a high affinity to the membrane and will be linked to a high rate of absorption (however, an extremely lipophilic drug will possibly tend to remain "trapped" in the membrane, which is termed *membrane retention*). Beside this issue, also consider that a drug must be in solution to be absorbed: as it is discussed in other chapters, drugs with low solubility will tend to have dissolution problems. Accordingly, adequate hydrophilic-lipophilic balance is often pursued when designing new drugs intended for oral administration. D and P vary from drug to drug. For a given drug and at a given moment, D, P, S, δ, and V are constants and can then be grouped together and replaced with a single constant that we will refer as the absorption rate constant through the lipid bilayer $K_{a,memb}$. $K_{a,memb}$ has units of s^{-1}:

$$\frac{dA}{dt} = -K_{a,memb} \times (A - B) \tag{2.6}$$

▶ **Important** The lipid bilayer of the cell membrane is often considered as the fundamental barrier that a drug must overcome to reach the intra-vascular space. Such assumption is in general well-founded, though the number of diffusional barriers that the drug must actually overcome greatly exceeds the bilayer. For instance, if we think of the intestinal epithelium, it is covered by a mucus layer rich in mucins, with gel-like properties (Johansson et al. 2013). After crossing the apical membrane, the drug must also cross the cytosol, the basolateral membrane, the connective tissue, the interstitial tissue, and the endothelium of the capillary bed. The drug will then be carried to the heart, and it will finally reach general circulation.

It can be observed that the aforementioned (sub)processes are serial: if any of them is slower than the others, it will be the limiting-step and govern the kinetics of the whole absorption process. In general, it is safe to assume that crossing the cell membranes is such a rate-limiting step (the hydrophobic core of the cell membrane is 100–1000 times more viscous than water!).

2.5.1 Sink Condition and Absorption

The diffusion processes linked to absorption will conclude once the drug has reached the capillary beds irrigating the site of absorption. As blood is circulating and removing drug molecules from the site of absorption, and taking into consideration that a substantial amount of the drug in plasma is complexed with plasma proteins (as further discussed in the next chapter), the term B in Eq. (2.6) will in general tend to be small in comparison with A, and, in most scenarios, it can be neglected. This condition is known as *sink condition* and allows simplifying Eq. (2.6):

$$\frac{dA}{dt} = -K_{a,\text{memb}} \times A \qquad (2.7)$$

This simple step has very important implications. The resulting expression (2.7) clearly represents first-order kinetics: the instantaneous rate at which the concentration of drug diminishes in the site of absorption is at any moment proportional to the remaining concentration of the drug in the site of absorption. The sink condition positively impacts on the extent of absorption, since it helps sustaining a concentration gradient (driving force of the diffusion process) and thus favors a complete absorption of the administered dose.

There are some conditions for which the sink condition may not apply, consequently reducing the extent of absorption. For instance, during physical exercise, the viscera perfusion is reduced, negatively impacting on the bioavailability of some drugs (the less permeable ones).

Figure 2.5 represents the instantaneous rate of absorption versus the concentration of drug in the site of absorptions: absorption by simple diffusion can be regarded, doubtlessly, as a linear process. A similar behavior would be found experimentally when performing permeability models if charting the initial rate of absorption versus initial concentration of the drug.

▶   **Important** Simple diffusion is a linear and non-saturable transport mechanism.

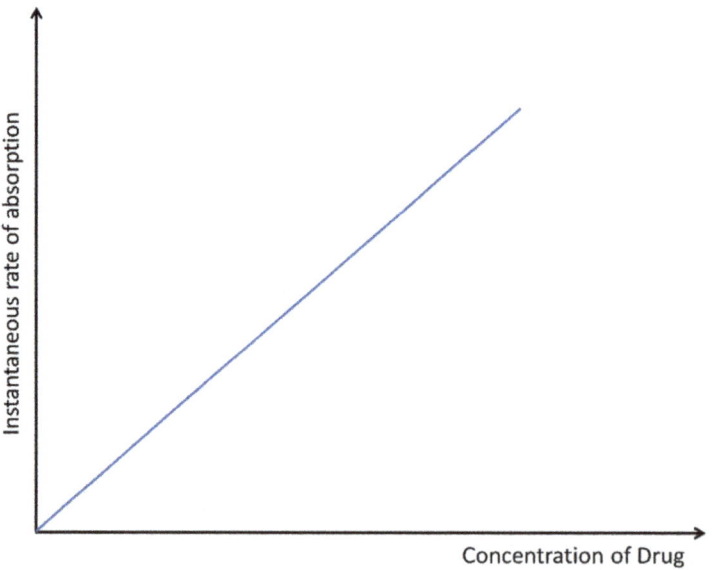

Fig. 2.5 Chart of instantaneous rate of absorption versus the (instantaneous) concentration of drug in the site of absorption, for a drug being absorbed through a linear process

2.5.2 The Impact of pH on Drug Absorption

As said, most of the known drugs are weak acids or bases. For a weak acid (for instance, aspirin, $pK_a = 3.5$), the ionization equilibrium can be written as

$$HA(aq) + H_2O(l) \rightleftharpoons H_3O^+(aq) + A^-(aq)$$

Notice that, following Le Chatelier's principle, in an alkaline media that consumes H_3O^+, the equilibrium will shift to the right, thus increasing the concentration of the charged species A^-. Conversely, in acid media the equilibrium will shift to the left, increasing the concentration of the non-charged species HA. Interestingly, while the solubility in an aqueous media will be higher for charged species, the permeability of charged species through the cell membrane by passive diffusion will be low. Moreover, the electrical potential gradient will also impact on the transport of charged species, in which case the Nernst-Planck equation will be useful to determine the flux across the barrier in question.

From the previous equilibrium, the corresponding acid ionization constant is:

$$K_a = \frac{[H_3O^+][A^-]}{[HA]} \qquad (2.8)$$

Taking the logarithm of both sides of the equation easily leads to the Henderson-Hasselbalch equation:

$$\log_{10} K_a = \log_{10} [H_3O^+] + \log_{10} \frac{[A^-]}{[HA]} \qquad (2.9)$$

Then:

$$-pK_a = -pH + \log_{10} \frac{[A^-]}{[HA]} \qquad (2.10)$$

Finally:

$$pH = pK_a + \log_{10} \frac{[A^-]}{[HA]} \qquad (2.11)$$

A very similar development can be achieved for a base:

$$B(aq) + H_2O(l) \rightleftharpoons BH^+(aq) + OH^-(aq)$$

Here, the base ionization constant is:

$$K_b = \frac{[OH^-][BH^+]}{[B]} \qquad (2.12)$$

This leads to:

Table 2.2 Negative effect of lansoprazole coadministration on atazanavir bioavailability

	AUC_{0-24} (μM · h)	C_{max} (μM)	T_{max} (h)
Atazanavir administered alone	16.3 ± 9.0	3.2 ± 1.7	2.8 ± 0.75
Atazanavir administered after lansoprazole	0.85 ± 1.8	0.13 ± 0.19	3.1 ± 1.0

Mean $\pm$ SD

$$pOH = pK_b + \log_{10} \frac{[BH^+]}{[B]} \tag{2.13}$$

Have in mind that pH + pOH equals 14.

Whereas the effects of pH variations on solubility and permeability are relatively straightforward for drugs having a single ionizable group within the physiologically relevant range of pH, the analysis gets more complicated when addressing drugs with many ionizable functions, where such ionizable groups will gain protons in a serial manner when pH gradually drops and vice versa. Accordingly, when a drug has many ionizable groups, the number of ionization states gets multiplied, and even zwitterionic species with no net charge can appear.

Example

Clinically, significant drug-drug interactions involving antiretroviral medications occur in up to 40% of HIV patients on treatment; the consequences of such interactions include fluctuations in antiretroviral plasma levels leading to toxicity, diminished efficacy, and increased risk of drug resistance (Lewis et al. 2016). Acid-reducing agents are commonly used co-medications by HIV-1-infected patients receiving antiretroviral treatment. Furthermore, acid-reducing therapy medications are widely used, and many of them are available over the counter in most countries.

Several reports demonstrate significant drug-drug interactions between acid-reducing medications and antiretrovirals. For instance, Tomilo et al. (2006) have reported a mean reduction of 94% in atazanavir bioavailability in nine healthy volunteers receiving such antiretroviral drug coadministered with lansoprazole (Table 2.2). The loss in solubility for atazanavir at increased pH values is considered responsible for this effect.

Studies suggest that some of the interactions between antiretroviral medications and acid-reducing agents may be mitigated by temporal separation of dose administration (Falcon and Kakuda 2008).

2.6   Facilitated Diffusion

If we pay close attention to expression (2.4), it is clear that simple diffusion will not be favored for certain chemical compounds. That is the case with large or highly hydrophilic molecules. How does the cell manage to transport such molecules (e.g., glucose from the intestinal lumen) through the cell membrane, if required?

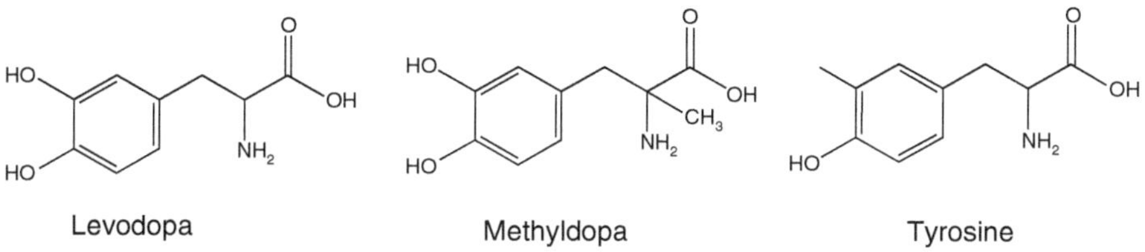

Levodopa   Methyldopa   Tyrosine

Fig. 2.6 The chemical similarity between the therapeutic agents levodopa and methyldopa explains why these drugs can use amino acid or peptide carriers for their absorption

Facilitated diffusion (also known as passive-mediated transport) is the process of spontaneous passive transport of molecules or ions across a biological membrane via specific transmembrane integral proteins.

There are essentially two types of protein structures facilitating diffusion. *Channel proteins* form open pores in the membrane, allowing small molecules of the appropriate size and charge to pass freely through the lipid bilayer. As previously stated, channel proteins also permit the passage of molecules between adjacent cells connected at tight junctions. In contrast, *carrier proteins* bind specific molecules to be transported and then undergo conformational changes that allow the molecule to pass through the membrane and be released on the other side. They are responsible for the facilitated diffusion of sugars, polar amino acids, and nucleosides across the plasma membranes of most cells.

▶   **Important** Since uptake carriers have evolved to transport very specific substrates with physiologic role through the membrane, only drug resembling such substrates will be able to exploit carriers as transport mechanism. Some classical examples of drugs that use uptake carriers are 5-fluorouracil, levodopa (antiparkinsonian), and methyldopa (antihypertensive). Notice the resemblance of levodopa and methyldopa to the amino acid tyrosine (Fig. 2.6).

A very important difference between facilitated diffusion and free diffusion is that the number of substrate molecules that can be transported by time unit depends on the number of channel or carrier copies available per cell. Accordingly, this kind of transport is saturable. The temperature dependence of facilitated transport is substantially different due to the presence of an activated binding event, as compared to free diffusion where the dependence on temperature is mild. Usually, variation of transport rate upon temperature changes will be used as experimental evidence of mediated transport.

The kinetics of mediated transport can be described by the Michaelis-Menten equation:

$$\frac{dA}{dt} = -\frac{V_m \times A}{K_m + A} \tag{2.14}$$

where V_m is the maximum rate of transport, which depends on the level of expression and the turnover number of the transporter, and K_m stands for the Michaelis-Menten constant, which is inversely related to the binding affinity of the substrate (note that it equals the substrate concentration at which half of the V_m is observed). At very low substrate concentrations (far from the saturation condition), A can be neglected versus K_m, thus resulting in apparent first-order kinetics (the *apparent* term denotes that the kinetics is not strictly linear over the entire concentration range but only at small substrate concentrations):

$$\frac{dA}{dt} = -\frac{V_m \times A}{K_m} \tag{2.15}$$

Apparent first-order kinetics will be observed under therapeutic conditions whenever diffusion is facilitated by channels, since the number of available channels is usually very large.

Above saturation conditions (where, at a given time point, every copy of transporter will be interacting with a substrate molecule and the system will be working at maximum velocity), K_m could be neglected, and the transport process assumes apparent zero-order kinetics:

$$\frac{dA}{dt} = -V_m \tag{2.16}$$

Both parts of the Michaelis-Menten saturation curve are shown in Fig. 2.7. It should be emphasized that, strictly speaking, Michaelis-Menten equation does not fully represent facilitated diffusion, which is intrinsically dependent on substrate concentrations on both sides of the plasma membrane. The approach is, however, useful to describe apparent transport properties under specific conditions (e.g., initial uptake rates) (Panitchob et al. 2015). More complex mechanistic models are required to capture transporter behavior under diverse physiological conditions.

2.7   Active Transport

Generally speaking and from a mathematical perspective, active transport does not differ of the previously discussed description for facilitated diffusion, since it may be adequately modeled through a Michaelis-Menten kinetics. Some points should be considered, though.

First, active transport is mediated by carrier proteins that undergo important conformational changes to move its substrate(s) across membranes.

Second, it consumes energy in a direct or indirect manner to move its substrate (s) against the concentration gradient. Primary active transport utilizes energy in the form of ATP to transport molecules across a membrane against their concentration gradient. ATP-powered pumps contain one or more ATP-binding sites, which are present on the cytosolic face of the membrane. Conversely, secondary active

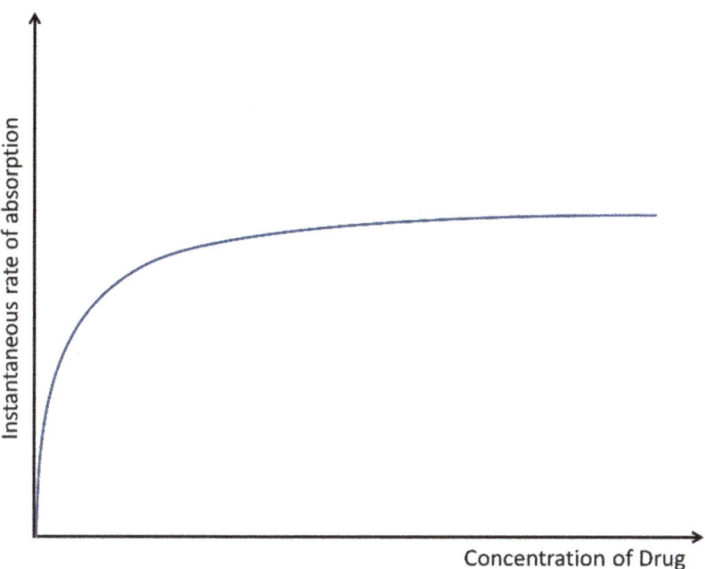

Fig. 2.7 Michaelis-Menten saturation curve

transport couples the movement of an ion (typically Na^+ or H^+) or molecule down its electrochemical gradient to the uphill movement of another molecule or ion against a concentration/electrochemical gradient. Thus, energy stored in the electrochemical gradient is used to drive the transport of another solute against a concentration or electrochemical gradient. The cotransported species may move in the same direction (symporters) or in opposite direction (antiporters). For example, transport of amino acids makes use of sodium-dependent symporters, of proton-motive force, and also of the gradient of other amino acids (Broër 2008).

Third, there are transporters that collaborate with drug absorption, whereas others oppose it. The former tends to display narrow substrate specificity and the latter, in contrast, wide substrate specificity. Consider that uptake transporters have evolved to help move specific molecules of physiologic value through membranes; conversely, efflux transporters have evolved to protect cells (or the organism) from strange, nonphysiologic compounds. We will discuss uptake and efflux transporters in a specific chapter of this volume.

Also, note that drug absorption might occur by a combination of transport mechanisms acting in parallel, e.g., active transport plus simple diffusion. In such case, the instantaneous rate of absorption will result from the addition of the mathematical equations explaining each of the individual processes involved in drug absorption.

Figure 2.8 illustrates the instantaneous rate of absorption versus concentration at the site of absorption, whenever the transport occurs by a combination of simple diffusion plus active transport or facilitated diffusion. Note that, as absorption proceeds, the concentration of substrate at the site of absorption will diminish, and

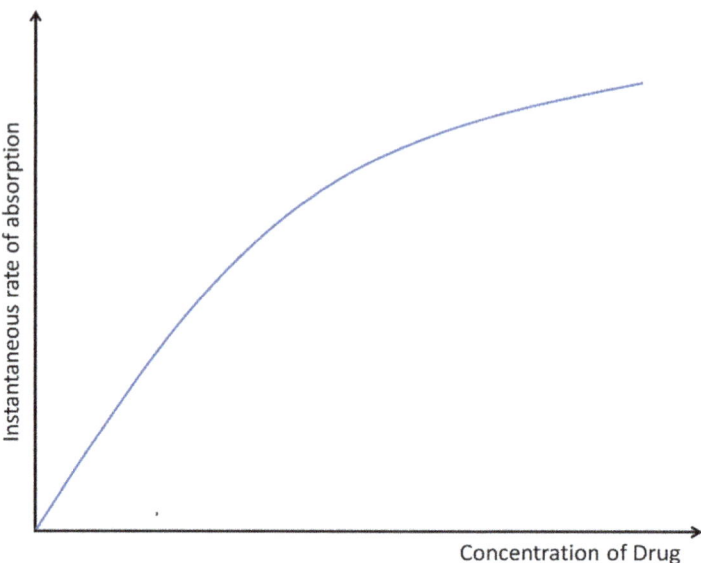

Fig. 2.8 Relationship between the instantaneous rate of absorption and drug concentration at the absorption site, for drug transport occurring by simple diffusion plus active transport or facilitated diffusion

the instantaneous rate of absorption will also go down. In other words, the rate of absorption will fall as the remaining (not yet absorbed) substrate at the site of absorption goes down.

Example

Parkinson's disease is one of the most common neurodegenerative diseases, characterized by motor and non-motor symptoms. It is characterized by the loss of dopaminergic neurons, with the consequent impairment of dopaminergic neurotransmission. Levodopa, an oral dopamine precursor, is used to restore dopamine levels at the central nervous systems. It is, so far, the most effective treatment for Parkinson's disease patients. Levodopa is rapidly metabolized to dopamine in the gastrointestinal tract by amino acid decarboxylase. This is not desired for two reasons: one, dopamine cannot penetrate the blood-brain barrier; two, peripheral conversion of levodopa to dopamine causes significant gastrointestinal side effects, such as nausea and vomiting. For these reasons, levodopa is commonly coadministered with peripherally acting amino acid decarboxylase inhibitors such as carbidopa and benserazide, which limit levodopa conversion to dopamine to the central nervous system.

The patients are likely to show reduced response to levodopa and develop motor complications such as dyskinesia and motor fluctuations (abrupt shifts between "on" and "off" states in which the symptoms are controlled and uncontrolled, respectively). The motor fluctuations in patients in advance stages become closely connected with the rapid rise and fall in levodopa concentration after each dose. Fluctuating motor performance is the major source of disability in Parkinson's disease patients and drastically impairs their quality of life. Therefore, maximizing the therapeutic efficacy of levodopa is of utmost importance.

Levodopa, which closely resembles some dietary amino acids, is absorbed by a saturable facilitated large neutral amino acid transport system. Such transport is also responsible for the penetration of levodopa to the central nervous system, where it exerts its therapeutic action. The presence of large neutral amino acids from the diet either in the intestinal lumen or in blood may reduce levodopa absorption and distribution to the brain, in that order, since they compete with levodopa for the transporter.

Accordingly, protein-restricted diets have shown effective in ameliorating motor fluctuations (Wang et al. 2017; Barichella et al. 2017). Such diets include low-protein diets (in which protein consumptions are restricted to 0.5–0.8 g protein/per kilogram of body weight per day) and protein-redistribution diet (in which the patient consumes low-protein food such as cereal products, fruits, and vegetables for breakfast and lunch and are allowed high-protein food including eggs, legumes, fish, and meat for dinner). This maximizes daytime motor function, since it is considered that diurnal fluctuations limit the quality of life more than fluctuations at night.

Case Study

A 67-year-old patient with a 23-year history of Parkinson's disease and type 2 diabetes presented to an emergency department on New Year's Day. She had hosted a large New Year's Eve party the preceding evening. She was on levodopa/carbidopa therapy as well as two dopamine agonists (pramipexole and amantadine) and metformin. She had taken her medication at 10 pm on New Year's Eve, although it was unclear whether she had taken her tablets at 11 am that morning. A large protein-rich Chinese meal is consumed shortly before her last confirmed dose of treatment. On New Year's Day, she had gone to sleep at 11 am after cooking lunch. At 2 pm her son could not rouse her and called an ambulance. Blood sugar was 9.0 mmol/l. Paramedics arranged transfer to the hospital. She had a Glasgow coma scale score of 5/15. Her blood pressure was 130/78 mmHg, pulse 84 beats per minute, and blood oxygen saturation 98% on room air. There was no focal neurological deficit. A tomographic scan revealed a mild degree of cerebral atrophy but was otherwise normal. Routine hematology and biochemistry investigations were normal. Peripheral white cell count was normal. Treatment with intravenous acyclovir was commenced to cover possible viral encephalitis. Shortly prior to admission, she had experienced an increased frequency of falls. The patient

(continued)

recovered very rapidly following reinstitution of dopaminergic therapy via nasogastric tube.

Considering that severe motor "wearing off" phenomena can be life-threatening, in extreme cases resembling neuroleptic malignant syndrome with confusion, hyperthermia, and rhabdomyolysis discusses possible causes to the patient's condition. Take into consideration the results of hematology and biochemistry investigations and other tests performed on the patient. Based on a case report by Arulanantham et al. (2014).

2.8 Conclusions

- Drug molecules must be in solution to be absorbed (with the exception of drugs included in some last-generation pharmaceutical carriers, such as nanocarriers).
- The vast majority of the drugs are absorbed by simple diffusion and the use the transcellular way to cross epithelia and/or endothelia.
- The same mass transfer mechanisms studied in this chapter will explain the drug distribution process.
- For a drug to be absorbed by carrier-mediated transport, it must closely resemble the natural susbtrates of the involved carrier, i.e., compounds of physiological value (e.g., amino acids, sugars, nucleotides). The number of therapeutically relevant small molecules that use uptake carriers is relatively small.
- Simple difussion can be modeled using Fick's first law. Facilitated transport, in contrast, can be modeled by Michaelis-Menten kinetics, though in some cases more complex models will be required.
- Simple difussion is a non-saturable (linear) process. Facilitated transport and active transport require transmembrane proteins and are saturable processes. In many therapeutic situations, though, facilitated transport will be regarded as an apparent first-order kinetic process.
- Since most drugs are either weak acids or weak bases, their solubility and permeability will be highly dependent on the pH of the absorption site. Moreover, the electrochemical gradient should also be taken into consideration to study the absorption of charged species.
- Ionic species with net charge tend to be more soluble than neutral species; conversely, charged species are less permeable.

References

Al-Mohizea A, Zawaneh F, Alam MA et al (2014) Effect of pharmaceutical excipients on the permeability of p-glycoprotein substrate. J Drug Deliv Sci Technol 24:491–495

Alberts B, Johnson A, Lewis J et al (2002) In: Alberts B, Johnson A, Lewis J et al (eds) Membrane structure. Garland Science, New York

Arulanantham N, Lee RW, Hayton T (2014) Lesson of the month 2: a case of coma in a Parkinson's patient: a combination of fatigue, dehydration and high protein diet over the new year period? Clin Med (Lond) 14:449–451

Barichella M, Cereda E, Cassani E et al (2017) Dietary habits and neurological features of Parkinson's disease patients: implications for practice. Clin Nutr 36:1054–1061

Bazzoni G (2006) Endothelial tight junctions: permeable barriers of the vessel wall. Thromb Haemost 95:36–42

Broër S (2008) Amino acid transport across mammalian intestinal and renal epithelia. Physiol Rev 88:249–286

Falcon RW, Kakuda TN (2008) Drug interactions between HIV protease inhibitors and acid-reducing agents. Clin Pharmacokinet 47:75–89

González-Mariscal L, Betanzos A, Nava P et al (2003) Tight junction proteins. Prog Biophys Mol Biol 81:1–44

González-Mariscal L, Posadas Y, Miranda J et al (2017) Strategies that target tight junctions for enhanced drug delivery. Curr Pharm Des 22:5313–5346

Hodaei D, Baradaran B, Valizadeh H et al (2014) The effect of tween excipients on expression and activity of p-glycoprotein in Caco-2 cells. Pharm Ind 76:788–794

Johansson MEV, Sjövall H, Hansson GC et al (2013) The gastrointestinal mucus system in health and disease. Nat Rev Gastroenterol Hepatol 10:5352–5361

Lewis JM, Stott KE, Monnery D et al (2016) Managing potential drug-drug interactions between gastric acid-reducing agents and antiretroviral therapy: experience from a large HIV-positive cohort. Int J STD AIDS 27:105–109

Panitchob N, Widdows IP, Crocker MA et al (2015) Computational modelling of amino acid exchange and facilitated transport in placental membrane vesicles. J Theor Biol 365:352–364

Sobue S, Sekiguchi K, Kikkawa H et al (2006) Comparison of nicotine pharmacokinetics in healthy Japanese male smokers following application of the transdermal nicotine patch and cigarette smoking. Biol Pharm Bull 29:1068–1073

Singer SJ, Nicolson GL (1972) The fluid mosaic model of the structure of cell membranes. Science 175:720–731

Sultatos L (2011) In: Enna SJ, Bylund DB (eds) Drug absorption from the gastrointestinal tract. Elsevier, Boston, pp 1–2

Tompkins L, Lynch C, Haidar S et al (2010) Effects of commonly used excipients on the expression of CYP3A4 in colon and liver cells. Pharm Res 27:1703–1712

Tomilo DL, Smith PF, Ogundele AB et al (2006) Inhibition of atazanavir oral absorption by lansoprazole gastric acid suppression in healthy volunteers. Pharmacotherapy 26:341–346

Van Itallie CM, Anderson JM (2004) The molecular physiology of tight junction pores. Physiology (Bethesda) 19:331–338

Wang L, Xiong N, Huang J et al (2017) Protein-restricted diets for ameliorating motor fluctuations in Parkinson's disease. Front Aging Neurosci 9:206

Further Reading

The reader is referred to the wonderful volume by Carsten Ehrhardt and Kwan-Jin Kim for additional insight on the absorption process with a focus on experimental and computational models of absorption (*Drug Absorption Studies. In Situ, In Vitro and In Silico models*, Springer, 2008). The *Molecular Biopharmaceutics* volume by Bente Steffansen et al. is also highly recommended to study experimental models of drug absorption (*Molecular Biopharmaceutics*, Pharmaceutical Press, 2010)

Drug Distribution

3

Alan Talevi and Carolina Leticia Bellera

3.1 Introduction

▶ **Definition** The term drug distribution describes the reversible mass transfer of a drug from one location to another within the body. Once drug molecules have been absorbed (i.e., when they have entered systemic circulation), they might extravasate to the interstitial fluid and, subsequently, to the intracellular space. Each organ or tissue can receive different levels of the drug, and a drug can remain in the different organs or tissues for different amounts of time. The mass transfer of drug from the vascular to the extravascular space will occur through the endothelial lining of blood capillaries. Accordingly, the rate of such mass transference will be strongly dependent on the features of the capillary bed under consideration. The possible drug transfer mechanisms across endothelia and cell membranes are essentially similar to those described for drug absorption. However, distribution will be also influenced by unspecific, reversible binding events to plasma proteins (which will conspire against drug extravasation) and tissue elements (which will promote drug extravasation). A key concept to understand distribution is that, if considering conventional drug dosage forms, only free, unbound drug molecules can efficiently cross endothelia and thus engage in the pseudo-equilibrium of distribution in a direct manner.

A. Talevi (✉)
Laboratory of Bioactive Research and Development (LIDeB), Department of Biological Sciences, Faculty of Exact Sciences, University of La Plata (UNLP), La Plata, Buenos Aires, Argentina

Consejo Nacional de Investigaciones Científicas y Técnicas (CONICET), La Plata, Buenos Aires, Argentina
e-mail: atalevi@biol.unlp.edu.ar

C. L. Bellera
Medicinal Chemistry/Laboratory of Bioactive Research and Development (LIDeB), Faculty of Exact Sciences, Universidad Nacional de La Plata (UNLP), La Plata, Argentina

Consejo Nacional de Investigaciones Científicas y Técnicas (CONICET), Buenos Aires, Argentina

© Springer Nature Switzerland AG 2018
A. Talevi, P. A. M. Quiroga (eds.), *ADME Processes in Pharmaceutical Sciences*,
https://doi.org/10.1007/978-3-319-99593-9_3

In the previous chapter, we have discussed the mass transfer of a drug from the site of administration/absorption to systemic circulation. We will use the term drug distribution to describe those phenomena and processes involved in the exchange of drug between the blood and the rest of the tissues (extravascular tissues). As absorption occurs, mass units of the drug arrive to the blood, resulting in a concentration gradient across blood vessel walls. This gradient explains the diffusion of drug molecules from the vascular space to the interstitial fluid, in the first place, and to the intracellular fluid, later. Exchange is facilitated at the level of blood capillaries, the smallest blood vessels, having a few micrometers in diameter and a wall one endothelial cell thick. Besides the facilitated diffusion provided by this rather narrow diffusion barrier, capillaries are part of a capillary bed, an interweaving network of capillaries supplying tissues and organs and presenting an impressive surface area. Drug distribution is thus an eminently passive process, except for some drugs that resemble physiologic compounds and may use active transport to extravasate and/or enter the cells. Exchange is a keyword here, reflecting the reversible nature of the process. While it is true that the initial sense of the drug transfer will be from the blood to the extravascular tissues, if occasionally the free drug levels in the blood drop, a net return of drug molecules from the extravascular tissues to the vascular space will be observed.

Underlying the previous description is the idea of distribution equilibrium, a concept which may frequently be found in the literature when approaching the topic of drug distribution. However, can a true distribution equilibrium be achieved? Can any equilibrium be achieved within living beings? If we think of living organism from a thermodynamic perspective, we will recognize an open system subjected to constant exchange of heat and matter with its environment. Living organisms are thus highly dynamic systems, exposed to permanent change and investing high amounts of energy in maintaining balance (homeostasis being a thermodynamic stationary state of non-equilibrium) (Recordati and Bellini 2004). Biological systems reach a state of thermal and mechanical equilibrium with the external environment only when dead.

Besides strictly thermodynamic considerations, there are also pharmacokinetic reasons to regard distribution equilibrium as an impossibility. Whereas once the drug has been absorbed it will be continuously subjected to elimination, the delivery of the drug (i.e., drug administration) is typically performed in a discontinuous, intermittent manner (through administration of successive dosing units). Only in those occasions where the drug enters the body in a continuous manner and following a zero-order kinetics (i.e., at a constant rate) will we approach a true state of distribution equilibrium (Fig. 3.1). Under the light of the previous discussion on the thermodynamics of living systems, the resulting steady state will frequently and more appropriately be referred as a distribution pseudo-equilibrium. Note that, under such circumstances, at some point after starting administration of the drug, a steady state will be reached in which the rate of drug inflow to the blood equals the rate of drug elimination from the blood and no net change is observed on drug levels as long as the inflow and elimination rates remain unchanged. This will likely happen in just a few therapeutic settings: (a) continuous intravenous infusion of a drug and (b) when modified-release drug delivery systems that achieve a sustained release are used (sustained-release drug delivery systems).

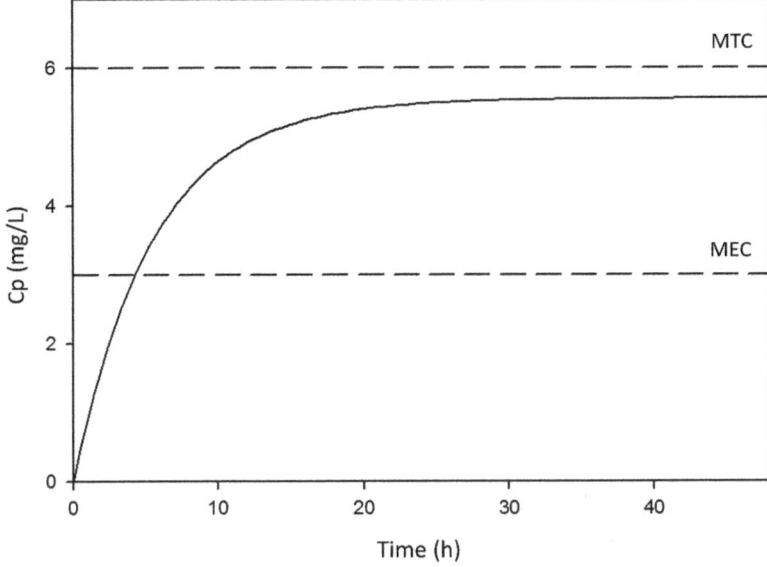

Fig. 3.1 Plasma concentration-time profile curve for a drug entering the body following a zero-order kinetics (i.e., at constant rate). The so obtained steady state is probably the closest state to a distribution equilibrium that one may achieve. MTC is the minimum toxic concentration, and MEC is the minimum effective concentration

▶   **Important** The (pseudo)-equilibrium of distribution is a state to which the drug will tend, but that will only be reached and maintained when using drug delivery systems that deliver the drug in a continuous and constant manner and as long as the clearance of the patient remains unchanged.

With immediate-release dosage forms, the steady state will only be met through administration of multiple and identical doses at regular time intervals. In such cases the steady state results from a balance between the mean absorption rate and the mean elimination rate during a dosing interval and will consist in fluctuations of the plasma drug levels between practically constant maximal and minimal concentrations. It then consists in a (pseudo) steady state in which plasma levels will oscillate within an approximately fixed range of concentrations (Fig. 3.2).

It is convenient to make clear that different pharmacokinetic models will describe the distribution process in different manners, both conceptually and mathematically. For instance, classical compartmental models depict the body as one (one-compartment model, Fig. 3.3), two (two-compartment model, Fig. 3.4), or more compartments and assume that the drug distributes uniformly and instantaneously within each of the compartments, though movement from one compartment to another is not instantaneous.

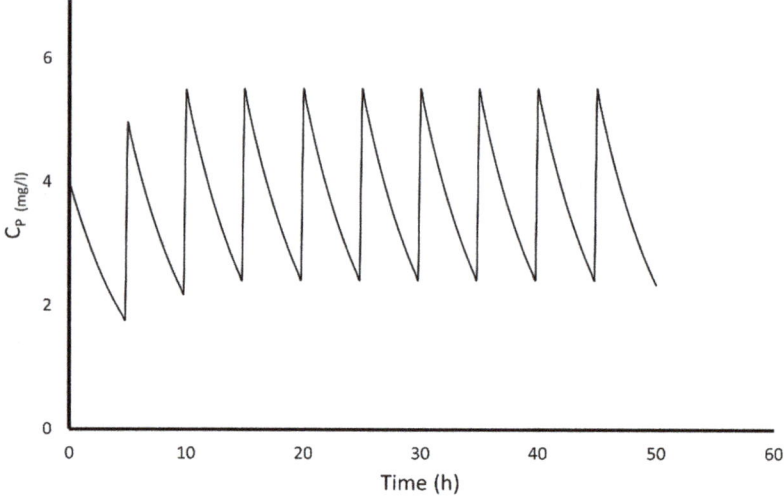

Fig. 3.2 Plasma concentration-time profile curve following multiple dosages and immediate-release dosage form

Fig. 3.3 Classic
one-compartment
pharmacokinetic model. K_a
is the absorption rate constant
and K_e is the elimination rate
constant

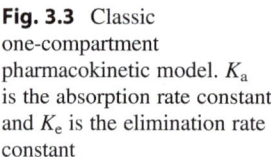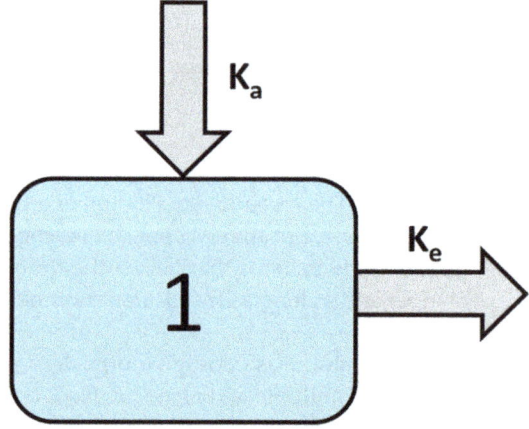

▶ **Definition** In the framework of classical compartmental modeling, a pharmacokinetic compartment is an abstract space (with no direct relationship with anatomical spaces, tissues, or fluids) in which the drug is assumed to distribute homogeneously and instantaneously. Whereas distribution *within* a compartment is assumed to be instantaneous, the drug exchange *between* compartments is not. Organs and tissues in which drug distribution is similar are grouped into one compartment. Whereas it is hard to imagine that any drug will distribute uniformly among different organs or tissues and it is even harder to believe that a mass transfer process could be instantaneous, such assumptions or simplifications lead to models which are relatively simple from a mathematical perspective.

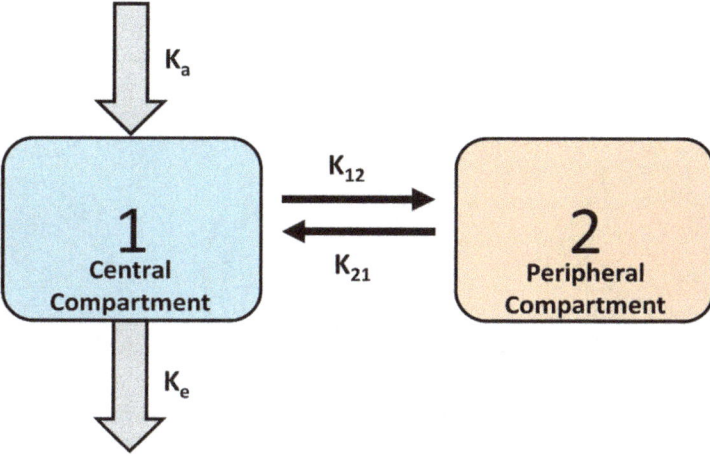

Fig. 3.4 Classic two-compartment pharmacokinetic model with first-order absorption and elimination. 1 and 2 are the central compartment (including plasma) and the peripheral compartment, respectively. K_a, K_e, K_{12}, and K_{21} represent, in that order, the kinetics constants linked to absorption, elimination, drug transfer from compartment 1 to compartment 2, and drug transfer from compartment 2 to compartment 1

3.2 Factors Impacting on [the Extent and Kinetics of] Drug Distribution

There are many factors that influence drug distribution. First, we may consider some physicochemical properties of the drug that can directly or indirectly affect its capacity to move across biological barriers. Among them, we could mention *molecular weight* (the larger a molecule, the more difficult its passive diffusion) and *polarity* (polar molecules will face difficulties to move through simple diffusion), and their movement will also be reduced in epithelia or endothelia where the paracellular route is precluded, i.e., at the blood-brain barrier. The *degree of ionization* will also be important, since the net charge of the molecule will depend on both the ionization constant/s of the drug and the pH of the physiological environment. The *affinity of the drug for drug transporters* should also be considered when studying drug distribution.

On the other hand, a group of physiologic factors impacting on drug distribution can be listed, which are in some cases subjected to variation in an individual in both health and disease states. These factors include the identity, expression levels, and subcellular localization of *drug transporters* throughout the body; the *characteristics of the capillary beds* that supply different tissues and organs, including the permeability of tight junctions between adjacent endothelial cells and the presence of pores and discontinuities; and the *tissue or organ perfusion* (blood flow to a given organ, i.e., rate at which the blood is delivered to a given tissue or organ, usually measured as volume of blood per unit of time per unit of tissue mass and

Continuous   **Fenestrated**   **Sinusoid**

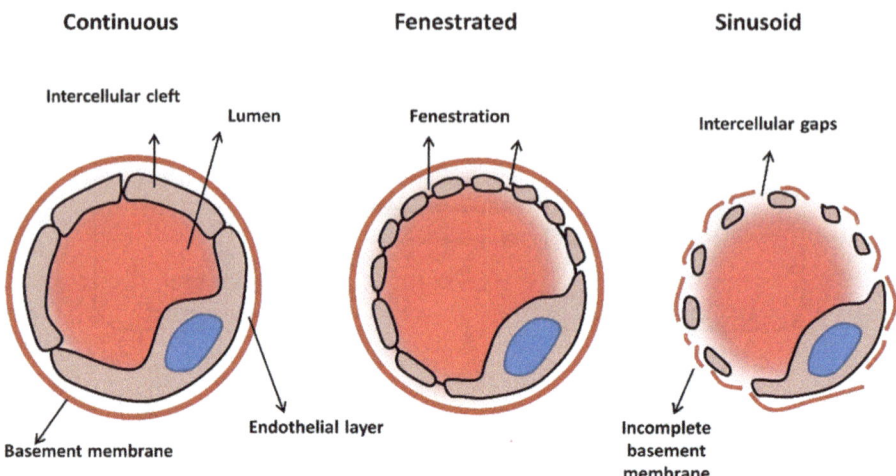

Fig. 3.5 Types of capillaries

related to the capillary density at a given tissue or organ). It should be remembered that the perfusion of a given organ or tissue will dynamically change depending on its metabolic demand. For instance, the blood flow to different organs will change during exercise, at resting state, at postprandial state, etc. *Affinity between the drug and the tissue* (which depends on the unspecific interactions that occur between the drug and tissue elements) should also be taken into consideration. Remember that there exist different types of capillary beds, as can be seen in the following definitions and in Fig. 3.5.

▶ **Definition** Continuous capillaries: The most common type of capillary, it is found in almost all vascularized tissues. They are characterized by a complete endothelial lining with tight junctions between endothelial cells. Such tight junctions usually display intercellular clefts that allow for exchange of water and other small, polar molecules (e.g., water, glucose) between the blood plasma and the interstitial fluid. Continuous capillaries not associated with the brain are rich in transport vesicles, contributing to either endocytosis or exocytosis. Those in the brain are part of the blood-brain barrier.

Fenestrated capillaries: This type of capillaries has pores (or fenestrations, from Latin *fenestra*, i.e., "window"). These make the capillary permeable to larger molecules. They are common in the small intestine as well as in the kidneys. They are also found in the choroid plexus of the brain and many endocrine structures.

Sinusoid capillaries (or simply sinusoids): the least common type of capillary, they are flattened, and they have extensive intercellular gaps and incomplete basement membranes, in addition to intercellular clefts and fenestrations. This allows for the passage of large molecules and structures, including plasma proteins and even cells. Blood flow through sinusoids is very slow, allowing more time for exchange of substances. Sinusoids are found in the liver and spleen, bone marrow, lymph nodes, and some endocrine glands.

Blood-brain barrier: the endothelial lining of capillary beds that perfuse the brain presents especially tight junctions with no intercellular clefts, plus a thick basement membrane and astrocyte extensions called end feet. In combination with high expression of efflux transporters, these structures pose a highly selective permeability barrier that provides precise control over which substances can enter or exit the brain.

In the previous section, we have (very) briefly discussed classical compartmental models. As empirical models, they are useful for data description and interpolation; they may also admit some (rough) degree of physiological parametrization. However, they do not have a direct physiological interpretation which makes it difficult to predict kinetic profiles with them when the underlying physiology changes (Aarons 2005; Espié et al. 2009). In contrast, physiologically based pharmacokinetic models strive to be mechanistic by mathematically transcribing anatomical, physiological, physical, and biochemical descriptions of the phenomena involved in the complex ADME processes. They intend to describe the pharmacokinetics of drugs within the body in relation to blood flows, tissue volumes, routes of administration, interactions with the tissue or organ, and biotransformation pathways. They are, in essence, compartmental models (in fact, multi-compartment models) but differ from classical pharmacokinetic models in that the compartments represent actual tissue and organ spaces and their volumes are the physical volumes of those organs and tissues. The biological and mechanistic bases of physiologically based models allow a better extrapolation of the kinetic behavior of drugs with regard to dose, route, species, individuals (males to females, adults to children, nonpregnant women to pregnant women, health conditions), and even in vitro to in vivo (Bouvier d'Yvoire et al. 2007). An example of a physiologically based model is presented in Fig. 3.6.

▶ **Important** In contrast with classic compartmental models, physiologically based pharmacokinetic models represent the body based on physiological data. They present an extended domain of applicability and facilitate transpositions and extrapolations. Some of these extrapolations are parametric: only changes in input or parameter values are needed to achieve the extrapolation (e.g., dose extrapolations). Others are nonparametric: a change in the model structure itself is needed (e.g., when extrapolating to a pregnant female, equations for the fetus should be added).

In the simplest physiologically based models, the compartments are assumed to be homogeneous. Such assumption may be relaxed if data at a lower level, for example, the cellular level, are available.

From the previous discussion, it can be appreciated that physiologically based models better capture the factors that influence drug distribution. Depending on the

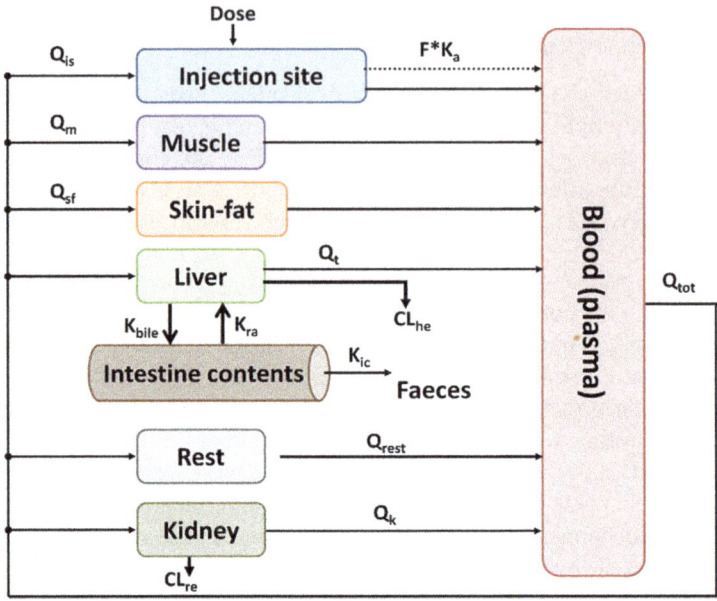

Fig. 3.6 Physiologically based model describing the pharmacokinetics of doxycycline in swine after IM administration. Q represents blood flow through tissue. Cl_{re} and Cl_{he} are the renal and hepatic clearances, respectively. F is bioavailability of doxycycline after IM administration, and Ka is absorption rate constant. K_{bile}, K_{ra}, and K_{ic} are biliary excretion rate constant, intestinal reabsorption rate constant, and intestinal elimination rate constant, respectively. Q_{tot} is cardiac output. (Extracted from Yang 2012)

rate of drug uptake by the tissues or organs, we may classify their distribution as (a) *perfusion rate-limited* or blood flow-limited or *permeability rate-limited* or membrane-limited (Espié et al. 2009). In the first situation, it is assumed that the drug distributes instantaneously from capillaries into interstitial fluid and tissue cells so that the rate-limiting step governing the movement of drug molecules into and out of the organ is the blood flow to the organ and no concentration gradients are likely within such physiological compartment. This is usually the case for small, lipophilic drugs. Such drugs will rapidly distribute to highly perfused organs such as the brain, the kidneys, the lungs, or the liver. Table 3.1 presents a list of blood flows to different organs. In the second situation (membrane-limited distribution), membrane permeability limits the distribution within the tissue/organ, and the tissue may be divided in subcompartments with a permeability rate-limited transfer between them (Nestorov 2003). The use of a membrane-limited organ model is indicated if tissue-drug concentrations do not decline in parallel with drug concentration plasma. Some examples are when the affinity of a tissue to the compound is high, when large and physically distributed tissues are modeled (e.g., muscle or skin), and when large molecules are modeled where diffusion and transport impediments are anticipated. Figure 3.7 schematically illustrates the difference between perfusion-limited and permeability-limited distribution.

Table 3.1 Regional blood flow distribution in a 70 kg healthy subject at rest

Organ	Blood flow (L/kg/min)
Kidneys	4.0
Liver	0.9
Heart	0.8
Brain	0.5
Muscle	0.08
Skin	0.03

Data extracted from Calzia et al. (2005)

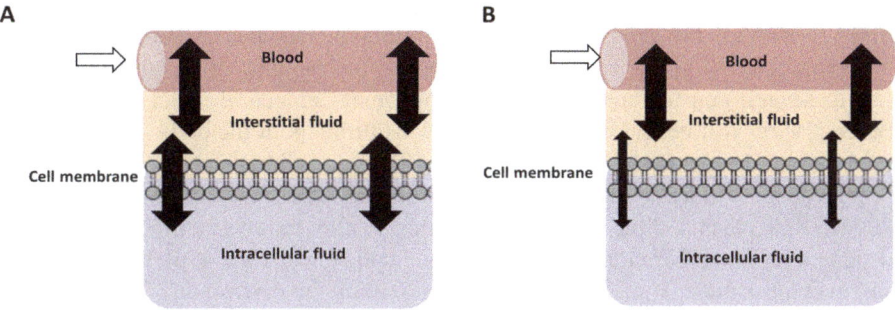

Fig. 3.7 Perfusion-limited (**a**) and permeability-limited (**b**) distribution. A thick double-headed arrow denotes a fat drug exchange, whereas a narrow arrow denotes a limited, slow exchange resulting in a lasting concentration gradient

3.3   Real and Apparent Distribution Volumes

The apparent volume of distribution (V_d) is a proportionality constant between the total amount of drug in the body and drug plasma concentration at a given time. In the framework of the simplest pharmacokinetic model (the one-compartment model), it can be defined as the theoretical volume that would be necessary to contain the total amount of a drug in the body at the same concentration that it is observed in the blood plasma. In the framework of more complex classic compartmental models where distribution is not assumed as an instantaneous process, other volumes of distribution may be computed, as snapshot plasma drug concentrations may be assessed in different conditions: before steady state, at steady state, and after steady state (at pseudo-distribution equilibrium) (Toutain and Bousquet-Mélou 2004). Though out of the scope of this book, the apparent volumes of distribution at steady state (V_{ss}) or beyond reaching steady state and at pseudo-equilibrium (the so-called V_{area}), as computed in the framework of the two-compartment model, are probably the more frequently reported volumes of distribution and the ones that find more applications in a clinical scenario.

▶ **Definition** The apparent volume of distribution V_d, in the context of the one-compartment model, is computed as:

$$V_d = \frac{Q}{C_p} \tag{3.1}$$

where Q stands for the total amount of drug in the body and C_p represents the total drug plasma concentration; both are time-dependent since they are subjected to constant changes (as soon as the drug enters the body, its elimination begins). However, as in the one-compartment model distribution is assumed to be instantaneous, the ratio between both variables (i.e, V_d) will be constant (independent of time). Note that the drug plasma concentration in the denominator is the total concentration, including free drug and plasma protein-bound drug, which will give rise to several considerations. The term "apparent" reflects the fact that the drug concentration will never be truly homogeneous throughout the body or identical to total plasma drug concentration. Consequently, V_d is an imaginary volume.

V_d can be estimated only after intravenous administration of the drug (kwon 2002). If a plasma drug concentration-time profile follows a monoexponential decline after administration of an intravenous bolus, i.e., if the body behaves like a single compartment, V_d can be estimated by dividing the intravenous dose (D) by an estimated drug concentration in plasma at time 0, determined by back-extrapolating the first two drug concentrations after intravenous injection:

$$V_d = \frac{D}{C_p(0)} \tag{3.2}$$

The apparent volume of distribution is an abstract entity that does not refer to any identifiable compartment in the body. Therefore, its "physical" interpretation is not unequivocal. Two drugs with different distribution patterns might have similar V_d values.

Frequently, the apparent volume of distribution will be simply referred as "volume of distribution" in the specialized literature. It is then important to establish the difference between the apparent volume of distribution of a drug and its real volume of distribution.

The real volume of distribution has physiological meaning and is related to the volume of body water/body fluids in which the drug is dissolved. Table 3.2 shows some typical physiological volumes for a person of 70 kg, which will be useful to further discuss the meaning of the real volume of distribution. The table shows that a person of 70 kg will have a total body water content of about 42 L (i.e., 0.6 L/kg). In other words, 42 L will approximately be the upper bound of the real volume of distribution for a subject weighting 70 kg. No drug can have, for that individual, a real volume of distribution above that approximate value. What is more, only a drug that gains access to all body volume (i.e., the entire interstitial and intracellular fluids) will have a real volume of distribution of 42 L. Since after an intravenous

Table 3.2 Approximate physiological volumes for a 70 kg healthy subject

Fluid	Volume (L)
Plasma	3
Blood	5
Interstitial fluid	12
Extracellular water	15
Intracellular water	27
Total body water	42

injection the drug molecules will almost instantaneously disperse within the plasma pool, V_d cannot be smaller than the actual volume of plasma in the body (approximately 3 L for a 70 kg person). 3 L thus represents the lower bound of both the apparent and real volumes of distribution. A drug that cannot extravasate nor enter blood cells would constitute one of the few examples in which the apparent volume of distribution and the real volume of distribution would be identical.

3.4 The Source of Discord: Drug Binding to Plasma Proteins and Tissue Components

Table 3.3 presents the approximate apparent volumes of distribution for different drugs. It is obvious from the table that V_d has no anatomical sense: note that the V_d of some drugs greatly exceeds the total body water. The origin of such behavior (discrepancy between the real and apparent volumes) lies in the fact that drug levels across the body are not uniform. In turn, the origin of such lack of uniformity in the drug concentrations once achieved the pseudo-equilibrium of distribution essentially lies in two factors: the affinity (if existent) of the drug for drug transporters and the nonspecific interactions between the drug and tissue elements (e.g., proteins).

The effect of active transport on drug concentrations is understandable: cells invest energy in maintaining concentration/electrochemical gradients, that is, in moving chemical entities uphill. Accordingly, if a drug behaves as a substrate of a drug transporter, the free drug hypothesis (see Chap. 1 of the present volume) will not apply, and free (unbound) drug concentrations at both sides of a biological barrier will not be the same at the pseudo-equilibrium of distribution.

Regarding unspecific interactions with tissue components, such components may be located within or outside the intravascular space. In the first case, the most relevant interaction is the one established between the drug and plasma proteins. In the second, drugs often interact with lipids or proteins, though the particular pattern of extravascular tissue binding depends on the drug. For instance, chloroquine and related compounds (e.g., hydroxychloroquine) have been shown to be differentially sequestered by different tissues (e.g., its concentration in the liver, spleen, and adrenal gland is 6000–80,000 times that of plasma) (Browning 2014), and it has been demonstrated that this drug binds to melanin, DNA, ferriprotoporphyrin, and phospholipids and also accumulates in lysosomes (MacIntyre and Cutler 1986). The influence of drug binding to plasma protein and

Table 3.3 Approximate apparent volumes of distribution of a selection of drugs

Drug	V_d (L/70 kg)	V_d (L/kg)
Chloroquine	15,000	214
Chlorpromazine	1400	20
Digoxin	490	7
Diazepam	63	0.90
Phenytoin	49	0.70
Phenobarbital	42	0.60
Aspirin	10.5	0.15
Ibuprofen	8.4	0.12
Warfarin	8	0.11
Tolbutamide	5.6	0.08

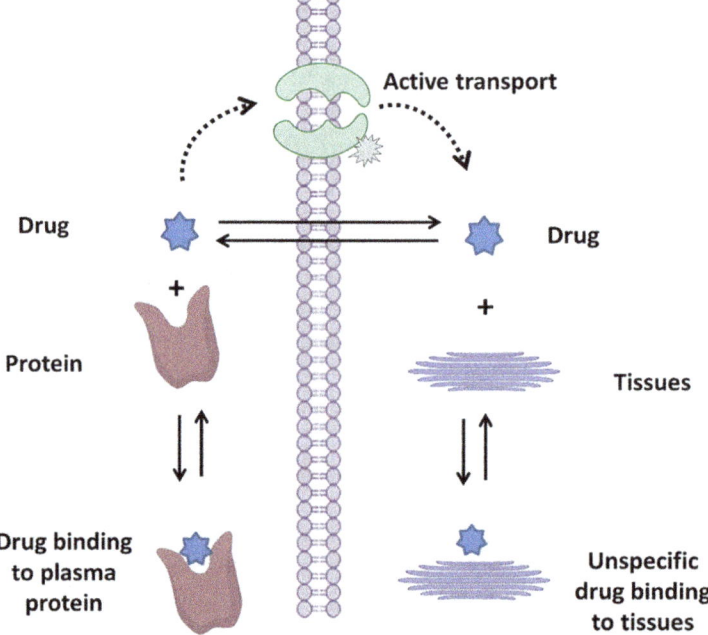

Fig. 3.8 A series of equilibria can be used to understand the influence of binding to plasma protein and tissue sites on distribution, along with the consequences of displacement

tissue components can be examined through the series of equilibria displayed in Fig. 3.8. A key concept worth remembering to gain a basic understanding of drug distribution is that, following the free drug hypothesis already discussed in the first chapter of this volume, only free (unbound) drug concentrations will tend to equilibrate at steady state. Le Chatelier's principle is also useful to anticipate the effect of a change in the biologic system (i.e., drug displacement) on drug distribution: when a system at equilibrium is subjected to change (e.g., in concentration of the drug or the binding partner), the system readjusts itself to counteract the effect of the applied change, evolving to a new equilibrium state.

   The disagreement between the real and apparent volumes is not the only consequence of the drug nonspecific interactions with vascular and extravascular
elements. Usually, the missing link between pharmacokinetics and pharmacodynamics (or the discrepancy between in vitro and in vivo results) is, precisely, the free
drug concentration. There is a wide consensus regarding the importance of using free
drug concentrations when developing pharmacokinetic-pharmacodynamic
relationships (see, for instance, Maurer et al. 2005; Di et al. 2013; Smith et al. 2010).

3.4.1   Binding to Plasma Proteins and Blood Cells

The high concentration of plasma proteins along with the propensity of practically
all drugs to bind to them to some extent explains the great interest in understanding
the impact of this process on pharmacokinetics and pharmacodynamics. The major
drug-binding components in plasma are albumin (which in a healthy individual is
present at a concentration similar to 600 μM), α-acid glycoprotein (AAG)
(12–30 μM), and lipoproteins such as γ-globulin (Bohnert and Gan 2013). For the
ease of clinical analysis, plasma has been historically the most common fluid
analyzed (Rowland and Tozer 1994), and binding to blood cells (in particular,
erythrocytes) has been frequently neglected in pharmacokinetic studies (Hinderling
1997), despite some drugs display high affinity for red blood cells (Kalamaridis and
DiLoreto 2014).
   Albumin is a 66 kDa protein with multiple binding sites (although two major sites
are primarily involved in ligand binding); whereas it can bind basic drugs, it displays
a preference for acidic ones (Trainor 2007; Bohnert and Gan 2013). In spite of its
large molecular size, it is not exclusively confined to the plasma compartment, as it
can extravasate. Nevertheless, the rate at which the albumin-drug complex
extravasates is much lower than that of the free drug, and extravasation of bound
drugs is frequently disregarded in pharmacokinetic analysis. Lower serum albumin
concentrations can be precipitated by several factors, such as burns, neoplastic
disease, or age (see Table 3.4 for some physiological and pathological states that
modify binding to plasma proteins).
   On the other hand, the other major plasma protein involved in drug binding,
AAG, is an acute-phase protein, synthesized in the liver, which shows preference to
bind basic and neutral drugs and whose levels could significantly change in disease
and inflammation states. The lower basal level of endogenous AAG (in comparison
with albumin), as well as its large relative fluctuations under various physiological
and pathological conditions, results in a higher variation from its basal level, and it
has been implicated in causing clinical relevant drug interactions (Bohnert and Gan
2013). The clinical impact of interactions involving drug displacement from plasma
proteins will be separately addressed in a subsequent subsection. By analyzing the
equilibria in Fig. 3.8, it can be seen that, since the free drug in plasma establishes a
comparatively rapid equilibrium with the extravascular free drug but the bound drug
does not, drug binding to plasma proteins will result in total plasma drug levels that
overestimate the extravascular drug levels. Accordingly, V_d of drugs with high-

Table 3.4 Some physiological and disease states that may alter plasma protein concentrations and thus drug binding to different plasma proteins

Decrease	Increase
Albumin	
Hepatic cirrhosis	Exercise
Age (neonate or elderly)	Schizophrenia
Pregnancy	Hypothyroidism
End-stage renal disease	Neurotic disorders
Cystic fibrosis	Psychosis
Nephrotic syndrome	
Acute pancreatitis	
α1-acid glycoprotein	
Nephrotic syndrome	Myocardial infarction
Oral contraceptives	Trauma
	Crohn's disease
	Surgery
	Rheumatoid arthritis
	Celiac disease
	Renal insufficiency
Lipoprotein	
Hyperthyroidism	Diabetes
Trauma	Hypothyroidism
	Nephrotic syndrome

bound fractions will tend to be small, especially if such affinity for plasma proteins is not compensated by a similar or higher affinity for tissue elements. In most cases, small V_{ds} will speak of high affinity of the drug for plasma proteins and low affinity to extravascular tissues.

Consider the free (or unbound) fraction of a drug in plasma:

$$f_u = \frac{C_{free}}{C_{tot}} \tag{3.3}$$

where C_{free} is the free drug plasma concentration and C_{tot} is the total drug plasma concentration (free plasma drug concentration plus bound plasma drug concentration, C_{bound}). For a drug having a single family of binding sites (binding sites characterized by the same affinity constant), C_{bound} is given by the following general equation (Toutain and Bousquet-Mélou 2002) (note its similarity to the Michaelis-Menten equation):

$$C_{bound} = \frac{B_{max} \times C_{free}}{K_D + C_{free}} \tag{3.4}$$

B_{max} denotes the maximal binding capacity (directly related to the molar concentration of the binding protein), and K_D is the equilibrium dissociation constant, which is the inverse of the affinity constant K_A, which is:

$$K_A = \frac{C_{\text{bound}}}{C_{\text{free}} \times P} \tag{3.5}$$

P stands for the unoccupied plasma protein concentration (which, in most therapeutic scenarios, will be similar to the total plasma protein concentration, if we have in mind the high concentrations stated in previous paragraphs). Now consider the total drug concentration that following Equation 3.4 might be expressed as:

$$C_{\text{tot}} = C_{\text{free}} + \frac{B_{\text{max}} \times C_{\text{free}}}{K_D + C_{\text{free}}} \tag{3.6}$$

After factorization and taking the inverse of the previous equation:

$$f_u = \frac{K_D + C_{\text{free}}}{B_{\text{max}} + K_D + C_{\text{free}}} \tag{3.7}$$

When a drug binds to more than one type of binding site (with different affinities), several equilibria are involved, each of them characterized by distinctive dissociation constants (Koch-Weser and Sellers 1976). Naturally, when the drug binds to only one binding site per binding partner molecule, the maximum molar concentration of drug that may bound the correspondent protein equals the molar concentration of the binding partner. The maximum possible binding capacity of a plasma protein (in relation to a given drug) equals the number of binding sites for that drug per protein molecule times the molar concentration of the protein.

▶ **Important** In most therapeutic situations, the free fraction will be relatively constant, at a given protein concentration, i.e., independent of drug concentration (Rowland and Tozer 1994; Toutain and Bousquet-Mélou 2002). This results from the rather high K_D values associated with the drug-protein complex formation equilibrium (have in mind that an appreciable free fraction exists for drugs even in the presence of a significant molar excess of plasma proteins, which speaks of a high dissociation constant). In such circumstances, Eq. 3.7 takes the following approximate form:

$$f_u = \frac{K_D}{B_{\text{max}} + K_D} \tag{3.8}$$

This is especially true for the case of drugs binding to albumin, due to the molar excess of albumin in comparison with the total plasma concentration range customary for most drugs (e.g., <25 μM). On the other hand, the free fraction for drugs binding to AAG will generally be observed to increase with increasing total drug concentration in the range achieved by many drugs (e.g., 1–10 μM) (Trainor 2007). Furthermore, drugs that bind to AAG are more prone to significant interindividual and intraindividual variations in their free fraction.

Drugs whose free fraction is independent of drug concentration in the therapeutic range are said to exhibit a linear binding behavior. There are, however, drugs that display nonlinear behavior in the therapeutic range. For instance, the antiepileptic drug valproic acid (which is also used to prevent migraines and to treat manic or mixed episodes associated with bipolar disorder) may achieve total plasma levels above the albumin concentration in plasma. Table 3.5 displays therapeutic plasma levels for different drugs, along with their approximate bound fraction and their binding partner. It can be observed that only those drugs whose therapeutic levels are similar to those of their binding partner are subjected to nonlinear binding behavior. If the reader notes that in other cases appreciable free plasma levels of the drug are observed in spite of the comparatively high levels of plasma proteins, high K_D values of the drug-protein complex can also be envisioned.

Table 3.5 Some examples of drugs with linear and nonlinear behavior in their binding to plasma proteins. Note that whether the binding follows a linear or nonlinear behavior essentially depends on the therapeutic range of the drug and the physiological concentration of its binding partner. Those drugs that bind to AAG tend to exhibit a nonlinear behavior at lower therapeutic concentrations. Note that the information in the table may significantly vary in physiological and pathological conditions where the plasma protein levels are modified or increased concentrations of endogenous displacing agents (e.g., fatty acids, bilirubin) are produced

Drug	Therapeutic range (μM)[a]	Bound fraction[b]	Binding partner(s)	Nonlinear behavior?
Carbamazepine	21.2–50.8	0.76	Albumin (70–80%) AAG (20–30%)	Yes (AAG) (MacKichan and Zola 1984)
Digoxin	0.001–0.003	0.25	Albumin	No (Evered 1972; Tillement et al. 1980)
Lidocaine	6.4–21.3	0.60–0.80	AAG (40–60%), albumin (20%)	No (Routledge et al. 1985)
Tacrolimus	0.004–0.025	0.99	Albumin (44–57%) AAG (39%)	No[b]
Theophylline	55.5–111.0	0.40	Albumin	No (Buss et al. 1985)
Valproic acid	346.7–693.4	0.80–0.90	Albumin	Yes, only significant for the highest therapeutic concentrations (Gugler and Mueller 1978; Gómez Bellver et al. 1993)

[a]Data extracted from MedlinePlus. Therapeutic drug levels. Reviewed on May 21, 2017. Available at http://www.nlm.nih.gov/medlineplus/ency/article/003430.htm. Accessed: May 21, 2018. Interpretation of reported drug concentrations must take into account the timing of the serum sample obtained for drug concentration determination
[b]Data extracted from DrugBank (https://www.drugbank.ca). Accessed: May 21, 2018

3.4.2   Binding to Plasma Proteins and Tissues: Their Relationship with Drug Kinetics

Plasma protein and tissue binding of drugs may impact on drug kinetics by acting at three levels: absorption, distribution, and elimination (McElnay 1996).

Regarding absorption, via their influence on maintaining a high concentration gradient of free drug between the absorption site and plasma, both plasma and tissue binding will reinforce the sink condition, thus favoring a more rapid absorption of drugs.

Plasma and tissue protein binding have opposing effects on drug distribution. Since plasma proteins are largely retained within the plasma compartment, drugs which are highly bound to plasma proteins tend to have low V_d (warfarin being one archetypical example). In contrast, drugs which are highly bound to tissues tend to have high V_{ds} (especially if the tissue to which the drug is bound is abundant in the body). For drugs which are highly bound to both plasma and tissues, the V_d will depend on the relative binding in both sites. In the case of tricyclic antidepressants, for instance, nearly all the drug in plasma is bound to albumin, but, due to extensive tissue binding, the circulating drug represents only a small fraction of the total drug in the body (Koch-Weser and Sellers 1976).

Let us now discuss the influence of tissue and plasma protein binding on elimination kinetics. If a drug is highly bound to tissues, it will be protected from elimination, since the drug outside the vascular compartment is not being delivered to the elimination organs (mainly, the liver and the kidneys). Thus, tissue-bound drug acts as a drug reservoir that will replenish plasma as drug in the plasma compartment is eliminated, and dues binding to extravascular tissue tends to enhance elimination half-lives. What about the impact of plasma protein binding on drug elimination? For those drugs displaying low extraction ratios in the elimination organs, plasma protein binding will protect the drug from elimination since only the unbound drug will be disposed by such organs during a single passage through them. This situation is mostly observed for those drugs that are moved from the serum to the elimination organs only through passive processes (e.g., glomerular filtration or passive diffusion to the hepatic biotransformation sites). For such drugs, the duration of action directly correlates with the bound fraction. In contrast, for drugs with high extraction in any of the elimination organs, protein binding has just the opposite effect (Koch-Weser and Sellers 1976; McElnay 1996): by keeping more drugs in the blood and increasing its delivery to the (highly perfused) elimination organs, protein binding will tend to shorten the duration of drug action. This is typically the case for drugs that exhibit high affinity for elimination systems, including metabolizing enzymes and drug transporters (e.g., efflux transporters that promote active tubular secretion of the drug or transporters that promote active uptake into the hepatocytes). In such cases, eliminated free drug molecules are rapidly replaced by additional free drug molecules that result from the dissociation of the drug-protein complex, even within a single passage through the elimination organ. This topic will be resumed in those chapters dedicated to elimination processes and also in the chapter dealing with drug transporters.

▶ **Important** For drugs with low extraction ratios in elimination organs, both plasma protein binding and tissue binding will protect the drug from elimination processes, contributing to a long duration of action. Conversely, for those drugs with high extraction rations in either the liver or the kidneys (or both), plasma protein binding will conspire against the duration of action, promoting fast elimination.

3.4.3 Some Therapeutic Considerations Related to Displacement from Binding Sites and Plasma Protein Binding

The displacement of a drug from its binding sites (either in plasma proteins or tissue components) by competition with other drugs or endogenous compounds has long been a concern among the healthcare community, due to the risk of adverse drug reactions related to drug-drug interactions. The equilibria involved in such competition are represented in Fig. 3.9, and the effect of the displacement on total plasma drug concentration and the free drug concentration in the plasma and the extravascular compartments can be easily predicted by once again resorting to the equilibrium law. Displacement of a drug from its binding partner due to noncompetitive mechanisms is also possible (Christensen et al. 2006). The potential importance of such interactions is vividly illustrated by the well-known interactions between warfarin and phenylbutazone or tolbutamide and sulfonamides. However, it should

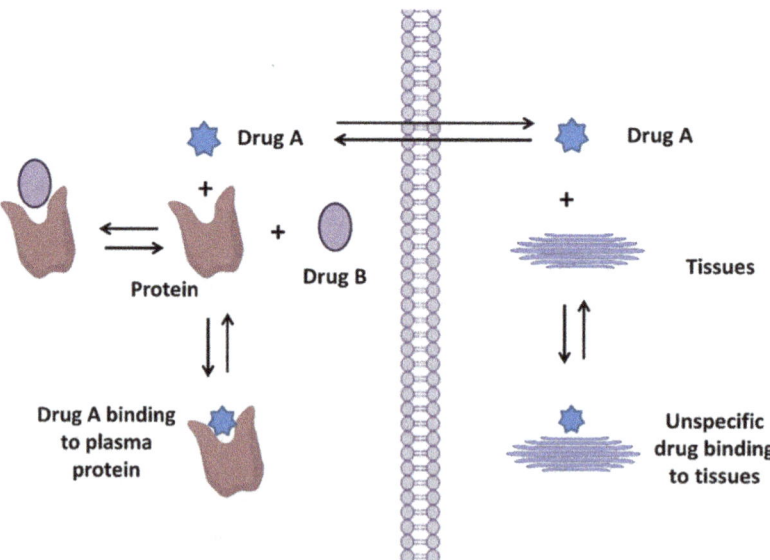

Fig. 3.9 Equilibria describing displacement of a drug from a binding protein by another exogenous (e.g., a drug) or endogenous compound

be highlighted that most drug-drug interactions involving displacement from plasma proteins will not be clinically significant (Rolan 1994; McElnay 1996).

▶   **Important** Most drug-drug interactions involving drug displacement from plasma proteins will not be clinically relevant.

Two reasons explain the previous statement. First, the transient increase on free drug concentrations due to displacement will usually be mitigated by diffusion of part of the free drug sudden excess to the extravascular compartment, coupled with the generally large sink capacity of the tissues. Second, immediately after displacement, higher amounts of free drug will be presented to the elimination organs. If linear elimination is observed, the excess of free drug will be rapidly compensated by an increased elimination rate. In other words, elimination and distribution to the extravascular compartment buffer or compensate the initial excess of free drug molecules in the vascular compartment. Accordingly, drug displacement from plasma proteins will likely be a concern only when several conditions are simultaneously met (Rolan 1994), namely, the displaced drug has low V_d (which speaks of limited binding to tissues); the displaced agent shows low/impaired elimination rate; the displaced agent has a narrow therapeutic window; and the displaced agent has a very high bound fraction, so that the displacement results in a relatively high increase in free drug levels (Table 3.6 depicts such effect). All these conditions are verified in the aforementioned case of warfarin, which has a very low V_d (around 9 L) and high bound fraction (about 0.99) and whose metabolism is inhibited by the same drug that displaces it from plasma proteins (phenylbutazone).

In contrast, adverse sequels are more likely to be observed if the displacement occurs at the level of extravascular tissues, since the low volume of plasma limits its capacity to buffer the sudden increases in tissue free drug concentrations. The subsequent decrease in V_d will result in a decrease in the displaced drug half-life. The best known example of a significant interaction of this type is the interaction between digoxin and quinidine (Bigger and Leahey 1982).

Special attention should be paid when administering drugs that are highly bound to plasma proteins by intravenous bolus. Rapid injection of the dose of such drugs

Table 3.6 Effect of drug displacement on the transient free drug levels, depending on the bound fraction. Displacement of highly bound drugs is more relevant from the clinical viewpoint, since it results in a more significant relative increase in the transient free drug levels, which would be later compensated by distribution to the extravascular compartment and elimination mechanisms

	% drug before displacement	% drug after displacement	% increase in free drug concentration
Drug A			
Bound	95	90	+100
Free	5	10	
Drug B			
Bound	50	45	+10
Free	50	55	

may surpass the binding capacity of the plasma proteins in the limited volume where the drug is being initially mixed, resulting in very high transient and local free drug concentrations to which highly perfused organs could be exposed (Koch-Weser and Sellers 1976). Caution must also be taken when administering a displacing agent intravenously (McElnay 1996). In both cases, slow administration is preferred to favor the dilution of the drug in the whole volume of the plasma compartment.

References

Aarons L (2005) Physiologically based pharmacokinetic modelling: a sound mechanistic basis is needed. Br J Clin Pharmacol 60:581–583

Bigger JT Jr, Leahey EB Jr (1982) Quinidine and digoxin. An important interaction. Drugs 24:229–239

Bouvier d'Yvoire M, Prieto P, Blaauboer BJ et al (2007) Physiologically-based kinetic modelling (PBK modelling): meeting the 3Rs agenda. The report and recommendations of ECVAM workshop 63. Altern Lab Anim 35:661–671

Bohnert T, Gan LS (2013) Plasma protein binding: from discovery to development. J Pharm Sci 102:2953–2994

Buss DC, Ebden P, Baker S, Routledge PA (1985) Protein binding of theophylline. Br J Clin Pharmacol 19:529–531

Browning DJ (2014) Hydroxychloroquine and chloroquine retinopathy. Springer, New York

Calzia E, Iványi ZD, Radermacher P (2005) In: Pinsky MR, Payen D (eds) Determinants of blood flow and organ perfusion. Springer-Verlag, Heidelberg

Christensen H, Baker M, Rucker GT, Rostami-Hodjegan A (2006) Prediction of plasma protein binding displacement and its implications for quantitative assessment of metabolic drug-drug interactions from in vitro data. J Pharm Sci 95:2778–2787

Di L, Rong H, Feng B (2013) Demystifying brain penetration in central nervous system drug discovery. J Med Chem 56:2–12

Espié P, Tytgat D, Sargentini-Maier ML et al (2009) Physiologically based pharmacokinetics (PBPK). Drug Metab Rev 41:391–407

Evered DC (1972) The binding of digoxin by the serum proteins. Eur J Pharmacol 18:236–244

Gómez Bellver MJ, García Sánchez MJ, Alonso González AC, Santos Buelga D, Dominguez-Gil A (1993) Plasma protein binding kinetics of valproic acid over a broad dosage range: therapeutic implications. J Clin Pharm Ther 18:191–197

Gugler R, Mueller G (1978) Plasma protein binding of valproic acid in healthy subjects and in patients with renal disease. Br J Clin Pharmacol 4:441–446

Hinderling PH (1997) Red blood cells: a neglected compartment in pharmacokinetics and pharmacodynamics. Pharmacol Rev 48:279–295

Kalamaridis D, DiLoretto K (2014) In: Caldwell GW, Yan Z (eds) Drug partition in red blood cells. Springer, New York

Koch-Weser J, Sellers EM (1976) Binding of drugs to serum albumin (first of two parts). N Engl J Med 294:311–316

Kwon Y (2002) Handbook of essentials pharmacokinetics, pharmacodynamics, and drug metabolism for industrial scientists. Kluwer Academic Publishers, New York

MacIntyre AC, Cutler DJ (1986) In vitro binding of chloroquine to rat muscle preparations. J Pharm Sci 75:1068–1070

MacKichan JJ, Zola EM (1984) Determinants of carbamazepine and carbamazepine 10,11-epoxide binding to serum protein, albumin and alpha 1-acid glycoprotein. Br J Clin Pharmacol 18:487–493

Maurer TS, Debartolo DB, DA T et al (2005) Relationship between exposure and nonspecific binding of thirty-three central nervous system drugs in mice. Drug Metab Dispos 33:175–181

McElnay JC (1996) In: D'Arcy PF, McElnay JC, Welling PG (eds) Drug interactions at plasma and tissue binding sites. Springer, Berlin

Nestorov I (2003) Whole body pharmacokinetic models. Clin Pharmacokinet 42:883–908

Recordati G, Bellini TG (2004) A definition of internal constancy and homeostasis in the context of non-equilibrium thermodynamics. Exp Physiol 89:27–38

Rolan PE (1994) Plasma protein binding displacement interactions--why are they still regarded as clinically important? Br J Clin Pharmacol 37:125–128

Rowland M, Tozer TN (1994) Clinical pharmacokinetics. In: Concepts and applications. Lippincott Williams & Wilkins, Philadelphia

Routledge PA, Lazar JD, Barchowsky A, Satrgel WW, Wagner GS, Shand DG (1985) A free lignocaine index as a guide to unbound drug concentration. Br J Clin Pharmacol 20:695–698

Smith DA, Di L, Kerns EH (2010) The effect of plasma protein binding on *in vivo* efficacy: misconceptions in drug discovery. Nat Rev Drug Discov 9:929–939

Tillement JP, Zini R, Lecompte M, d'Athis P (1980) Binding of digitoxin, digoxin and gitoxin to human serum albumin. Eur J Durg Metab Pharmacokinet 5:129–134

Toutain PL, Bousquet-Mélou A (2002) Volumes of distribution. J Vet Pharmacol Ther 25:460–463

Toutain PL, Bousquet-Mélou A (2004) Volumes of distribution. J Vet Pharmacol Therap 27:441–453

Trainor GL (2007) The importance of plasma protein binding in drug discovery. Expert Opin Drug Discov 2:51–64

Yang F, Liu HW, Li M, Ding HZ, Huang XH, Zeng ZL (2012) Use of a Monte Carlo analysis within a physiologically based pharmacokinetic model to predict doxycycline residue withdrawal time in edible tissues in swine. Food Additives & Contaminants: Part A 29 (1):73–84

Further Reading

The reader is advised to complement this chapter with those that focus on drug elimination (Chaps. 4 and 5) and drug-drug interactions (Chap. 12), in this same volume. If interested in drug interactions related to the distribution process, the reader is strongly advised to read the chapter by McElnay, which is listed among the references.

Drug Metabolism

4

Alan Talevi and Carolina Leticia Bellera

4.1 Introduction

Through evolution, the body has developed different strategies to dispose metabolic waste and protect itself from exposure to potential toxic chemicals. It is useful to differentiate chemical compounds with a well-defined physiologic role from those that do not have any. The latter will be generically called *xenobiotics* and include most drugs, natural and industrial contaminants, other industrial compounds used for technical purposes (i.e., agrochemicals), cosmetics, food additives, food components with no physiological function, and recreational and social drugs (Testa and Krämer 2006), along with their biotransformation products without physiologic role. There is also a tendency to view as xenobiotics those endogenously produced compounds which are administered at relatively high doses (for medical or nonmedical reasons) (Jenner et al. 1981; Testa and Krämer 2006) and are thus found within the body outside their physiological range.

A. Talevi (✉)
Laboratory of Bioactive Research and Development (LIDeB), Department of Biological Sciences, Faculty of Exact Sciences, University of La Plata (UNLP), La Plata, Buenos Aires, Argentina

Consejo Nacional de Investigaciones Científicas y Técnicas (CONICET), La Plata, Buenos Aires, Argentina
e-mail: atalevi@biol.unlp.edu.ar

C. L. Bellera
Medicinal Chemistry/Laboratory of Bioactive Research and Development (LIDeB), Faculty of Exact Sciences, Universidad Nacional de La Plata (UNLP), La Plata, Buenos Aires, Argentina

Consejo Nacional de Investigaciones Científicas y Técnicas (CONICET), Buenos Aires, Argentina

© Springer Nature Switzerland AG 2018   55
A. Talevi, P. A. M. Quiroga (eds.), *ADME Processes in Pharmaceutical Sciences*,
https://doi.org/10.1007/978-3-319-99593-9_4

▶ **Definition** The prefix *xeno-* comes from the Ancient Greek and means "alien." As in *xenophobia*, it is used to denote strangeness. Accordingly, the word xenobiotic describes a chemical entity which is found within a living organism but has no physiological function. Xenobiotics can be regarded as potentially toxic, and the body has mechanisms to dispose them.

In contrast, *physiological compounds* are chemicals having essential biological functions, indispensable to the survival or well-being of the body. Among them, we can mention water, nutrients, oxygen, mineral, dietary fiber, or vitamins.

Those mechanisms available within the body to reduce its exposure to nonphysiological compounds will also limit the bioavailability of that particular category of xenobiotic which are therapeutic drugs. Essentially, these mechanisms can be broadly classified as (a) *preventing xenobiotics from entering the blood-stream or sensitive organs* (e.g., the blood–brain barrier limits the distribution of many xenobiotics to the brain); (b) *physically removing the xenobiotic molecules* (mainly through the bile and the urine but also, sometimes, through perspiration and respiration); and (c) *metabolizing xenobiotics (biochemical transformation or bio-transformation) to produce drug metabolites whose excretion is faster* than that of the parent (unchanged) compound.

▶   **Important** Although drug metabolism and drug excretion are presented
      separately, they are indeed highly integrated phenomena. As will be
      extensively discussed throughout this and the next chapter, metabolism
      and excretion work in a concerted manner to provide a more efficient
      disposal of xenobiotics.
         The ultimate objective of all bodily drug elimination systems is to
      promote excretion of the xenobiotic molecules (either unchanged or as
      metabolites). The major routes of excretion are bile (those drugs or drug
      derivatives excreted though bile are eventually, if everything goes well,
      excreted in feces) and urine. Taken globally, biotransformation processes
      transform the drug into a considerably more polar biochemical product.
      Such polar metabolite will not be (re)absorbed in either the intestine or
      the renal tubules. Furthermore, they can be occasionally subjected to
      active secretion into the bile or the tubular content.
         Some textbooks point to the enhanced water solubility of drug
      metabolites as the origin of a more efficient excretion. This is only
      indirectly true, since commonly the drug concentrations in biologic fluids
      are way below its solubility. Improved hydrophilicity increases the excre-
      tion rate of the metabolite (in comparison with the parent compound)
      because highly polar metabolites are neither efficiently reabsorbed nor
      distributed (they experience much greater diffusional barriers) and in
      some occasions are actively secreted into the intestinal, bile canalicular,
      or tubular lumen.

It emerges from the previous two chapters that to be properly absorbed and distributed most drugs (with the possible exception of those that experience facilitated diffusion or active uptake) require an adequate lipophilic-hydrophilic balance. They must display a certain minimal solubility in physiological media (only soluble molecules are transferred through biological barriers), but they must also possess some lipophilicity so that they can passively permeate to the bloodstream and then move forward to the remaining bodily compartments. Furthermore, lipophilic drugs tend to display higher affinity for their molecular targets. On the one hand, the formation of the drug-target complex implies desolvation; desolvation free energies, which will be higher for polar drugs, oppose the binding event (Kólar et al. 2011). On the other, lipophilic substituents in the drug are often wanted to exploit hydrophobic binding pockets in the target.

The same requisites that a drug must generally fulfill to be absorbed and distributed, and interact with its molecular target conspire against excretion, since lipophilic drugs tend to be extensively reabsorbed. Accordingly, biotransformation to polar metabolites is often required to enhance the excretion of a drug.

4.2   Metabolism: Biotransformation Reactions

▶ **Definition** Drug biotransformation involves the enzyme-mediated conversion of drugs into drug metabolites (the products of a biotransformation reaction).

A drug is usually metabolized in a sequential manner. The *parent* (unchanged) *drug* (i.e., the pharmacologically active chemical entity that has been administered through a drug delivery system) is thus initially converted to a *primary metabolite*, which may be substrate for a second biotransformation that will produce a *secondary (or sequential) metabolite* (Smith 2008).

Biotransformation reactions can be broadly classified into functionalization reactions (historically known as phase I reactions) and conjugation or synthetic reactions (also known as phase II reactions). Functionalization reactions imply the creation of a functional group or the modification of an existing one. This type of biotransformation introduces a "chemical handle" or anchoring point in its substrate, so that the resulting metabolite is more prone to experience a synthetic reaction (Testa and Krämer 2007; Talevi 2016). They include redox reactions and hydrolyses/hydrations (Testa and Krämer 2006). Synthetic reactions imply coupling the drug or one of its phase I metabolites with a diversity of (endogenous) moieties (e.g., glucuronic acid, glutathione, sulfate, phosphate, amino acids, acetyl, etc.) generally resulting in metabolites with a drastic increase in polarity and also a (moderately) higher molecular weight, which are not likely to be reabsorbed.

Sequential metabolism formed the basis for the common nomenclature of phase I and phase II metabolism, although a phase II reaction may occur to a drug without a precedent phase I reaction if the drug has a suitable anchoring site to which the correspondent conjugation moiety is transferred. In other words, conjugation reactions are able to produce first-generation and later-generation metabolites

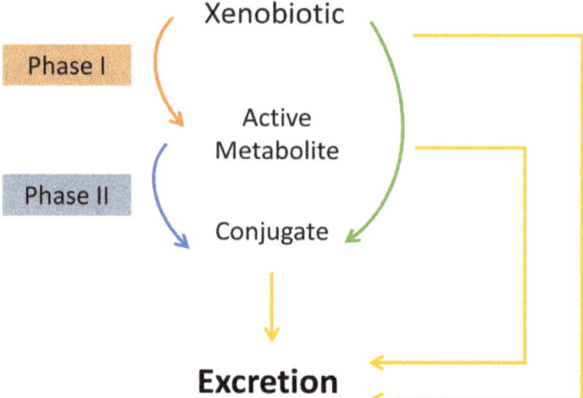

Fig. 4.1 The fate of xenobiotic within the body may depend on different elimination pathways. The unchanged xenobiotic molecule might be directly excreted. Alternatively, it may be excreted as a phase I or phase II metabolite. Synthetic reactions could occur directly on the parent compound or may require a previous phase I reaction to occur and predispose the drug to a conjugation. Often, all these possibilities take place in parallel and contribute to the overall elimination of the drug

(Testa and Krämer 2008). Accordingly, phase I and phase II nomenclature may be misleading in the sense that it may be *wrongly* understood that phase II biotransformations always occur on phase I metabolites. A meta-analysis by Testa et al. on the biotransformation reactions of more than 1100 xenobiotics (Testa et al. 2012) shows that first-generation metabolites are formed mainly (almost 70%) by redox reactions, but about 22% are formed by conjugations. In the second generation, the contribution of redox reactions decreases to about 50%, whereas conjugations increase to 37%. In third and later generations, both redox and conjugation reactions accounted for the same proportion (46%) of metabolites. The proportion of metabolites generated by hydrolysis does not vary significantly from one generation to another, remaining in the 8–12% range. The previous analysis demonstrates the inadequacy of the phase I/phase II classification, which assumes the metabolism of xenobiotics to begin with redox or hydrolysis reactions, followed in subsequent metabolic steps by conjugations.

The possible fate of a xenobiotic after entering the body is illustrated in Fig. 4.1. Acetaminophen metabolism is used in Fig. 4.2 to provide the reader with a concrete idea of the complexity of elimination processes and how phase I and phase II reactions may be or may be not coupled.

Drug metabolism not only introduces physical-chemical modifications to the drug. Aside from the changes in molecular weight and polarity, the resulting metabolites are usually inactivated from a pharmacological viewpoint. Occasionally, however, they can be pharmacologically active (and even more active than the parent compound), or, if excessively reactive from a chemical perspective, they may be toxic.

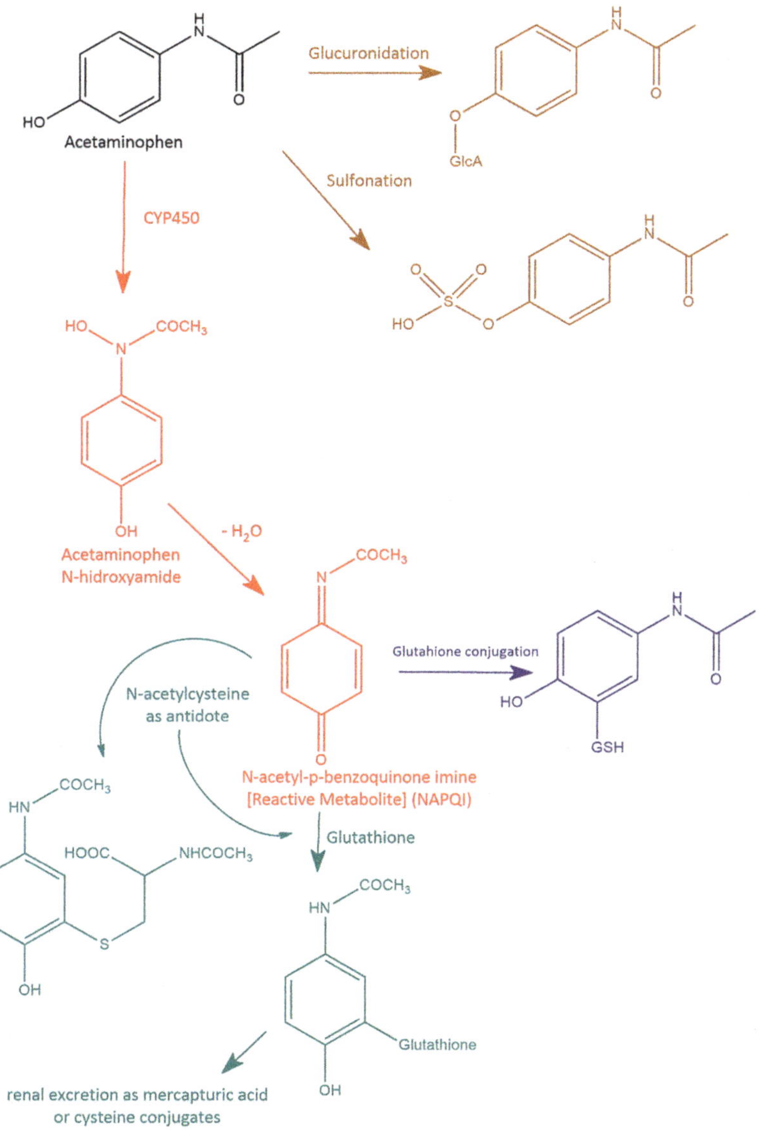

Fig. 4.2 Elimination pathways for acetaminophen/paracetamol. A fraction of the dose is eliminated with no changes in urine. The drug undergoes direct conjugation with glucuronide and sulfate or sequential CYP450-mediated biotransformation to the highly toxic N-acetyl-p-benzoquinone imine (NAPQI) which is inactivated by conjugation with glutathione. If not inactivated, NAPQI reacts with hepatic proteins resulting in acute hepatic failure

4.2.1 Functionalization (Phase I) Reactions

Functionalization reactions expose, as in hydrolysis reactions, or introduce, usually through oxidation reactions, chemically reactive functional groups such as -OH, -NH$_2$, or -SH. The enzymes that catalyze these reactions may be oxygenases/oxidases (such as cytochrome P450 (CYP450) isozymes, flavin-containing monooxygenases, monoamine oxidases, peroxidases, xanthine oxidases, alcohol oxidases, and others), reductases (such as aldo-keto reductases), or hydrolytic enzymes (e.g., esterases, amidases). It is worth mentioning that some of the phase I enzymes can catalyze both oxidations and reductions (Testa and Krämer 2007). It has been estimated that redox reactions account for about 60% of total xenobiotic metabolic reactions (Testa et al. 2012).

Regarding subcellular localization of phase I reactions, a significant fraction of the correspondent enzymes are membrane proteins located in the smooth endoplasmic reticulum (e.g., CYP450, flavin-containing monooxygenases, some esterases, and epoxide hydrolases) (Cribb et al. 2005). The lumen of the endoplasmic reticulum has an oxidizing environment relative to the cytosol (the glutathione/glutathione disulfide ratio is around 3, compared to 100 in the cytosol). This balance seems to be crucial to the process of oxidative protein folding within the reticulum and may contribute to the generation of oxidative stress through the use of oxygen as the terminal electron acceptor in protein folding (Tu and Weissman 2004). It also creates an environment that favors further oxidation of reactive intermediates that enter the endoplasmic reticulum lumen. To a lesser extent, some phase I enzymes are also located in the mitochondria (e.g., monoamine oxidase) and the cytosol (e.g., xanthine oxidase or alcohol dehydrogenase).

▶ **Important** The liver is, by far, the organ that expresses the highest levels of metabolizing enzymes. However, some metabolism enzymes are also extensively expressed in other organs (and, in some particular cases, at higher levels than in the liver itself) (Jennen et al. 2010). Aside from the liver, other organs that significantly contribute to xenobiotic metabolism are the small intestine, the lungs, and the kidneys. Note that all of them are involved in matter exchange with the environment; accordingly, their involvement in biotransformation of xenobiotics does not lack evolutionary logic. It is also important to note that gut microbiota can also significantly contribute to drug metabolism and have an impact on drug elimination kinetics (Swanson 2015).

Some examples of functionalization reactions can be found in Table 4.1.

A key feature of phase I reactions is that, while intending to contribute to xenobiotic detoxification, they may play a significant role in the etiology of some diseases and metabolic-mediated toxification processes by resulting in the formation of electrophilic intermediates capable of covalently modifying biological

Table 4.1 Some examples of phase I biotransformation reactions

Reaction	Examples
Hydroxylation	Aromatic hydrocarbons (R-Ar)
Sulfo-oxidation	Disulfides (R-S-R)
Dehydrogenation	Alcohols (R-OH)
Reduction	Nitro compounds (R-NO$_2$)
Hydrolysis	Esters (R-COO-R')

Table 4.2 Examples of bioactivation of halogenated solvents and their associated toxicity

Halogenated solvent	Toxicity
Bromobenzene	Hepatic necrosis
Vinyl chloride	Liver cancer
Carbon tetrachloride	Hepatic necrosis, renal necrosis
Chloroform	Hepatic necrosis, renal necrosis

macromolecules (adduct formation), typically proteins or DNA, a process known as drug bioactivation (Tang and Lu 2010; Matias et al. 2014; Gan et al. 2016) (note, though, that mechanistic pathways of toxicity pathways for drugs that form adducts are no fully understood yet and that adduct formation may coexist with adduct-independent pathways). A relevant example of drug bioactivation will be later discussed when describing the molecular basis of paracetamol toxicity. Furthermore, the responsible moiety for toxification by some halogenated solvents (after oxidation) is presented in Table 4.2. After scrutinizing xenobiotic metabolites, Testa et al. concluded that about 7% are toxic, mostly due to strong electrophilicity (Testa et al. 2012). Toxic compounds usually arise from Csp^2 and Csp^3 oxidations (e.g., resulting in epoxide formation), N- and S-oxidations, and, above all, generation of quinones and analogues (quinonimines, quinonimides, and quinone diimines), which account for around 40% of all toxic and/or reactive metabolites. The major enzyme systems involved in bioactivation are the CYP450 and peroxidase systems (Walsh and Miwa 2011).

Furthermore, it is relatively frequent for phase I metabolites to retain pharmacological activity (occasionally, a metabolite could display more potency than the unchanged drug!) (Talevi 2016). C-hydroxylation and hydrolysis reactions play a predominant role in the generation of pharmacologically active metabolites (Testa et al. 2012). Pharmacologically active metabolites are majorly first-generation ones. The role of ester hydrolyses is not surprising having in mind the current interest in prodrugs (which are defined later). As for the pharmacological significance of alkyl and aryl hydroxylations, this is majorly owing to the hydroxylated metabolite retaining the target affinity and thus pharmacological activity of the parent compound. Note that while phase I metabolites are chemically distinct from the parent drug, they are still chemically similar, and a similar activity profile to the one of the parent compound might be attained if the modification is introduced in a non-pharmacophoric group or whenever it leads to optimization in binding to the molecular target (Fura 2006).

▶ **Definition** A prodrug is a compound that, after administration, is metabolized into a pharmacologically active drug within the body. Prodrugs are often designed to improve bioavailability (enhanced oral absorption, differential distribution) or improve safety and, sometimes, to mask unpleasant organoleptic properties.

▶   **Important** Sometimes, phase I metabolites will retain the pharmacological activity of the parent compound. In some occasions, they may even display more activity than the parent compound. Bioactivation to toxic metabolites by phase I reactions is also possible.

4.2.1.1 Cytochrome P450

CYP450 comprises a superfamily of hemoproteins (i.e., they contain the organic cofactor *heme*, a prosthetic group crucial for their catalytic activity). This enzymatic system is arguably the most important phase I xenobiotic metabolizing system, participating in the elimination of 70–85% of known drugs. Furthermore, CYP450 is also involved in the metabolism of endogenous compounds, including the biosynthesis, bioactivation, and, sometimes, breakdown of steroids (sex hormones, neurosteroids, cholesterol, vitamin D) and fatty acids (Gibbons 2002; Jones et al. 2014; Westphal et al. 2015). Fifty-seven individual P450s have been characterized in humans (Lewis 2004), distributed across 18 families. In mammals, CYP450 members are located either in the endoplasmic reticulum or in the inner mitochondrial membrane. Mutations or other defects in genes encoding CYP450 members (especially, from those families characterized by low redundancy and narrow substrate specificity) result in P450-mediated diseases including those caused by aberrant steroidogenesis; defects in fatty acid, cholesterol, and bile acid pathways; and vitamin D and retinoid dysregulation (Nebert et al. 2013). Furthermore, CYP members from microorganisms provide the basis for the development of anti-infective therapies (e.g., antifungal and antiprotozoal drugs) (Lepesheva and Waterman 2011; Lepesheva et al. 2018).

▶   **Important** Monooxygenation is beyond a doubt the most common reaction catalyzed by CYP450, i.e., insertion of one atom of oxygen into an organic substrate (RH), while the remaining oxygen atom from an oxygen molecule is reduced to water:

$$RH + O_2 + NADPH + H^+ \rightarrow ROH + H_2O + NADP^+ \qquad (4.1)$$

However, other less common reactions are also catalyzed by CYP450, including reductions, ester cleavage, and ring expansions (Guengerich 2001).

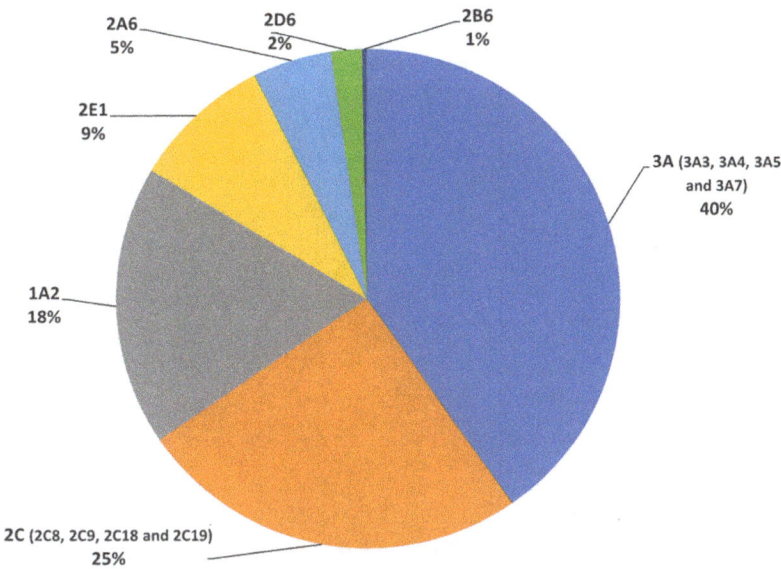

Fig. 4.3 Comparison of expression levels of individual P450 isoforms in the total P450 in human liver microsomes. (Adapted from Kwon 2002)

The liver, kidneys, intestine, lungs, and skin are the organs with the highest expression levels of CYP450 members. Since each CYP450 member has its own substrate specificity, it is not hard to imagine that, as a whole, the superfamily confers great versatility for the biotransformation of a wide range of xenobiotics. Furthermore, some of its members display a wide substrate specificity (i.e., they are polyspecific): in particular, CYP3A4 is the most promiscuous member of the superfamily, catalyzing the phase I biotransformation of hundreds of known drugs (Bu 2006).

Figure 4.3 displays the relative abundance in the liver of different CYP450 isozymes. Figure 4.4 shows the approximate percentage of drugs in the market whose metabolism is mediated by those CYP450 members. Note that the rate at which a given member catalyzes the conversion of a substrate depends on *the turnover number* (maximum number of chemical conversions of substrate molecules per second that a single catalytic site will execute), *the number of available enzyme copies*, and *the affinity of the substrate for the enzyme*. Consequently, there exist isoforms (e.g., CYP2D6) which have a remarkable role in xenobiotic metabolism despite being expressed at comparatively low expression levels.

Besides the monooxygenases, the CYP450 includes "auxiliary" enzymes that contribute with the electrons required to the redox reaction: NADPH-CYP450 reductase and cytochrome b5, which in turn may be reduced by NADPH-cytochrome reductase and NADH-cytochrome-b5 reductase (Gan et al. 2009).

The most common proposed mechanism of the catalysis involves the following (Meunier et al 2004; Guengerich 2007) (Fig. 4.5): (a) binding or the substrate in the vicinity of the heme group, which induces a change in the conformation of the active

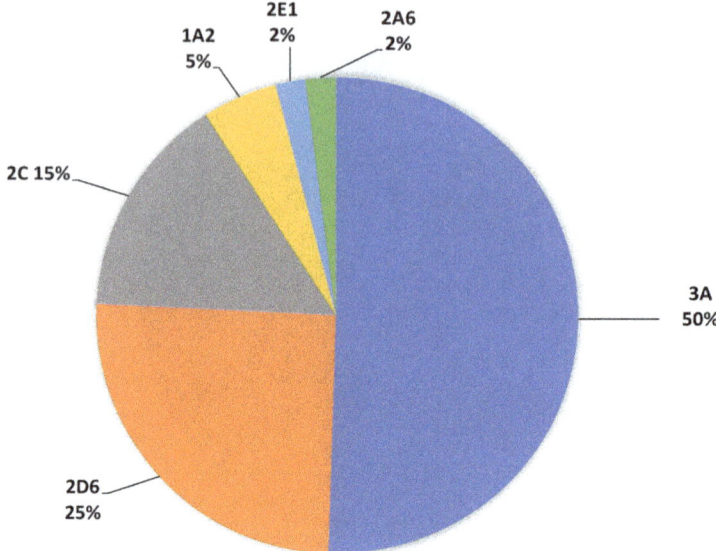

Fig. 4.4 Estimated percent of drugs on the market metabolized by various cytochrome P450 isoforms. (Adapted from Kwon 2002)

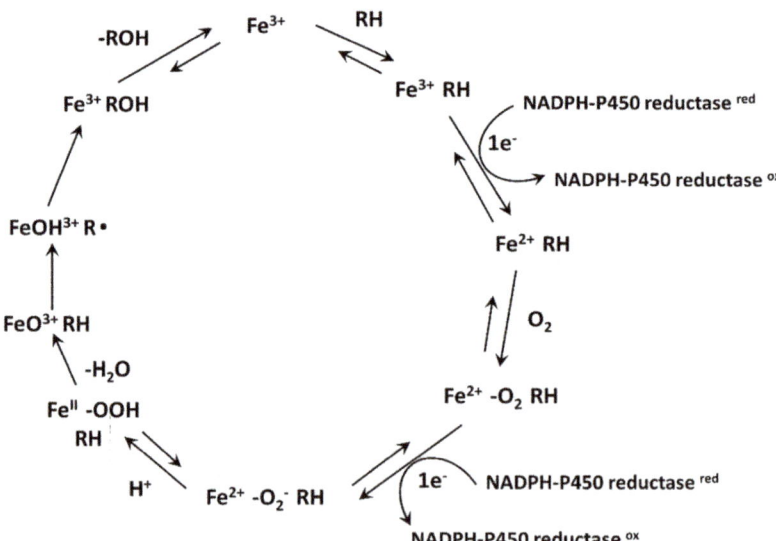

Fig. 4.5 Generalized catalytic cycle for CYP450 reactions. RH denotes the parent compound (substrate for the biotransformation), whereas ROH represents the metabolite (i.e., product of the mono-oxygenation reaction)

site, often displacing a water molecule from the distal axial coordination position of the heme iron; (b) substrate binding induces electron transfer from NADPH via CYP450 reductase or another associated reductase; (c) molecular oxygen binds to the resulting ferrous heme center at the distal axial coordination position, initially giving a dioxygen adduct; (d) a second electron is transferred from either cytochrome P450 reductase, cytochrome b5, or ferredoxins, reducing the $Fe-O_2$ adduct to give a short-lived peroxo state; (e) the peroxo group formed in step d is rapidly protonated twice, releasing one molecule of water and forming highly reactive species (an iron(IV) oxo or ferryl species with an additional oxidizing equivalent; the remaining oxygen atom is transferred to the substrate. After the product is released from the active site, the enzyme returns to its original state, with a water molecule returning to occupy the distal coordination position of the iron nucleus.

The names of CYP450 enzymes are given with a number-letter-number designation (e.g., CYP3A4) with the CYP abbreviation indicating CYP450 membership, the first number indicating the P450 family based on 40% or greater sequence identity, the letter indicating the P450 subfamily based on 55% or greater sequence identity, and the final number representing the individual P450 within that family/subfamily. In humans, CYP3A4, CYP2D6, CYP2C9, CYP1A2, and CYP2C19 perform the majority of drug metabolism, with some contribution from CYP2E1, CYP2A6, CYP2C8, and CYP2B6 (Furge & Guenguerich 2006).

CYP450 activity is frequently studied using microsomes.

▶ **Definition** Microsomes are vesicle-like artifacts reformed from pieces of the endoplasmic reticulum when eukaryotic cells (frequently, hepatocytes) are broken up in the laboratory. They can be concentrated and separated from other cellular debris by differential centrifugation. Unbroken cells, nuclei, and mitochondria sediment out at 9000–10,000 g, whereas soluble enzymes and fragmented ER, which contains CYP450, remain in solution. At 100,000 g, achieved by faster centrifuge rotation, endoplasmic reticulum components sediment out, and the soluble enzymes remain in the supernatant (called cytosolic fraction). The so-called S9 fraction can be obtained after the first centrifugation step at about 9000 g, hence the name. It contains both cytosolic and microsomal enzymes.

Incubation of test compounds with hepatic microsomal preparations is the primary mean by which phase I biotransformations are determined. For microsomal stability determination, the compound is typically incubated for 30 min in about 1.0 mg/mL microsomal protein, phosphate buffer pH 7.4 at 37 C. Compound stability is then determined by an appropriate analytical method.

Note, however, that microsomes cannot be used to study enzyme induction phenomena (see later in this chapter), since the later requires the whole cell machinery functional.

4.2.2 Synthetic (Phase II) Reactions

In phase II reactions, a chemical compound is conjugated with (generally polar) endogenous moieties (the endogenous conjugating moiety, sometimes abbreviated as the *endocon*), such as glucuronic acid, glutathione, sulfate, or acetate, among others. A marked difference between phase I and phase II reactions is that the latter tend to produce pharmacologically inactive and chemically nonreactive products. Only about 10% of the known toxic metabolites originate in conjugation reactions (Testa et al. 2012). Only exceptionally phase II metabolites are pharmacologically active (a typical counterexample is morphine 6-O-glucuronide). We will then say that, generally, phase II reactions *inactivate* their substrates.

Besides detoxifying reactive molecules, the addition of such polar groups through conjugation produces more hydrophilic and larger metabolites which are not able to easily diffuse across cell membranes, conditioning their reabsorption and distribution. Furthermore, many of the products from phase II metabolism are weak acids which are (a) nearly completely ionized at physiological pH range and (b) prone to interact with albumin, which also conspires against their extravasation (Table 4.3). As we will discuss in Chap. 5, some of the phase II metabolites are so polar that they will need the help of efflux transporters to leave the metabolizing cell.

Conjugations are catalyzed by a series of transferases. As previously mentioned, the substrate must have an appropriate "anchoring point" for the conjugation to occur. Sites on drugs where conjugation reactions occur include carboxyl, hydroxyl, amino, and sulfhydryl groups. An important aspect of phase II reactions is that they require a co-substrate (often called cofactor) to take place (the transferase is sometimes figuratively described as a "nuptial bed"). The co-substrate carries the endogenous conjugating moiety, with the chemical bond linking the cofactor and the endocon being a high-energy one such that the Gibbs energy released upon its cleavage drives the transfer of the endocon to the substrate (Testa and Krämer 2008). The molecular structures of the cofactors of the main phase II reactions are shown in Fig. 4.6. Since there is a limited supply of the co-substrate, its availability (and the rate at which the body can replenish it) will determine the capacity of the reaction. For example, sulfonation is a high affinity but low capacity reaction: it has a fast initial turnover rate that decreases as the limited amount of the co-substrate (3'-phosphoadenosine-5'-phosphosulfate, PAPS) is depleted. In contrast, glucuronidation has comparatively low affinity but high capacity (the correspondent cofactor, uridine-5'-diphospho-a-d-glucuronic acid, is produced endogenously by the C(6) oxidation of UDP-a-d-glucose, with about 5 g synthesized daily in adults). At low doses, for a xenobiotic which is subjected both to sulfonation and glucuronidation, sulfonation is faster and its metabolite predominates; at high doses, the sulfonation capacity is exceeded and glucuronidation predominates. Metaphorically, sulfonation can be thought as a sprinter, whereas glucuronidation may be regarded as a marathon runner.

Some of the phase II conjugates may experience deconjugation. Sulfates and acyl glucuronides can hydrolyze within physiological pH range. Glucuronides are susceptible to β-glucuronidase cleavage in the gut, as will be described in the next

Table 4.3 Increase in molecular weight and pKa values of the resulting products from some common phase II reactions. The capacity and affinity of the correspondent transferases are also listed

Reaction	Transferase	ΔMW	pKa	Affinity	Capacity
Glucuronidation	UDP glucuronosyltransferases (UGTs)	176	3.0–3.5	Low	High
Glycine conjugation	Glycine transferases	57	3.5–4.0	Intermediate	Intermediate
Sulfonation	Sulfotransferases	81	<1	High	Low
Glutathione conjugation	Glutathione S-transferases (GSTs)	289	2.1, 3.5	High	Low
Acetylation	N-acetyl transferases	42	Neutral	Variable	Variable

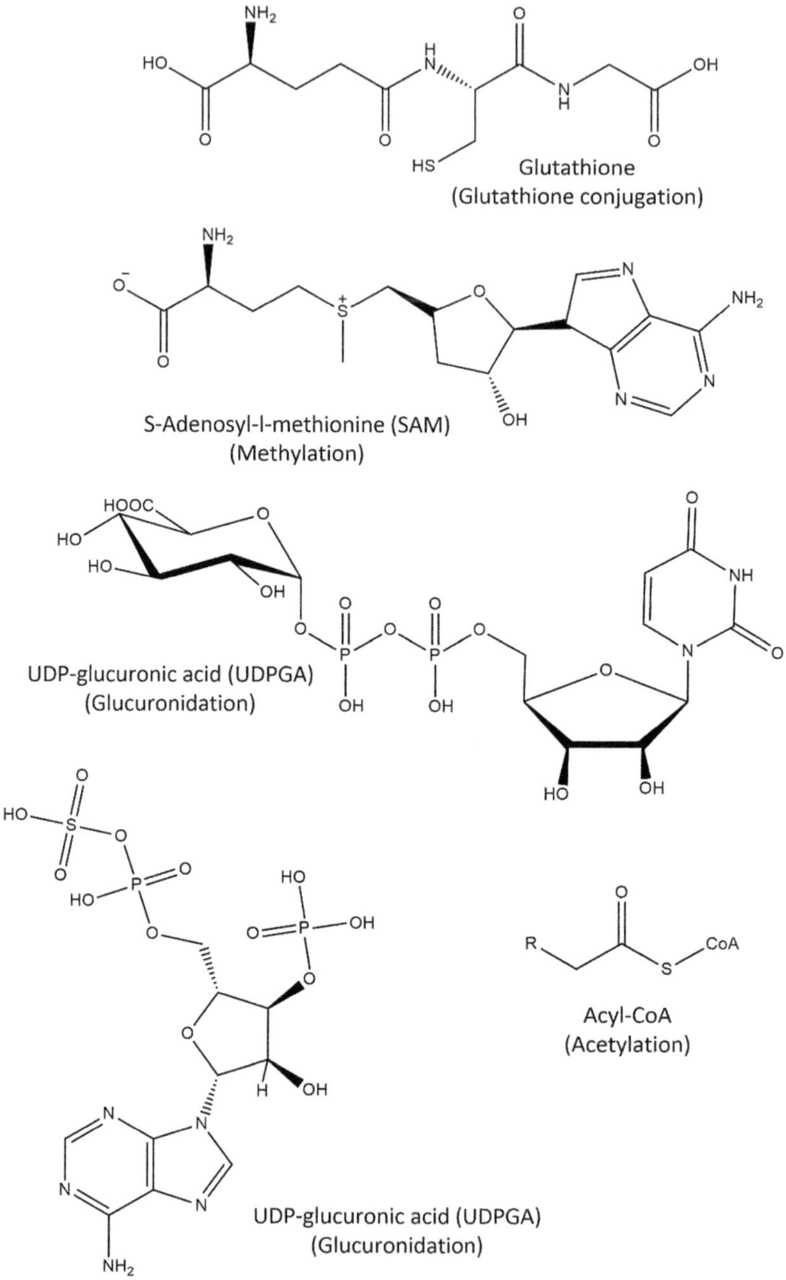

Fig. 4.6 Molecular structures of co-substrates for common phase II reactions

chapter. Glycine conjugates and acetylation products are subject to possible in vivo cleavage by hydrolases/esterases.

Regarding the subcellular localization of the different phase I reactions, sulfonation and N-acetylation are cytosolic reactions; conjugation with amino acids takes place in the cytosol and the mitochondria; glutathione conjugation occurs in the cytosol, the endoplasmic reticulum, mitochondria, and peroxisomes (though glutathione S-transferases are often found in higher levels in the cytosol); and glucuronidation locates in the endoplasmic reticulum (Testa and Krämer 2008; Badenhorst et al. 2013). By far, the most frequent phase II reaction is glucuronidation, followed by glutathione conjugation and sulfonation (Williams et al. 2004; Testa et al. 2012). Note how convenient is that the enzymes that catalyze the most frequent phase II reaction colocalize with those catalyzing the most common phase I biotransformation, increasing the efficiency of the overall process and limiting the exposure of the metabolizing cell and tissue to reactive phase I metabolites (the bioactivated phase I metabolite is rapidly converted to the inactivated phase II product).

Regarding organ distribution, UDP glucuronosyltransferases are the most ubiquitously distributed, having significant expression levels in the liver and bile ducts, kidneys, gastrointestinal tract, reproductive organs, and skin; sulfate conjugation primarily takes place in the liver, the kidneys, and the intestine; glutathione S-transferases are highly expressed in the liver and kidneys; high levels of N-acetyltransferases are found in the liver, kidneys, gastrointestinal tract, lungs, and skeletal muscle (Liston et al. 2001; Testa and Krämer 2008).

Example

Paracetamol/acetaminophen is a widely used over-the-counter analgesic and antipyretic drug. When properly used it is generally safe (in fact, it is regarded as the safest analgesic and antipyretic for pregnant women); it is, though, the leading cause of acute liver failure in the Western world, including developed countries (Lauretti 2012) and large doses (such as those observed in intentional or unintentional overdose) may lead to severe hepatic necrosis and fatal hepatic failure (Mahadevan et al. 2006).

Now we have enough tools to perform a deeper analysis of Fig. 4.2 and understand the mechanism of paracetamol toxicity, risk factors, and its antidote. Note that the elimination of paracetamol follows different parallel competing processes: a small proportion of the unchanged drug is excreted in urine, whereas most of the ingested drug is metabolized before excretion. The drug might be directly biotransformed through sulfonation and glucuronidation (thus generating a first-generation sulfonate or glucuronide conjugate, respectively). Alternatively, it may be oxidized through CYP450 (CYP2E1) resulting in the toxic N-acetyl-p-benzoquinone imine, which attacks cellular components in the main metabolizing organ and produces the liver failure. At therapeutic doses, though, this toxic metabolite is rapidly inactivated by glutathione conjugation. In contrast, overdose leads to depletion of the co-substrates of the sulfonation and glutathione conjugation reactions. Depletion of the co-substrate in sulfonation reactions, 3'-phosphoadenosine-5'-phosphosulfate (PAPS) implies overloading the remaining biotansformation pathways (including the CYP450-mediated one, thus leading to higher NAPQI levels). Depletion of glutathione impedes NAPQI inactivation.

Several variables (such as chronic ethanol misuse, concomitant use of CYP450 inducing drugs or UDP glucuronosyltransferase inhibitors, and prolonged fasting or malnutrition) may increase the risk of hepatic injury (Bray et al. 1992; Castellano et al. 2001; Daly et al. 2008; Liu et al. 2011), though the influence of such factors is source of controversy and some authors claim that none of them have so far been unequivocally linked to paracetamol toxicity (Caparrotta et al. 2018). Aside from several case reports suggesting the previously mentioned factors increase the risk of toxicity from paracetamol (sometimes, even at therapeutic doses), there are reasonable grounds to suspect from them. Both CYP450 inducers and glucuronosyltransferase inhibitors would favor the generation of NAPQI. On the other hand, a malnourished state could deplete hepatic glutathione reserves and thus reduce the ability to detoxify NAPQI. In fact, many guidelines for the management of paracetamol poisoning have taken the risk factors into consideration (Daly et al. 2008).

Activated charcoal administered within 1–2 h of ingestion reduces the absorbed paracetamol dose and the likelihood of hepatic failure. If activated charcoal cannot be administered within that time frame, infusion of N-acetylcysteine is an effective antidote which guarantees survival if administered within 8 h of paracetamol ingestion. Beyond 8–10 h after ingestion, efficacy decreases. N-acetylcysteine is a glutathione precursor that acts to replenish depleted glutathione reserves in the liver.

Case Study

A 53-year-old woman with concurrent squamous cell carcinoma of the anus and renal cell carcinoma developed severe hepatotoxicity while receiving paracetamol at recommended dosage (4 g daily) under medical supervision and without concomitant administration of enzyme-inducing agents. Her appetite had been poor for several months, during which time she received chemotherapy and radiotherapy and lost 8 kg in bodyweight. She was hospitalized for insertion of brachytherapy rods to treat the anal malignancy. On admission, she was severely malnourished based on standards of weight for height. Liver span, enzyme levels, and standard functional parameters were normal. Brachytherapy rods were inserted, and she received morphine for postoperative analgesia. Administration of paracetamol started postoperatively, after a period of fasting of about 20 h. She remained fasting for other 18 h after starting administration of this drug. Four days later, the alanine

(continued)

aminotransferase level had increased to 269 IU/L (normal range 4–51 IU/L) and the aspartate aminotransferase level to 252 IU/L (normal range 15–45 IU/L). The next day, she developed severe nausea, right upper quadrant discomfort, and tender hepatomegaly. Severe paracetamol-related hepatotoxicity was diagnosed. Paracetamol was suspended. N-acetylcysteine and glucose were administered via intravenously. Serological and virological testing excluded acute infection with hepatitis viruses A, B, C, and E, cytomegalovirus, Epstein–Barr virus, herpes simplex virus, varicella–zoster virus, enteroviruses, and adenovirus. Testing for metabolic, immune-mediated, and metastatic liver disease proved negative. The blood glucose and bilirubin levels and prothrombin time normalized after 2 days of treatment with N-acetylcysteine. Liver enzyme and albumin levels normalized after 4 weeks. Over the next 3 months, the patient underwent nephrectomy and abdominoperineal resection of the anal tumor. Her appetite had remained poor, and she had lost additional 4 kg in bodyweight. She received paracetamol again, ranging from 1.3 g to 2.6 g daily up to three times a week, over several weeks following the anal surgery. Nonetheless, liver enzymes, bilirubin, prothrombin time, and blood glucose values remained normal during this second period of exposure.

Do you concur with the diagnosis? Why do you think liver toxicity was not observed the second time the patient received paracetamol? Would you suggest limiting the daily dose of paracetamol to malnourished or fasting patients, or do you think you lack evidence to formulate such recommendation?

Based on a case report by Kurtovic and Riordan (2003).

4.2.3 Further Drug Metabolism: Phase III Reactions

After phase II reactions, the xenobiotic conjugates may be further biotransformed. For example, glutathione conjugates may be processed to acetylcysteine (mercapturic acid) conjugates (the glutamate and glycine residues in the glutathione molecule are removed by gamma-glutamyl transpeptidase and dipeptidases). In the final step, the cystine residue in the conjugate is acetylated.

Some authors also include within the phase III the transporter-mediated excretion of conjugates, with the anionic groups in most conjugates acting as affinity tags for a variety of membrane efflux pumps. Whereas this process is certainly articulated with metabolism, we will restrict the idea of metabolism to chemical modifications catalyzed by enzymes, and we will consider the previously mentioned transport processes separately in subsequent chapters (Homolya et al. 2003).

4.3   First-Pass Effect

▶ **Definition** We will define first-pass effect (or first-pass metabolism or pre-systemic metabolism) as any biotransformation suffered by drug molecules before reaching systemic circulation. While first-pass effect might be present in any administration route (except, maybe, intra-arterial administration), it will be considerably more significant for the oral route, since there the drug will face organs expressing high levels of biotransformation enzymes before reaching systemic circulation.

Orally administered drugs will be majorly absorbed in the small intestine and transported to the portal vein (which accounts for approximately 75% of the total liver blood flow) through the mesenteric vessels. Blood passes from branches of the portal vein through liver sinusoids between "plates" of hepatocytes. Blood also flows from branches of the hepatic artery and mixes in the sinusoids to supply the hepatocytes with oxygen. This mixture percolates through the sinusoids and collects in a central vein which drains into the hepatic vein. The hepatic vein subsequently drains into the inferior vena cava, which carries deoxygenated blood from the lower and middle body into the right atrium of the heart. This blood will later go to the lungs before finally reaching systemic circulation.
    A significant fraction of the absorbed drug amount might be subjected to pre-systemic loss due to biotransformation in the intestinal walls, biotransformation and/or biliary excretion in the liver, or biotransformation in the lungs (note that these three organs are highly relevant in terms of drug metabolism). This first-pass effect can be clinically relevant when the metabolized fraction is high or when it varies significantly from individual to individual or within the same individual over time, resulting in variable or erratic absorption. Note that all the substances absorbed in the stomach and the intestines will have to pass through the liver before reaching systemic circulation, with the exception of lipids, which form chylomicrons that are not absorbed directly into capillary blood but transported first into the lymphatic vessel that penetrates each intestinal villus. Chylomicron-rich lymph then drains into the lymphatic system and only then into blood, without participation of the portal system.
    Substances absorbed through the sublingual mucosa also evade hepatic first-pass effect, since the veins originating there do not join the portal system. For its part, about two thirds of the drug absorbed through the rectal route bypasses the hepatic first-pass metabolism as the rectum's venous drainage is two thirds systemic (middle and inferior rectal vein) and only one third hepatic portal system (superior rectal vein).

▶   **Important** Have in mind that, since enzymatic systems are saturable systems (and generally, but, not always, described through the Michaelis–Menten kinetics discussed in Chap. 2), the fraction of the dose whose absorption will be affected by the first-pass effect will largely depend on the drug flux to the metabolizing organ. If the metabolizing system is exposed to a large quantity of drug per unit of time, it could get saturated (this is particularly true for drugs administered in large doses).

The higher the amount of drug above the saturation condition, the higher the fraction of the dose that will survive (unchanged) the first-pass effect.

The fraction of the dose absorbed (F) when the drug is given by any route of administration can be estimated by comparing the area under the plasma concentration-time curve (AUC) after administering the drug for that route with the area under the plasma concentration-time curve after administering the drug intravenously. Remember that the AUC is proportional to the amount of drug that has reached systemic circulation.

For example, if one wants to estimate the fraction of the dose absorbed for a drug given orally, we should compute

$$F = \frac{\text{AUC}_{\text{oral}}}{\text{AUC}_{\text{iv}}} \times \frac{D_{\text{iv}}}{D_{\text{oral}}} \tag{4.2}$$

An F below 1 suggests incomplete absorption of the drug, due to inappropriate release of the drug from the dosage form, drug degradation in gastric media, intestinal low permeability, and/or first-pass metabolism in the gut and/or liver.

Example

Recently, Fathi et al. studied the bioavailability of paracetamol tablet, capsule, and effervescent dosage forms (each one containing 500 mg of the drug) in healthy subjects (Fathi et al. 2018). Thirty volunteers were divided into three gender-balanced groups of ten subjects. Participants displayed a mean ($\pm$SD) age of 21 ($\pm$2) years. They were prohibited to use acetaminophen or other analgesics 1 day before the experiment.

The effervescent dosage form had a quicker absorption (greater Cmax and lower Tmax), whereas the area under the plasma concentration curve was significantly higher for the effervescent dosage form than for the two other formulations (data is shown in Table 4.4). In other words, the effervescent dosage form displayed better bioavailability, both in quantitative and kinetic terms.

Table 4.4 Results obtained by Fathi et al. when comparing three paracetamol dosage forms. The table shows mean data for the ten volunteers in each group

Plasma concentration (μg/mL) – mean (SD)			
Tablet	Capsule	Effervescent	Time (min)
0	0	0	0
6.61 (2.41)	11.29 (3.94)	15.25 (2.54)	60
8.74 (2.49)	7.43 (1.37)	9.94 (1.63)	120
7.35 (2.80)	4.67 (1.49)	6.67 (1.81)	240
2.64 (0.85)	2.08 (1.70)	1.49 (0.64)	480
47.04	40.62	53.11	$\sum$AUC (μg min/mL)

Data extracted from Fathi et al. (2018)

Can you provide any possible explanations for the observations? Do you think that the results could have been confounded by other factors asides the differences between the dosage forms under comparison?

4.4   Inter- and Intra-Individual Factors Affecting Drug Metabolism: Intrinsic and Extrinsic Sources of Variability

Factors affecting drug metabolism are among the most important sources of inter- and intra-individual variability in drug pharmacokinetics and in the pharmacological response to medications.

▶ **Definition** Inter-individual factors are factors which differ between individuals but are constant throughout the life of an organism. This applies to factors dependent on the genetic makeup of an individual (Krämer and Testa 2008). Variations in metabolism which are not based on genetic differences will be called intra-individual factors and mostly depend on environmental factors and physiological or pathological states (in some cases it could be argued that some of these factors have mixed origin: e.g., disease conditions often impact on drug pharmacokinetics, and usually disease is the result of a combination of intrinsic and extrinsic factors).

It is worth underlining, though, that two individuals of the same species may differ, at a given time point, in their drug metabolism capacity due to either constitutive or environmental factors.

Regarding inter-individual factors, within a particular species there might be stable (i.e., at significant frequencies) genetic variations called polymorphisms. Genetic variations in metabolizing enzymes (usually arising from point mutations) can result in differences in enzyme expression levels, substrate specificity or activity. Carriers of variant alleles can be at risk of experimenting toxic effects or therapy failure when treated with a substrate of the respective enzyme. Other variations of genetic origin may result from gene duplication/multiplication events (leading to enhanced expression levels of the gene products and, in the long term, to divergence of duplicates). We may also include here sex-related differences in drug metabolism, though in general they have a minor impact in humans (Krämer and Testa 2008).

Personalized pharmacotherapy considering the genetic makeup of an individual will be specially required when the therapeutic range is narrow, and the altered enzyme represents the major elimination pathway of the administered drug. Although the relationship between pharmacokinetics and pharmacogenomics is abundantly addressed in Chaps. 8 and 9 of the present volume, here we will briefly discuss some key points on the matter.

▶ **Definition** Precision medicine is an emerging approach for disease treatment and prevention that considers individual variability in genes, environment, and lifestyle for each person. It is closely related to personalized and stratified medicine, and these terms are often used interchangeably, though subtle distinctions between them have been realized (Day et al. 2017; Talevi 2018).

Particular mutations may be unique for a certain population or show differences in their frequency when comparing ethnicities, leading to population-specific responses to xenobiotics. For instance, a frequent polymorphism in the aldehyde dehydrogenase ALDH2 gene in Asians is responsible for facial flushing and unpleasant sensations after alcohol intake, with this variant being infrequent in Caucasians. Realization of such population differences has even impacted on regulatory status. For instance, the Japanese regulatory authority requires clinical trials on Japanese volunteers to consider new drug applications and, for multi-regional clinical trials, confirmation of similarities in the dose-response curve and pharmacokinetic data between Japanese and non-Japanese subjects. Note that adaptation to different environments and differences in nutrition constitute additional reasons for ethnic differences in drug metabolism.

Before the genetic background of polymorphic behavior in drug metabolism was known, the consequences of polymorphism were described in a purely phenotypic manner. Patients were assigned to categories such as poor, intermediate, extensive, and ultrarapid metabolizers, according to the frequency distribution of individual metabolic rates. In the case of acetylation, the phenotypes are called slow and rapid acetylators. At present, patients are sometimes classified in these categories based on their genotype rather than on the observed metabolic rates (Krämer and Testa 2008). The drug metabolism enzymes with the clinically most relevant polymorphisms are CYP2D6, CYP2C19, CYP2C9, UDP glucuronosyltransferase 1A1, N-acetyltransferase, dihydropyrimidine dehydrogenase, cholinesterase, and thiopurine methyltransferase.

In general, people who cannot metabolize a drug will require a much lower dose than is recommended by the manufacturer; those who metabolize it quickly may require a higher dose. Note, however, that the pharmacological consequences of limited or enhanced drug metabolic rates depend on whether the metabolite is or is not more active than the parent compound (think, e.g., of a prodrug).

Regarding intra-individual factors influencing drug metabolism, the most common ones are physiological factors (diurnal cycle, age, pregnancy, sex hormones levels), pathological conditions (e.g., celiac disease, liver disease, diabetes, inflammation, infection, heart failure), and interactions of the drug with physiological compounds or with other xenobiotics (drugs, drinks, tobacco, pollutants) (Krämer and Testa 2009). The major mechanisms that mediate drug-drug interactions (or interaction between a drug and any other type of xenobiotic) are enzyme induction and enzyme inhibition.

▶ **Definition** In enzyme induction, a ligand (the *inducer*) activates a transcription factor, which consequently dimerizes with another transcription factor (usually belonging to the class of nuclear receptor), and the dimer binds to a regulatory sequence on the DNA. DNA binding of the transcription factor dimer generally results in an increase in transcription of the controlled gene and consequently increased levels of the transcribed enzyme. When a xenobiotic induces its own metabolism, we will speak of *auto-induction*; if the xenobiotic induces the metabolism of other xenobiotic, we will speak of *hetero-induction*. Some typical enzyme inducers include phenobarbital, phenytoin, rifampin, and dexamethasone.

Enzyme inhibition, on the other hand, involves the interaction of an inhibitor with a biotransformation enzyme, decreasing its activity in a reversible or irreversible manner. Typical examples of inhibitors include ketoconazole, ritonavir, and amiodarone.

Example

In the package insert of a well-known carbamazepine brand, it reads:

"Because carbamazepine induces its own metabolism, the half-life is also variable. Autoinduction is completed after 3–5 weeks of a fixed dosing regimen. Initial half-life values range from 25 h to 65 h, decreasing to 12–17 h on repeated doses. (...) Carbamazepine is a potent inducer of hepatic 3A4 and is also known to be an inducer of CYP1A2, 2B6, 2C9/19 and may therefore reduce plasma concentrations of co-medications mainly metabolized by CYP 1A2, 2B6, 2C9/19 and 3A4, through induction of their metabolism. When used concomitantly with carbamazepine, monitoring of concentrations or dosage adjustment of these agents may be necessary..."

Note that, due to the induction phenomenon, the pharmacokinetics of carbamazepine will be time-dependent, showing a two- to fivefold decrease in the half-life of the drug, whose pharmacokinetics will require weeks to stabilize. Also note that carbamazepine (auto) induces its metabolism and (hetero) induces the metabolism of other CYP450 substrates.

Example

Back in 2010, Xiang et al. studied the effect of CYP2D6 variants on risperidone bioavailability. Twenty-three healthy Chinese subjects took part in the study. Table 4.5 shows the AUC for CYP2D6 *1/*1, *1/*10, and *10/*10 carriers.

How would you characterize *1/*10 and *10/*10 carriers according to the results? Would you expect lower or higher risperidone pharmacological effect for these subjects (tricky question: some research required!)? In the light of the results obtained for *1/*1 and *10/*10 subjects, are those obtained for *1/*10 subjects reasonable?

It is interesting to note that the fold change in expression levels and activity for CYP450 enzymes (either due to polymorphism or induction) tends to be much more pronounced than for phase II enzymes (Krämer and Testa 2008; Krämer and Testa

Table 4.5 Mean risperidone AUC for Chinese subjects displaying different CYP2D6 genotypes. Data extracted from Xiang et al. 2010. CYP2D6*1 is the most common form, considered "fully functional," also known as wild type. Data for homozygous and heterozygous carriers of the *10 variant is shown

Genotype	n	Risperidone – mean (95% confidence interval) (ng h/mL)
1*/1*	5	30.6 (22.9, 38.3)
1*/10*	12	64.8 (45.2, 84.4)
*10/*10	6	248.2 (77.2, 419.2)

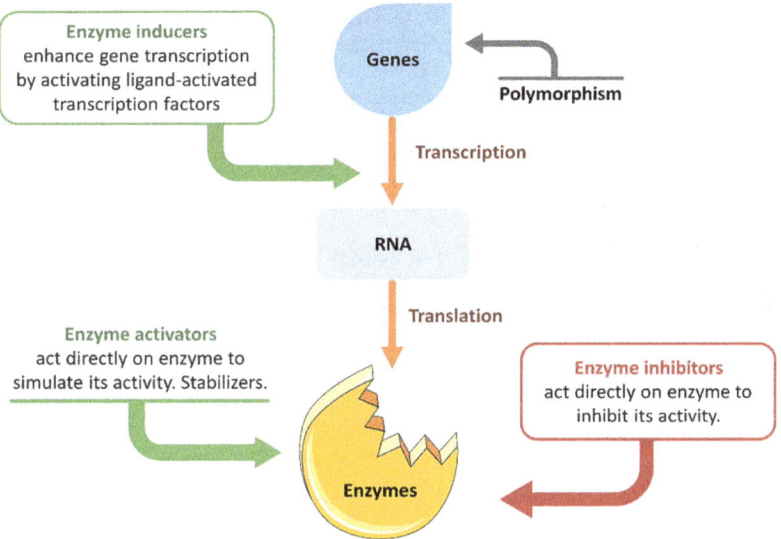

Fig. 4.7 Diagram of the different levels at which a factor of variability in drug metabolism might act

2009). Adjusting treatment with certain CYP450 substrates based on their genotype and concomitant medication would thus be desirable (Lynch and Price 2007). Genotype testing may predict people who are poor metabolizers or nonresponsive to drugs metabolized by CYP450 enzymes. CYP450 genetic variants should be considered when patients exhibit unusual sensitivity or resistance to drug effects at normal doses. Patients should be monitored carefully for the development of adverse drug effects or therapeutic failures when a potent CYP450 enzyme inhibitor or inducer is added to drugs metabolized by one or more CYP450 isozymes. Because they are known to cause clinically significant CYP450 drug interactions, caution must be used when adding the following substances: amiodarone, antiepileptic drugs, antidepressants, antitubercular drugs, grapefruit juice, macrolide and ketolide antibiotics, non-dihydropyridine calcium channel blockers, and protease inhibitors (the list is not exhaustive!). Further examples may be found in Chap. 12 of the present volume.

Fig. 4.7 shows at what level do some of the factors that influence drug metabolism impact

References

Badenhorst CP, van der Sluis R, Erasmus E, van Dijk AA (2013) Glycine conjugation: importance in metabolism, the role of glycine N-acyltransferase, and factors that influence interindividual variation. Expert Opin Drug Metab Toxicol 9:1139–1153

Bray GP, Harrison PM, O'Grady JG, Tredger JM, Williams R (1992) Long-term anticonvulsant therapy worsens outcome in paracetamol-induced fulminant hepatic failure. Hum Exp Toxicol 11:265–270

Bu H, (2006) A Literature Review of Enzyme Kinetic Parameters for CYP3A4-Mediated MetabolicReactions of 113 Drugs in Human Liver Microsomes: Structure- Kinetics Relationship Assessment. Current Drug Metabolism 7 (3):231-249

Castellano I, Novillo R, Gómez-Martino JR, Covarsi A, Herrero JL (2001) Fracaso renal agudo debido a intoxicación por paracetamol. Nefrologia (Madr) 21:592–595

Caparrotta TM, Antoine DJ, Dear JW (2018) Are some people at increased risk of paracetamol-induced liver injury? A critical review of the literature. Eur J Clin Pharmacol 74:147–160

Cribb AE, Peyrou M, Muruganandan S (2005) The endoplasmic reticulum in xenobiotic toxicity. Drug Metab Rev 37:405–442

Day S et al., (2017) Stratified, precision or personalised medicine? Cancer services in the 'real world' of a London hospital. Sociol Health Illn. 39(1):143–158. https://doi.org/10.1111/1467-9566.12457

Daly FFS, Fountain JS, Murray L, Graudins A, Buckley NA (2008) Guidelines for the management of paracetamol poisoning in Australia and New Zealand — explanation and elaboration. Med J Aust 188:296–302

Fathi M, Kazemi S, Zahedi F, Shiran MR, Moghadamnia AA (2018) Comparison of oral bioavailability of acetaminophen tablets, capsules and effervescent dosage forms in healthy volunteers. Curr Issues Pharm Med Sci 31:5–9

Fura A (2006) Role of pharmacologically active metabolites in drug discovery and development. Drug Discov Today 11:133–142

Furge LL, Guenguerich FP (2006) Cytochrome P450 enzymes in drug metabolism and chemical toxicology: an introduction. Biochem Mol Biol Educ 34:66–74

Gan L, von Moltke LL, Trepanier LA, Harmatz JS, Greenblatt DJ, Court MH (2009) Role of NADPH-cytochrome P450 reductase and cytochrome-b5/NADH-b5 reductase in variability of CYP3A activity in human liver microsomes. Drug Metab Dispos 37:90–96

Gan J, Zhang H, Humphreys WG (2016) Drug-protein adducts: chemistry, mechanisms of toxicity, and methods of characterization. Chem Res Toxicol 29:2040–2057

Gibbons GF (2002) The role of cytochrome P450 in the regulation of cholesterol biosynthesis. Lipids 37:1163–1170

Guengerich FP (2001) Common and uncommon cytochrome P450 reactions related to metabolism and chemical toxicity. Chem Res Toxicol 14:611–650

Guengerich FP (2007) Mechanisms of cytochrome P450 substrate oxidation: MiniReview. J Biochem Mol Toxicol 21:163–168

Homolya L, Váradi A, Sarkadi B (2003) Multidrug resistance-associated proteins: export pumps for conjugates with glutathione, glucuronate or sulfate. Biofactors 17:103–114

Jennen DGJ, Gaj S, Giestbertz PJ, van Delft JHM, Evelo CT, Kleinjans JCS (2010) Biotransformation pathway maps in WikiPathways enable direct visualization of drug metabolism related expression changes. Drug Discov Today 15:851–858

Jenner P, Testa B, Di Carlo FJ (1981) Xenobiotic and endobiotic metabolizing enzymes: an overstretched discrimination? Trends Pharmacol Sci 2:135–137

Jones G, Prosser DE, Kaufmann M (2014) Cytochrome P450-mediated metabolism of vitamin D. J Lipid Res 55:13–31

Kolár M, Fanfrlík J, Hobza P (2011) Ligand conformational and solvation/desolvation free energy in protein-ligand complex formation. J Phys Chem B 115:4718–4724

Krämer SD, Testa B (2008) The biochemistry of drug metabolism – an introduction. Part 6. Inter-individual factors affecting drug metabolism. Chem Biodivers 5:2465–2573

Krämer SD, Testa B (2009) The biochemistry of drug metabolism – an introduction. Part 6. Intra-individual factors affecting drug metabolism. Chem Biodivers 6:1477–1660

Kurtovic J, Riordan SM, (2003) Paracetamol-induced hepatotoxicity at recommended dosage. Journal of Internal Medicine 253 (2):240–243

Kwon Y (2002) Handbook of essentials pharmacokinetics, pharmacodynamics, and drug metabolism for industrial scientists. Kluwer Academic Publishers, New York

Lauretti WJ (2012) In: Gatterman MI (ed) The safety and effectiveness of common treatments for whiplash. Elsevier Mosby, St. Louis

Lepesheva GI, Waterman MR (2011) Sterol 14alpha-demethylase (CYP51) as a therapeutic target for human trypanosomiasis and leishmaniasis. Curr Top Med Chem 11:2060–2071

Lepesheva GI, Friggeri L, Waterman MR (2018) CYP51 as drug targets for fungi and protozoan parasites: past, present and future. Parasitology 12:1–17

Lewis DF (2004) 57 varieties: the human cytochromes P450. Pharmacogenomics 5:305–318

Liston HL, Markowitz JS, DeVane CL (2001) Drug glucuronidation in clinical psychopharmacology. J Clin Psycopharmacol 21:500–515

Liu Y, Ramírez J, Ratain MJ, (2011) Inhibition of paracetamol glucuronidation by tyrosine kinase inhibitors. British Journal of Clinical Pharmacology 71 (6):917–920

Lynch T, Price A (2007) The effect of cytochrome P450 metabolism on drug response, interactions, and adverse effects. Am Fam Physician 76:391–396

Mahadevan SBK, McKiernan PJ, Davies P, Kelly DA (2006) Paracetamol induced hepatotoxicity. Arch Dis Child 91:598–603

Matias M, Canário C, Silvestre S, Falcao A, Alves G (2014) In: Wu J (ed) Cytochrome P450-mediated toxicity of therapeutic drugs. Nova Science Publishers, New York

Meunier B, Samuël P, de Visser, Shaik S (2004) Mechanism of Oxidation Reactions Catalyzed by Cytochrome P450 Enzymes. Chemical Reviews 104 (9):3947–3980

Nebert DW, Wikvall K, Miller WL (2013) Human cytochromes P450 in health and disease. Philos Trans R Soc Lond Ser B Biol Sci 368:20120431

Smith FC (2008) In: Pearson PG, Wienkers LC (eds) Pharmacokinetics of drug metabolites. Informa Healthcare, New York

Swanson HI, (2015) Drug Metabolism by the Host and Gut Microbiota: A Partnership or Rivalry?. Drug Metabolism and Disposition 43 (10):1499–1504

Talevi A (2016) The importance of bioactivation in computer-guided drug repositioning. Why the parent drug is not always enough. Curr Top Med Chem 16:2078–2087

Talevi A (2018) Drug repositioning: current approaches and their implications in the precision medicine era. Expert Review of Precision Medicine and Drug Development 3 (1):49–61

Tang W, Lu AY (2010) Metabolic bioactivation and drug-related adverse effects: current status and future directions from a pharmaceutical research perspective. Drug Metab Rev 42:225–249

Testa B, Krämer SD (2006) The biochemistry of drug metabolism – an introduction. Part 1. Principles and overview. Chem Biodivers 3:1053–1101

Testa B, Krämer SD (2007) The biochemistry of drug metabolism – an introduction. Part 2. Redox reactions and their enzymes. Chem Biodivers 4:257–405

Testa B, Krämer SD (2008) The biochemistry of drug metabolism – an introduction. Part 4. Reactions of conjugation and their enzymes. Chem Biodivers 5:2171–2336

Testa B, Pedretti A, Vistoli G (2012) Foundation review: reactions and enzymes in the metabolism of drugs and other xenobiotics. Drug Discov Today 17:549–560

Tu BP, Weissman JS (2004) Oxidative protein folding in eukaryotes: mechanisms and consequences. J Cell Biol 164:341–346

Walsh JS, Miwa GT (2011) Bioactivation of drugs: risk and drug design. Annu Rev Pharmacol Toxicol 51:145–167

Westphal C, Konkel A, Schunck WH (2015) Cytochrome p450 enzymes in the bioactivation of polyunsaturated fatty acids and their role in cardiovascular disease. Adv Exp Med Biol 851:151–187

Williams JA, Hyland R, Jones BC, Smith DA, Gurst S, Goosen TC, Peterkin V, Koup JR, Ball SE (2004) Drug-drug interactions for UDP-glucuronosyltransferase substrates: a pharmacokinetic explanation for typically observed low exposure (AUCi/AUC) ratios. Drug Metab Dispos 32:1201–1208

Xiang Q, Zhao X, Zhou Y, Duan JL, Cui YM (2010) Effect of CYP2D6, CYP3A5, and MDR1 genetic polymorphisms on the pharmacokinetics of risperidone and its active moiety. J Clin Pharmacol 50:659–666

Further Reading

Drug metabolism is a vast topic, and entire books have been written about it. Here, we intended to provide a summary of the more relevant points of drug biotransformation, but the chapter is far from exhaustive. For a much deeper insight on the topic, the reader is advised to Pearson and Wienkers's *Handbook of Drug Metabolism* (currently in its 3rd edition by CRC Press); the *Drug Metabolism Handbook* edited by Nassar, Hollenberg, and Scatina (2008, Wiley & Sons); and the extensive and unbelievably comprehensive series of articles by Testa and Krämer, many of which have been included in the reference list of the present chapter.

Drug Excretion

5

Alan Talevi and Carolina Leticia Bellera

5.1 Introduction

Together, drug metabolism and excretion constitute the drug elimination strategies of the body. They might occur in parallel (a fraction of the administered dose might be excreted as unchanged drug molecules, while, in simultaneous, the other fraction is subjected to metabolism prior to excretion), or, in some cases (i.e., lipophilic drugs), drug metabolism will be a prerequisite for drug excretion to take place (Fig. 5.1).

▶ **Important** Without a shadow of a doubt, urine and bile are the major excretion routes (for both xenobiotic compounds and physiologic waste products). Compounds that have been excreted in bile and are not reabsorbed will be ultimately excreted in feces. Both the kidneys and liver are organs that, through evolution, have become specialized in removal of waste products and xenobiotics.

A. Talevi (✉)
Laboratory of Bioactive Research and Development (LIDeB), Department of Biological Sciences, Faculty of Exact Sciences, University of La Plata (UNLP), La Plata, Buenos Aires, Argentina

Consejo Nacional de Investigaciones Científicas y Técnicas (CONICET), La Plata, Buenos Aires, Argentina
e-mail: atalevi@biol.unlp.edu.ar

C. L. Bellera
Medicinal Chemistry/Laboratory of Bioactive Research and Development (LIDeB), Faculty of Exact Sciences, Universidad Nacional de La Plata (UNLP), La Plata, Buenos Aires, Argentina

Consejo Nacional de Investigaciones Científicas y Técnicas (CONICET), Buenos Aires, Argentina

© Springer Nature Switzerland AG 2018
A. Talevi, P. A. M. Quiroga (eds.), *ADME Processes in Pharmaceutical Sciences*,
https://doi.org/10.1007/978-3-319-99593-9_5

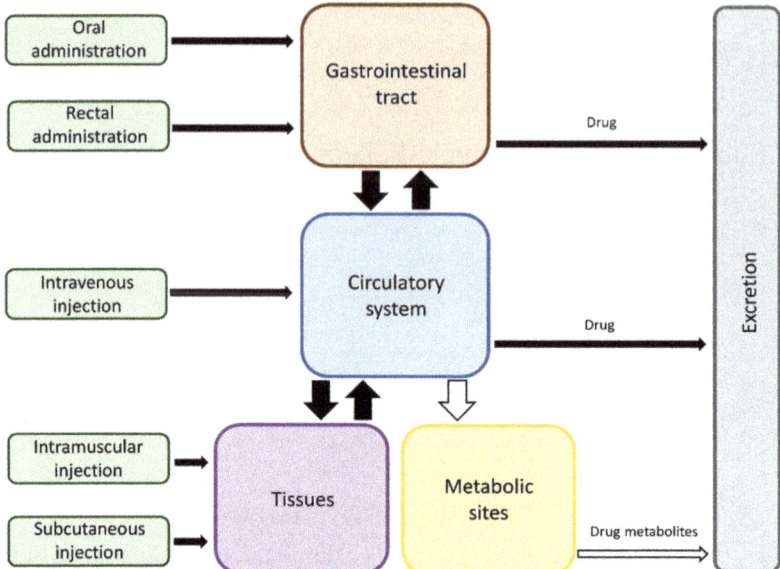

Fig. 5.1 Diagram representing different drug elimination routes, including direct intestinal and renal excretion and sequential metabolism and excretion

Secondary excretion routes include respiration, sweat, saliva, semen, hairs, milk, and tears (Testa and Krämer 2006; Kapusta 2008). In the case of poultry, drugs may also be excreted in eggs. Of course, many of these are physiologically unintended excretion ways, emerging from the fact that some drugs are extensively distributed and reach virtually every body fluid and tissue. However, they are also relevant for different reasons, such as undesired exposure to xenobiotics through farm and dairy products (Novaes et al. 2017) or breastfeeding (Sachs 2018) and bioanalytical (e.g., measuring drug bioavailability in saliva) (Ruiz et al. 2010) and forensic purposes (Franz et al. 2018).

Drug excretion depends on transport across various organs and tissues. Over the past decades, increasing knowledge has been generated on the role played by drug transporters (in particular, efflux transporters) in elimination and excretion processes. These systems will be opportunely (and deeply) discussed in Chap. 14, although we will briefly overview them here as well. We will dedicate a specific section to the presentation of the concept of clearance, which is central to all drug elimination phenomena, including drug metabolism and drug excretion. Next, the renal and intestinal routes of drug excretion will be specially emphasized.

5.2 Clearance

▶ **Definition** The most common definition of clearance (Cl) is a proportionality factor between the (instantaneous) rate of elimination of a drug from the entire body (*systemic clearance*) or an organ (*organ clearance*, i.e., hepatic or renal clearance) and its concentration in a reference body fluid, typically blood or plasma. The systemic plasma clearance is thus:

$$Cl = \frac{dQ_{el}(t)/dt}{C_p(t)} \tag{5.1}$$

where dQ_{el} represents the amount of drug eliminated in the body in a given infinitesimal time interval dt and C_p stands for the drug plasma concentration. Note that both the amount of drug eliminated and the drug plasma concentration are time-dependent. Also note that, if we are working in a postabsorptive phase (i.e., the whole dose has already been absorbed, and only elimination phenomena are impacting on the drug levels in the organism), Eq. 5.1 can be simply expressed as:

$$Cl = \frac{-dQ(t)/dt}{C_p(t)} \tag{5.2}$$

where Q is the amount of drug in the body and dQ/dt represents the instantaneous rate at which such quantity changes (due to elimination processes only!), which is naturally negative in the postabsorptive phase since drug levels can only be falling.

A physiologically more relevant definition of systemic clearance is the apparent volume of reference that is being cleared from drug per unit of time, that is, the apparent volume of reference fluid that contains the amount of drug that is being eliminated in the body per unit of time. The clearance terms depend on where the drug concentration is measured: blood (*blood clearance*), plasma (*plasma clearance*), and plasma water (*clearance based on unbound drug con*centration).

▶ **Important** Note that if we estimate the clearance in a given organ, if such organ eliminates drug by both metabolism and (unchanged drug) excretion, the estimated clearance will comprise both phenomena. Also note that if certain a drug molecule undergoes sequential biotransformation and excretion, the loss or elimination of such molecule will be only counted once (i.e., once the molecule has been biotransformed, it has already been eliminated; the excretion of the metabolite now counts as metabolite elimination or loss but not as unchanged drug loss). Put simply, a given molecule is not eliminated twice.

Let's return to Eq. 5.1. If elimination follows a first-order kinetics (which is usually the case in therapeutics, except for drugs with nonlinear kinetics), the instantaneous rate of elimination dQ_{el}/dt may be rewritten as $k_{el}Q$, which lead us to Eq. 5.3:

$$Cl = \frac{k_{el} \times Q(t)}{C_p(t)} \tag{5.3}$$

If we are working in the context of the one-compartment model, then Q equals $C_p \times V_d$ (V_d being the apparent volume of distribution); therefore, Eq. 5.3. can be transformed in:

$$Cl = k_{el} \times V_d \tag{5.4}$$

This is a very important result. It tells us that Cl is time-independent and, accordingly, constant if the elimination constant or the apparent volume of distribution of the person receiving the drug does not change (of course, both parameters can vary along time, but frequently they remain approximately constant during short time periods). The instantaneous rate of elimination changes continuously, but the apparent volume cleared from drug per unit of time remains constant no matter how much drug is within the body (as long as first-order elimination holds, i.e., no elimination system gets saturated).

Let's go once again to Eq. 5.1. After minimal Algebra:

$$Cl \times C_p(t) \times dt = dQ_{el} \tag{5.5}$$

Now we can integrate both sides of the equation from zero to infinity. According to Eq. 5.4, Cl is time-independent. Then:

$$Cl \int_0^\infty C_p(t)dt = \int_0^\infty dQ_{el} \tag{5.6}$$

The integral from zero to infinity of C_p dt is, by definition, the total area under the plasma concentration-time curve. On the other hand, if we integrate dQ_{el} from zero to infinity, we get the total amount of drug that has been eliminated (between zero and infinity). Since only what has been absorbed can be eliminated, then the total amount of eliminated drug must equal the total amount of absorbed drug, which can be expressed as the product of the absolute bioavailability F (fraction of the dose that gets absorbed) and the dose D, which leads us to a very important equation:

$$Cl \times AUC_{0-\infty} = F \times D \tag{5.7}$$

The total AUC is proportional to the amount of drug that reaches systemic circulation (*the quantitative component of drug bioavailability*)! And the proportionality constant is no other than the systemic clearance.

Expression 5.7 is also a good starting point to understand deviations from the expected total AUC due to nonlinear kinetics, but this is out of the scope of the present volume.

The unit of clearance is the same as flow rate, that is, volume per time unit. For instance, mL/min (or mL/min kg if normalized to body weight).

5.2.1 Organ Clearance

Organ clearance reflects the ability of a given eliminating organ to remove a drug from the blood and equals the apparent volume of blood cleared of drug per time

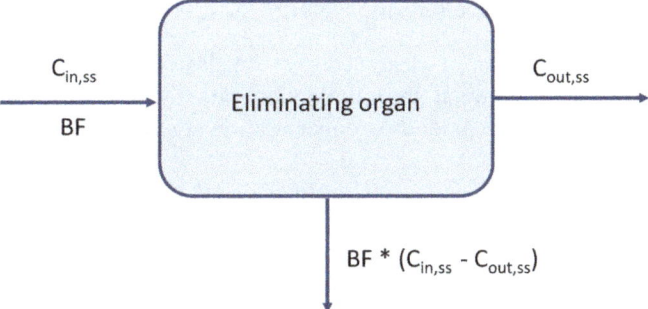

Fig. 5.2 Schematic representation of organ perfusion under steady state conditions: $C_{in,ss}$, drug concentration in blood entering the organ; $C_{out,ss}$, drug concentration in blood leaving the organ; and BF, eliminating organ blood flow

unit. It is important to underline that the reference fluid to measure organ clearance is the blood, not plasma (Kwon 2002). The notion of organ clearance can be best illustrated by considering elimination processes of a drug in a single organ using isolated organ perfusion under a steady state condition (Fig. 5.2). This physiologic approach to clearance allows predicting the effect of changes in the biological factors such as blood flow, the unbound fraction of the drug, or the activity of metabolizing enzymes (or drug transporters) on the elimination of a drug (Wilkinson and Shand 1975).

Among the various physiologic models characterizing organ clearance, the most widely applied is the venous equilibration (or well-stirred) model (Taft 2009). The model depicts an eliminating organ as a single well-stirred compartment through which the concentration of free drug in the existing blood is in equilibrium with the free drug within the organ. The model was originally used to characterize hepatic clearance but was subsequently applied to describe urinary excretion. Important assumptions for the well-stirred model include (Kwon 2002) as follows: (a) only unbound drug in blood is subject to elimination; (b) no membrane transport barrier; (c) no concentration gradient of the drug within the eliminating organ; (d) concentration of the drug within the organ equal to that in emergent venous blood; and (e) linear kinetics (in other words, instantaneous and complete mixing occurs within the organ, with no diffusion barrier).

Based on mass balance *at steady state*, the rate of elimination of a drug by the organ is equal to the difference between the input and output rates of the drug through the organ:

$$\text{Rate of elimination by the organ} = \text{Input rate} - \text{Output rate} \qquad (5.8)$$

The drug input rate to the organ can be expressed as the product of the eliminating organ blood flow (BF) and the drug concentration in the blood entering the organ at steady state, $C_{in,ss}$. Similarly, the drug output rate can be expressed as the product of BF and the concentration in the blood leaving the organ at steady state, $C_{out,ss}$. Then:

$$\text{Rate of elimination} = \text{BF} \left(C_{\text{in, ss}} - C_{\text{out, ss}} \right) \tag{5.9}$$

According to the definition of clearance, we can describe organ clearance as the proportionality factor between the elimination rate by the organ and the drug concentration in blood, i.e., inlet drug concentration at steady state:

$$\text{Organ clearance} = \frac{\text{BF} \left(C_{\text{in, ss}} - C_{\text{out, ss}} \right)}{C_{\text{in, ss}}} \tag{5.10}$$

The ratio $\frac{C_{\text{in, ss}} - C_{\text{out, ss}}}{C_{\text{in, ss}}}$ is called *the extraction ratio E*, representing the fraction of the amount of drug entering the organ that is extracted (eliminated) by the organ during perfusion. E is dimensionless and ranges from 0 to 1, with $E = 0$ meaning that the organ does not remove drug at all during perfusion, whereas $E = 1$ indicates complete elimination of the drug from the blood by the organ during perfusion.

▶ **Definition** On the basis of its extraction ratio by an organ, drugs can be classified as drugs with low ($E \leq 0.3$), intermediate ($0.3 < E < 0.7$), or high ($E \geq 0.7$) extraction, which allows predicting the dependence of organ clearance on the physiologic factors (BF, unbound fraction f_u, and intrinsic clearance Cl_{int}) (Taft 2009). Unbound fraction has already been discussed in Chap. 3. Cl_{int} represents the maximum organ capacity to clear the drug with no blood flow or binding limitations (Taft 2009; Duconge 2008).

Rowland et al. introduced organ blood flow and intrinsic clearance in their definition of the organ clearance of the unbound drug (Rowland et al. 1973); soon later, Wilkinson and Shand added the protein binding parameter to allow conversion for total drug clearance characterization (Wilkinson and Shand 1975). Organ clearance (with respect to blood concentrations) can then be expressed as:

$$\text{Organ clearance} = \frac{\text{BF} f_u \times \text{Cl}_{\text{int}}}{\text{BF} + f_u \, \text{Cl}_{\text{int}}} \tag{5.11}$$

Equation 5.11 can be simplified for drugs displaying either high (>0.7) or low (<0.3) extraction ratios. For the former (i.e., those with high $f_u \times \text{Cl}_{\text{int}}$), BF can be neglected in the denominator and thus:

$$\text{Organ clearance} = \text{BF} \tag{5.12}$$

These are considered perfusion- or flow-limited drugs, as the organ clearance will depend on the volume of blood entering the organ per time unit. The clearance of this kind of drugs is relatively independent of changes in the intrinsic clearance or the unbound fraction. Examples include propranolol or lidocaine (Duconge 2008). Whereas it is often assumed that only the unbound drug is available for extraction, there exist examples in which the avidity of the active processes taking place in the eliminating organ is sufficiently high (high intrinsic clearance) that drug molecules can be "stripped" from its binding sites in plasma proteins during passage through the organ (Wilkinson and Shand 1975; McElnay 1996). Two different types of extraction might then be defined:

▶ **Definition** *Restricted extraction* is limited to the circulating free drug; in contrast, in *nonrestrictive extraction*, both the bound drug and unbound drug are eliminated by the eliminating organ.

In contrast, for drugs with low extraction ratios, f_u Cl_{int} is the term neglected in the denominator of Eq. 5.11 and thus:

$$\text{Organ clearance} = f_u \times Cl_{int} \qquad (5.13)$$

Drugs in this second group are considered capacity-limited, because clearance is mainly determined by the enzymatic and/or secretory capacity of the eliminating organ. The clearance of this drug is highly dependent on the expression levels of those systems that participate in the elimination (which may vary, for instance, due to enzyme or transporter induction phenomena). It is also highly sensitive to drug-drug interactions. For these drugs, the eliminating organ will only be able to remove free drug from the blood during a single pass through the eliminating organ. This latter group could also be subclassified as binding sensitive ($\geq 90\%$ bound) or binding insensitive ($<90\%$ bound), depending on the percentage of drug bound to plasma proteins (for binding sensitive drugs, a change in the bound fraction will significantly impact on the drug clearance by the organ, since it would imply a high relative change in free drug levels and it is assumed that only free drug can be cleared). Examples of these drugs are phenytoin and warfarin (the reader might want to check Table 3.6 to refresh this concept).

5.3   Renal Excretion

The renal excretion of a drug comprises three relevant processes: *glomerular filtration*, *active secretion*, and *passive reabsorption*. Figure 5.3 displays a schematic representation of the nephron, the functional unit of the kidneys.

5.3.1   Glomerular Filtration

Urine formation begins with glomerular filtration (or *ultrafiltration*) that occurs in the *renal corpuscle* (Fig. 5.4), which consists of the *glomerulus* (a dense capillary network) and a glomerular capsule, the *Bowman's capsule*. The glomerulus receives its blood supply from an afferent arteriole of the renal arterial circulation. Unlike most capillary beds, the glomerular capillaries exit into efferent arterioles instead of venules. The resistance of the efferent arterioles results in enough hydrostatic pressure to provide the force for ultrafiltration. The filtering structure has three layers composed of (fenestrated) endothelial cells, a basement membrane, and podocytes.

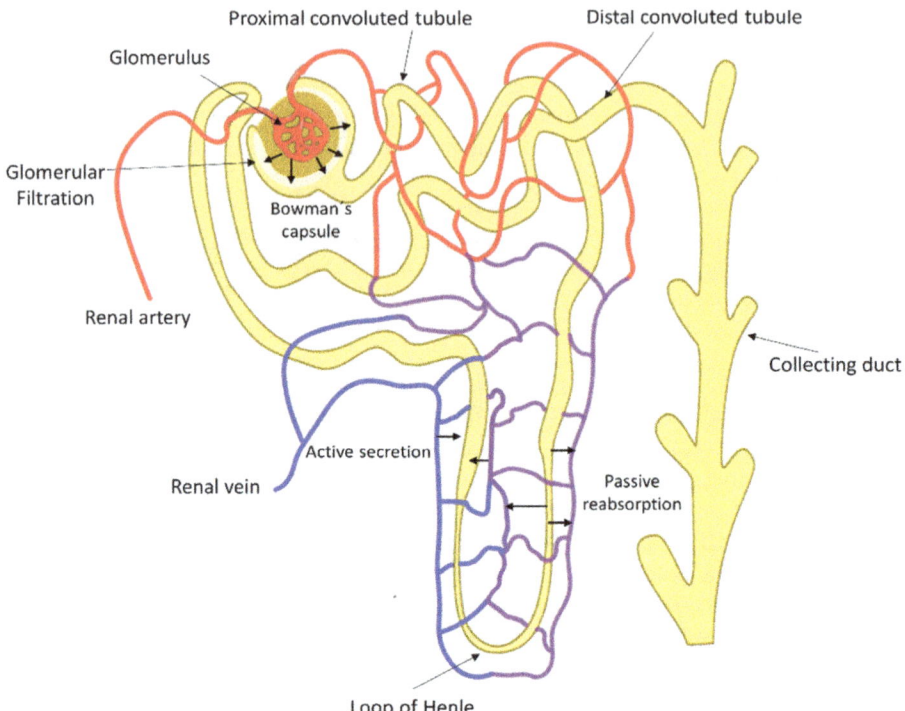

Fig. 5.3 Scheme of a nephron, indicating where do the processes impacting on drug excretion take place

Three important factors governing filtration are molecular size, molecular shape, and electrical charge (Haraldsson et al. 2008).

As molecular size increases, filterability becomes progressively smaller, with restricted filtration of molecules that are greater than 20 kDa in size and a molecular weight cutoff of around 60 kDa. Small molecules like the active pharmaceutical ingredient of conventional medications display nonrestricted passage through the glomerular barrier. However, the linear dimensions of a molecule are more directly relevant to its ability to move through a small pore. Size restriction is probably better reflected by the Stokes-Einstein radius of a solute. For globular proteins, for example, a hydrodynamic diameter below 5–6 nm is associated with the ability to be cleared rapidly from the body via renal filtration and urinary excretion (Choi et al. 2007).

▶ **Definition** The *Stokes-Einstein radius* of a solute is the radius of a rigid sphere that diffuses at the same rate as that solute. It factors in not only size but also solvent effects. A smaller ion with stronger hydration, due to solvation effects, for example, may have a greater Stokes radius than a larger ion with weaker hydration. The term

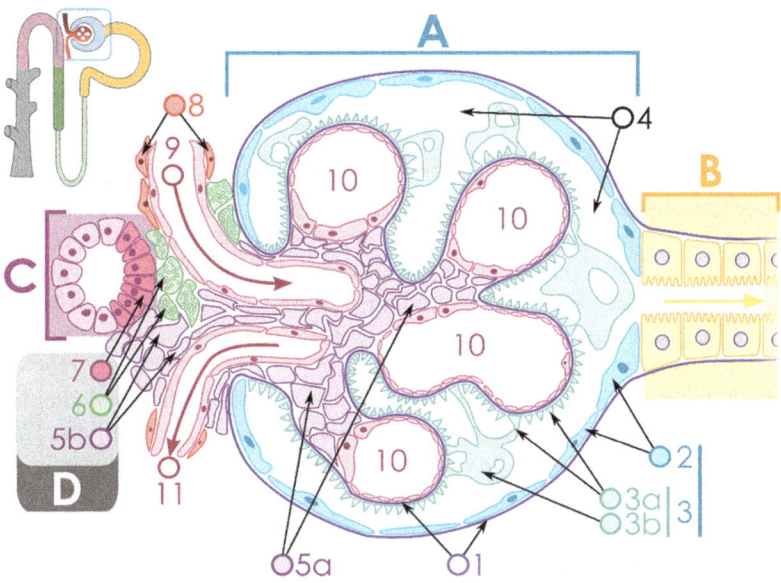

Fig. 5.4 Diagram of renal corpuscle structure: (**a**) Renal corpuscle. (**b**) Proximal tubule. (**c**) Distal convoluted tubule. (**d**) Juxtaglomerular apparatus. 1. Basement membrane (basal lamina). 2. Bowman's capsule – parietal layer. 3. Bowman's capsule – visceral layer. 3a. Pedicels (foot processes from podocytes). 3b. Podocyte. 4. Bowman's space (urinary space). 5a. Mesangium – intraglomerular cell. 5b. Mesangium – extraglomerular cell. 6. Granular cells (juxtaglomerular cells). 7. Macula densa. 8. Myocytes (smooth muscle). 9. Afferent arteriole. 10. Glomerulus capillaries. 11. Efferent arteriole. (The illustration belongs to Michał Komorniczak and is reproduced under Creative Commons)

hydrodynamic radius is often used as synonym of the *Stokes-Einstein radius* of a polymer or other macromolecule.

For neutral solutes with Stokes-Einstein radius in the 3–5 nM range, the sieving coefficient (the ratio of a specific solute concentration in the ultrafiltrate, removed only by a convective mechanism, divided by the mean plasma concentration in the filter) (Neri et al. 2016) is theoretically estimated to fall by one order of magnitude for every 1 nm increase in molecular radius (Haraldsson et al. 2008). For neutral solutes up to the Stokes-Einstein radius of 3 nM, the sieving coefficient only falls from 1.0 to 0.1. Increasing the molecular radius from 5 nM to 7 nM reduces the sieving coefficient by a factor of 5.

Electrical charge is the second variable determining filterability of macromolecules. For any given size (though seemingly more pronouncedly for Stokes-Einstein radius between 3 nm and 4 nm), negatively charged macromolecules are filtered to a lesser extent than neutral molecules, while positively charged molecules are filtered slightly faster than those neutral solutes of similar molecular weight. The charge selectivity probably results from the contributions of the endo-thelial cell surface layer (composed of negatively charged glycoproteins, glycosaminoglycans, and membrane-associated proteoglycans) and the podocytes (whose interdigitated foot processes are covered by a negatively charged glycocalyx facing the urinary space and containing sulfated molecules such as glycosaminoglycans and sialylated glycoconjugates) (Haraldsson et al. 2008; Resiser and Altintas 2016).

At last, molecular shape substantially affects solute transport across the glomerular barrier for solutes that are elongated or random coils, while minor deviations from a spherical shape have less influence on the rate of filtration (Haraldsson et al. 2008).

Filtration is a passive phenomenon, and, as a result from the previous restrictions, only unbound drug can be filtered. In contrast, filtration of plasma proteins such as albumin is negligible (as a result of its size, charge, and shape, its estimated sieving coefficient is usually in the 10^{-3}–10^{-4} order).

Glomerular filtration rate (GFR) is estimated as the clearance of a marker that can be filtered and has no (or negligible) tubular secretion and reabsorption. The most frequently used markers are creatinine and inulin. In adults, normal GFR is 100–125 mL/min, approximately 20% of the total renal blood flow.

5.3.2 Active Tubular Secretion

A second mechanism by which drug is extracted from the blood into the urine is proximal tubular secretion. It involves translocation of drug molecules from the plasma to the urine, which is mediated by membrane transporters which will be opportunely discussed in Chap. 14. Such transporters facilitate the transfer of drugs from the peritubular capillaries to the cytosol of the epithelial cells that line the tubules (i.e., blood to kidney drug transfer) and/or the movement of drug molecules from the cytosol of the tubular epithelium to the tubular lumen, where urine is being formed (i.e., kidney to urine drug transfer). Genetic polymorphisms or drug-drug interactions that affect the function of these transporters may explain differences in patient susceptibility to drug nephrotoxicity.

As explained in Sect. 5.2, drugs displaying high affinity for transporters involved in active tubular secretion may show high extraction ratios, and their elimination may become perfusion limited. An example is para-aminohippurate (PAH), which is used as a marker of renal plasma flow. The majority of excreted medications, however, are restrictively cleared.

Example

Many anticancer drugs (cisplatin, ifosfamide, methotrexate, etc.) are transported into tubular cells from the peritubular capillaries by organic anion and organic cation transporters and subsequently excreted into tubular lumens via multiple efflux transporters on the apical membrane of the epithelial cells (among them, multidrug resistance proteins and P-glycoprotein). Dysfunction or inhibition of such efflux drug transporters can result in abnormal drug accumulation within the tubular cells, resulting in drug-induced nephrotoxicity. For instance, the deposition of insoluble methotrexate crystals within the tubules can lead to crystalline nephropathy (Izzedine and Perazella 2017).

Nephrotoxic agents have been implicated as etiologic factors in 17–26% of inhospital acute kidney injury (Pazhayattil and Shirali 2014).

5.3.3 Tubular Reabsorption

Being mainly a passive process, tubular reabsorption of drug molecules occurs mostly at the distal tubule, where drug concentrations in the tubular lumen are high due to water reabsorption in the proximal tubule and, in occasions, following active secretion of the drug in the proximal tubule. Reabsorption also depends on the urine pH and the acid-base properties of the drug, which will impact on its degree of ionization (as always, the neutral molecular species will be more easily reabsorbed). On average, urinary pH is close to 6.3, but diet, drugs, and various disease states can alter urine pH. Under forced acidification and alkalization, urinary pH approximately ranges from 4.4 to 8.2 (Kwon 2002).

Whereas active secretion clearly contributes to renal excretion, reabsorption conspires against it.

5.3.4 Renal Clearance

Taking into consideration the previous three processes impacting on renal clearance, it may be computed as follows (Taft 2009):

$$\text{Renal clearance} = \left(\text{Cl}_{\text{filtration}} + \text{Cl}_{\text{secretion}} - \text{Cl}_{\text{reabsorption}}\right) \tag{5.14}$$

Filtration clearance can be estimated from the product of GFR and the unbound fraction of the drug, and the secretion clearance term may be calculated from the well-stirred model (Eq. 5.11). However, an accurate estimation of reabsorption is not easy to obtain. Reabsorption is best described in terms of the fraction of drug filtered and secreted that is reabsorbed. If we call such fraction R, the following general equation describing renal clearance can be written:

$$\text{Renal clearance} = \left[\text{GFR} \times f_{\text{u}} + \left(\frac{\text{BF}f_{\text{u}} \times \text{Cl}_{\text{int}}}{\text{BF} + f_{\text{u}}\,\text{Cl}_{\text{int}}}\right)\right] \times (1 - R) \tag{5.15}$$

If the drug is significantly metabolized at the kidneys, Eq. 5.15 can be expanded with a term Cl_{rm}, representing renal biotransformation:

$$\text{Renal clearance} = \left[\text{GFR} \times f_{\text{u}} + \left(\frac{\text{BF}f_{\text{u}} \times \text{Cl}_{\text{int}}}{\text{BF} + f_{\text{u}}\,\text{Cl}_{\text{int}}}\right)\right] \times (1 - R) + \text{Cl}_{\text{rm}} \tag{5.16}$$

Let's neglect Cl_{rm} and divide Eq. 5.15 by $\text{Cl}_{\text{filtration}}$. We will obtain the excretion ratio XR, which is simply the renal clearance corrected for filtration clearance:

$$\text{Renal clearance} = \left[1 + \left(\frac{\text{Cl}_{\text{secretion}}}{f_{\text{u}} \times \text{GFR}}\right)\right] \times (1 - R) \qquad (5.17)$$

XR provides a general indication of the mechanism of renal elimination for the compound of interest, and it can be useful for drug interaction assessment. An XR equal to 1 indicates net filtration of the drug (absence of secretion and reabsorption or compensation of secretion by reabsorption). XR > 1 denotes net secretion, and XR < 1 denotes net reabsorption.

Question

A patient is medicated with drug A, which has a XR of 2.2. After adding a medication B, the XR is measured again resulting in 1.5.

Discuss possible mechanisms behind A and B evident interaction.

5.4   Intestinal (Hepatobiliary) Excretion

Although generally identified with its primary role of drug metabolism, the liver also acts as an important excretory organ through biliary excretion. Hepatic clearance can be expressed as the sum of hepatic metabolic clearance (Cl_{hm}) and biliary clearance ($\text{Cl}_{\text{biliary}}$):

$$\text{Hepatic clearance} = \left(\text{Cl}_{\text{hm}} + \text{Cl}_{\text{biliary}}\right) \qquad (5.18)$$

Bile is formed in the canaliculus between adjacent hepatocytes following active secretion of bile acids and other components such as phospholipids and bilirubin across the canalicular membrane. These bile components are either synthesized by the liver or transported into it across the sinusoidal membrane. The resulting bile drains into the branches of intrahepatic bile ducts that later converge in the common hepatic bile duct. The bile plays an important role in the excretion of xenobiotics, including drugs and their metabolites.

The extraction ratio in the liver may be greatly boosted by the expression of drug transporters that contribute to the active uptake of drug molecules in the sinusoidal (basolateral) membrane. These are mostly members of the solute carrier (SLC) family, which will be discussed in detail in Chap. 14.

▶   **Important** The uptake transporters and the metabolizing enzymes work as highly articulated systems; the former help increase the rate at which xenobiotic molecules are presented to the biotransformation systems. The high expression levels of metabolizing enzymes in the liver would not be fully exploited if the diffusion into the hepatocyte became the limiting step of metabolic clearance. Sinusoidal uptake is then a determinant of net hepatic clearance: changes in it will affect overall hepatic clearance even for highly extracted drugs.

Once within the hepatocyte, drug molecules could be subjected to hepatic biotransformation or excreted to the bile. Note that, once metabolized, the compounds will have an enhanced polarity that would conspire against back diffusion into the blood or diffusion into the bile (this is especially true for Phase II metabolites). If not for the help of efflux transporters, such metabolites will tend to accumulate within the hepatocyte. Some efflux transporters are expressed in the sinusoidal membrane (transporting the metabolites back into the blood, so that they may be later excreted in urine), whereas others locate in the canalicular membrane, favoring biliary excretion.

Molecular size appears to be a general determinant of the final fate of xenobiotics or their metabolites, with the size threshold for biliary excretion varying among species. In humans, xenobiotics with molecular weights above 500 Da tend to have biliary excretion as primary excretion route, with renal excretion being the primary route of excretion for small molecules. A compound of intermediate molecular weight tends to be excreted in both the urine and the bile (Kwon 2002; Taft 2009). In any case, the ultimate molecular determinants to define if a compound returns to blood for subsequent renal excretion or goes to the bile for subsequent intestinal excretion (or both) is the affinity of such compound for the different transporters expressed in the sinusoidal and canalicular membranes. Figure 5.5 shows drug transport across the hepatocyte and its coordination with drug metabolism within the cell.

5.4.1 Enterohepatic Recycling

Enterohepatic recycling involves cyclic circulation of drug molecules resulting from the combined roles of the liver and the intestine in xenobiotic metabolism and excretion (Fig. 5.6). It begins with drug absorption across the gastrointestinal tract into the portal circulation, followed by uptake into the hepatocytes. Next, a fraction of the absorbed drug and/or its conjugated metabolites are secreted into the bile, returning to the intestine. Phase II metabolites (e.g., glucuronides) are poorly reabsorbed in the gut due to their high polarity; thus, one would expect them to be excreted in feces. However, gut microbiota produce enzymes capable of deconjugating (hydrolyzing) drug conjugates (e.g., β-glucuronidase), releasing the parent molecule, which can be (re)absorbed across the intestine wall and taken to the liver once again, restarting the cycle.

Thus, for drugs undergoing biliary excretion, enterohepatic recycling represents a secondary absorption phase for drug, prolonging the drug elimination half-life (in comparison to a hypothetical situation in which the microbiota were not able to perform deconjugation) and may also produce multiple peaks in the plasma concentration-time profile of the drug (Fig. 5.7) (Roberts et al. 2002). Many drugs undergo enterohepatic recycling, among them some antibiotics, many hormones,

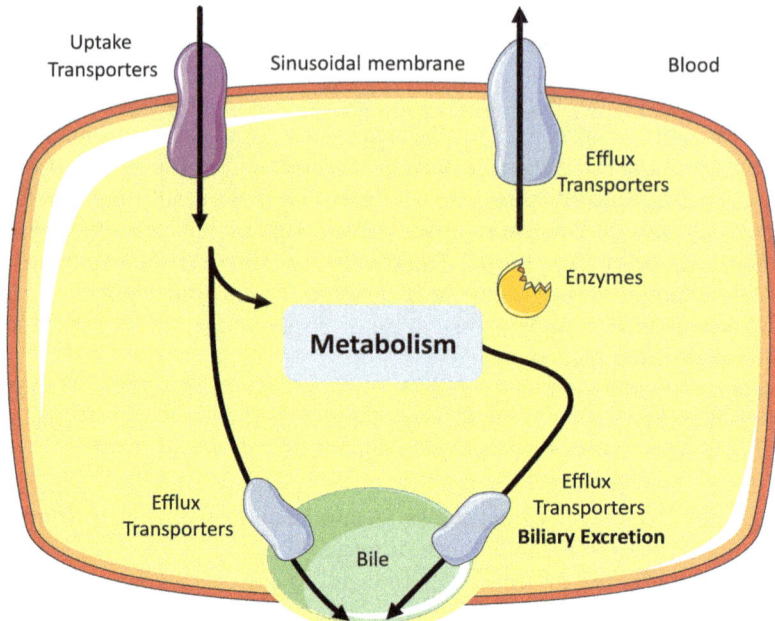

Fig. 5.5 Diagram depicting mechanisms of hepatobiliary clearance. Drug can be uptaken into hepatocytes across the sinusoidal membrane via transporters and/or passive diffusion, followed by metabolism and/or biliary excretion. Return of drug molecules or metabolites to the blood (for subsequent renal elimination) is also possible, with a prominent role of efflux transporters

opioids, and warfarin. Such agents might be subjected to clinically relevant drug-drug interactions involving oral antibiotic therapy, due to the effects of some antibiotics on the gut microbiota, which may impact on their capacity to metabolize drug conjugates. For instance, some oral antibiotic therapies are thought to limit the efficacy of oral contraceptives. The mechanism of this interaction is interruption of enterohepatic circulation by eliminating gut flora responsible for enzymatic deconjugation of contraceptives. Consequently, women taking oral contraceptives may be at increased risk of becoming pregnant. Although other mechanisms explaining increased risk of contraceptive failure due to interaction with antibiotics are possible (e.g., enzyme induction), a second contraceptive method should be recommended whenever antibiotics are prescribed.

Example

An 18-year-old woman with chronic granulomatous disease underwent a staphylococcal skin infection. She received semisynthetic penicillin (500 mg every 6 h for 6 weeks), and her skin infection was remarkably improved. However, she became pregnant in the meanwhile despite being on oral contraceptives. Careful review of her contraceptive package and antibiotic prescription suggested that she

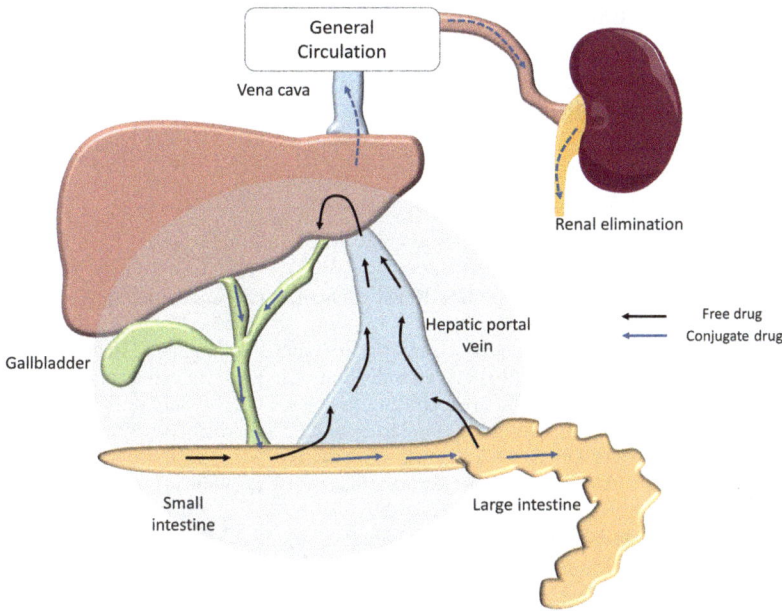

Fig. 5.6 Diagram of enterohepatic circulation

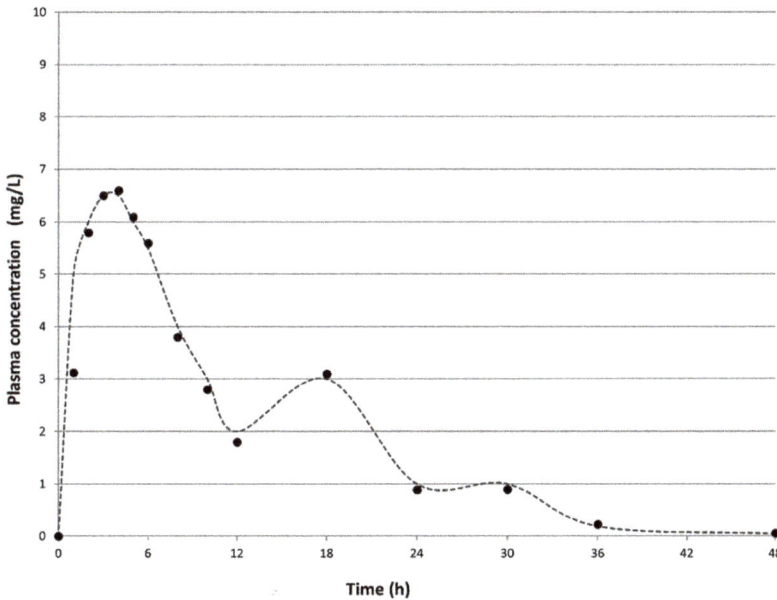

Fig. 5.7 Example of concentration-time profile for an orally administered drug undergoing enterohepatic circulation (note the multiple peaks of the profile)

had complied faithfully with both medications. She decided to undergo an abortion. The physicians considered interruption of enterohepatic circulation as a possible explanation for the unplanned pregnancy.

This case is taken from Silber (1983).

References

Choi HS, Liu W, Misra P et al (2007) Renal clearance of nanoparticles. Nat Biotechnol 25:1165–1170

Duconge J (2008) Applying organ clearance concepts in a clinical setting. Am J Pharm Educ 72:121

Franz T, Scheufler F, Stein K et al (2018) Determination of hydroxy metabolites of cocaine from hair samples and comparison with street cocaine samples. Forensic Sci Int 288:223–226

Haraldsson B, Nyström J, Deen WM (2008) Properties of the glomerular barrier and mechanisms of proteinuria. Physiol Rev 88:451–487

Izzedine H, Perazella MA (2017) Anticancer drug-induced acute kidney injury. Kidney Int Rep 2:504–514

Kapusta D (2008) In: Enna SJ, Bylund DB (eds) Drug excretion. Elsevier, Amsterdam

Kwon Y (2002) Handbook of essential pharmacokinetics, pharmacodynamics and drug metabolism for industrial scientists. Kluwer Academic Publishers, New York

McElnay JC (1996) Drug interactions at plasma and tissue binding sites. In: D'Arcy PF, McElnay JC, Welling PG (eds) Handbook of experimental pharmacology. Springer, Berlin

Neri M, Villa G, Garzotto F et al (2016) Nomenclature for renal replacement therapy in acute kidney injury: basic principles. Crit Care 20:318

Novaes SF, Schreiner LL, Pereira e Silva I et al (2017) Residues of veterinary drugs in milk in Brazil. Cienc Rural 47:e20170215

Pazhayattil GS, Shirali AC (2014) Drug-induced impairment of renal function. Int J Nephrol Renovasc Dis 7:457–468

Resiser J, Altintas MM (2016) Podocytes [version 1; referees: 2 approved]. F1000Research 5: 114

Roberts MS, Magnusson BM, Burczynski FJ et al (2002) Enterohepatic circulation: physiological, pharmacokinetic and clinical implications. Clin Pharmacokinet 41:751–790

Rowland M, Benet LZ, Graham GG (1973) Clearance concepts in pharmacokinetics. J Pharmacokinet Biopharm 1:123–135

Ruiz ME, Conforti P, al FP (2010) The use of saliva as a biological fluid in relative bioavailability studies: comparison and correlation with plasma results. Biopharm Drug Dispos 31:476–485

Sachs HC (2018) The transfer of drugs and therapeutics into human breast milk: an update on selected topics. Pediatrics 132:e796–e809

Silber TJ (1983) Apparent antibiotic oral contraceptive administration. J Adolesc Health Care 4:287–289

Taft DR (2009) Drug excretion. In: Hacker M, Bachmann K, Messer W (eds).Academic Press, Burlington

Testa B, Krämer SD (2006) The biochemistry of drug metabolism – an introduction. Part 1. Principles and overview. Chem Biodivers 3:1053–1101

Wilkinson GR, Shand DG (1975) Commentary: a physiological approach to hepatic drug clearance. Clin Pharmacol Ther 18:377–390

Further Reading

The reader is advised to complement this chapter with Chap. 14 of this volume

Routes of Drug Administration

6

María Esperanza Ruiz and Sebastián Scioli Montoto

6.1 Introduction

The route of administration of a medication directly affects the drug bioavailability, which determines both the start and the duration of the pharmacological effect. Some considerations must be taken into account when designing a drug dosage form: (1) the intended route of administration; (2) the amount or dose to be administered; (3) the anatomical and physiological characteristics of the site of absorption, such as membrane permeability and blood flow; (4) the physicochemical properties of the site, such as pH and osmotic pressure of physiological fluids; and (5) the potential effect of the medication over the site of administration.

When the systemic absorption of a drug is desired, medications are usually administered by two main routes: the *parenteral route* (through the skin by injection, avoiding the digestive system) and the *enteral route* (directly at some point of the gastrointestinal tract). To a lesser extent, the pulmonary (or respiratory) and nasal routes are employed. Other routes of administration, such as ophthalmic and vaginal, are not included here because their application is almost exclusive for local (not systemic) drug administration.

M. E. Ruiz (✉)
Quality Control of Pharmaceutical Products/Laboratory of Bioactive Research
and Development (LIDeB), College of Exact Sciences, National University of La Plata (UNLP),
La Plata, Buenos Aires, Argentina

Argentinean National Council of Scientific and Technical Research (CONICET), La Plata,
Buenos Aires, Argentina
e-mail: eruiz@biol.unlp.edu.ar

S. Scioli Montoto
Quality Control of Pharmaceutical Products/Laboratory of Bioactive Research
and Development (LIDeB), College of Exact Sciences, National University of La Plata (UNLP),
La Plata, Buenos Aires, Argentina

© Springer Nature Switzerland AG 2018
A. Talevi, P. A. M. Quiroga (eds.), *ADME Processes in Pharmaceutical Sciences*,
https://doi.org/10.1007/978-3-319-99593-9_6

6.2   Parenteral Routes of Administration

Parenteral drug administration is carried out directly through the skin, in or towards systemic circulation. It is the route of choice for drugs that cannot be absorbed orally and/or that are unstable in the gastrointestinal tract (e.g. insulin, heparin). These routes of administration are also used for the treatment of unconscious patients or under circumstances that require a rapid onset of action.

Parenteral routes of administration exhibit higher bioavailability than other routes and are not subjected neither to first-pass metabolism nor to the sometimes extreme conditions of the gastrointestinal environment, while offering the greatest control over the real drug amount that accesses systemic circulation. As main drawbacks, drug administration by these routes is irreversible and can cause fear, pain, tissue damage and/or infections (Florez 1998).

Parenteral administration can be performed by injection (small volumes), infusion (large volumes) or implant, and while its typical goal is to achieve systemic effects, it can also be used locally on a specific organ or tissue by injecting the pharmaceutical active ingredient directly on the site of action, in order to minimize systemic adverse effects (Øie and Benet 2002).

The three main parenteral routes are intravenous (IV), intramuscular (IM) and subcutaneous (SC).

6.2.1   Intravenous (IV)

Administering a drug through the IV route involves the introduction of a drug solution through a needle, directly into a vein. It is the best way to deliver a dose rapidly and accurately, as the drug enters directly into systemic circulation without the delay associated to absorption processes, achieving its therapeutic effect faster than by any other route. For the same reason, this route presents a bioavailability of 100%, since the pharmaceutical active ingredient usually reaches the site of action without suffering alterations due to pre-systemic effects.

There are three main methods to administer medications by IV route:

▶ **Definition** *Fast IV injection.* Also called 'IV bolus', is the administration of a single dose by direct injection into a vein, so it only supports small volumes (smaller than 10 mL). After an IV bolus injection, the drug diluted in the venous system reaches the heart, is pumped to the lung and is distributed to the rest of the body by the arterial system. According to the fraction of the arterial blood flow that reaches the site of action, therapeutic effects can be observed even 20–40 s after injection, and thus this is a very useful route for emergencies and pain management (Saxen 2016).

Slow IV infusion. It consists in the administration of the medication into a vein, during a prolonged period of time (large solution volumes). The amount of drug administered in a certain period is determined by the rate of infusion, and the

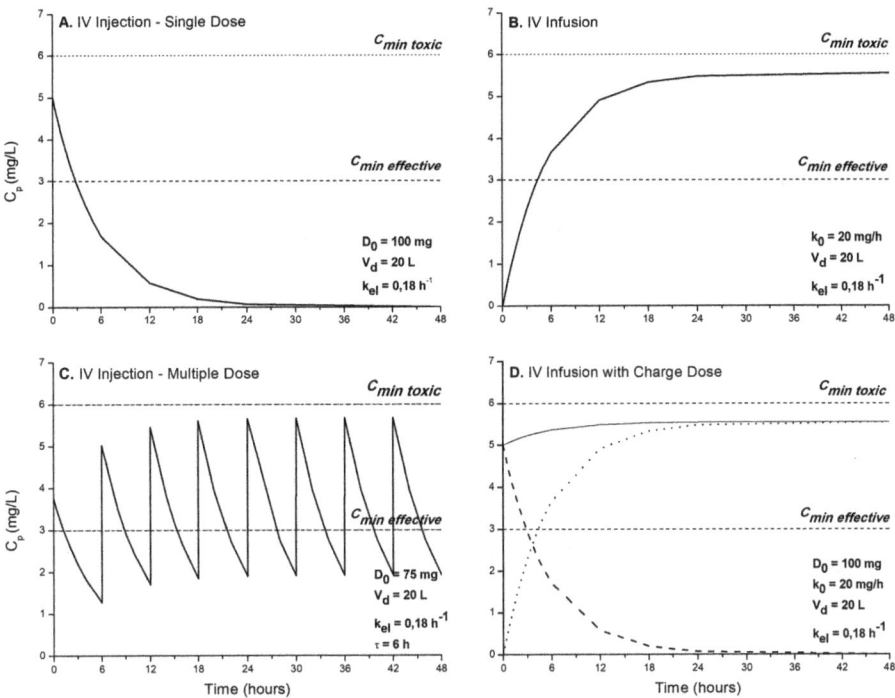

Fig. 6.1 Simulated pharmacokinetics profiles for (**a**) fast IV injection of a single dose; (**b**) slow IV infusion; (**c**) fast IV injection of multiple doses; (**d**) slow IV infusion one loading dose. For simulation, it was considered a single-compartment model and a drug with an apparent volume of distribution (V_d) of 25 L and an elimination constant (k_e) of 0.17 h^{-1}. C_p, plasma concentration (mg/L); D_0, dose (mg); k_0, infusion rate (mg/h); τ, dosing interval (h); $C_{\text{min toxic}}$, minimum concentration with toxic effects (mg/L); $C_{\text{min effective}}$, minimum concentration with therapeutic effect (mg/L)

medication enters the body by gravity or by an infusion pump, forcing the solution to pass through a plastic catheter inserted in a vein.

Although for patient comfort veins of the forearms (basal and middle cephalic veins) or in the wrists (cephalic accessory and medial antebrachial) are the most common sites, IV administration can be performed in any superficial vein. In certain cases, it may be necessary to resort to a central IV route, which consists in a catheter located at the outlet of the superior vena cava (right atrium). This is required when it is not possible to channel peripheral routes, for prolonged treatments or when hypertonic solutions must be administered (Boylan and Nail 2002).

6.2.1.1 Pharmacokinetic Characteristics of IV Administration

An example of a typical pharmacokinetic profile (drug plasma concentration vs. time) obtained after an IV bolus is shown in Fig. 6.1a. After the maximum concentration peak, an exponential decay is observed, and the drug concentration rapidly falls to subtherapeutic levels (i.e. below the minimum effective concentration). This is the reason why when longer times of effect are desired, slow IV infusion is used, and constant plasma drug levels are achieved (Fig. 6.1b). Additionally, slow infusions do not exhibit significant fluctuations of serum drug concentrations, as indeed occur with repeated IV bolus or IM administrations (Fig. 6.1c).

In the case of drugs with long systemic half-life, for which the initiation of the therapeutic effect by IV infusion could be slow, it is common to administer a loading dose by IV injection to rapidly achieve the therapeutic level, after which systemic concentrations are maintained within the therapeutic range by controlling the rate of infusion (Fig. 6.1d).

It should be noted that as IV injection allows obtaining high systemic concentrations in a very short time, highly vascularized and perfused organs (such as the heart, lungs, liver and kidneys) are also subjected to high concentrations in a short time, which may produce unwanted side effects. For this reason, this type of administration must be done slowly, monitoring for possible signs of intoxication. IV phenytoin solution, for example, must be administered at a speed of less than 50 mg/min, since faster administrations may cause hypotension, cardiovascular collapse or CNS depression (FDA and CDER 2014).

6.2.1.2 Physicochemical and Technological Considerations for IV Drug Administration

The IV route of administration is only applicable to aqueous solutions, since suspensions and oily solutions carry the risk of embolism and/or thrombophlebitis. Special care must be taken when selecting the vehicle used to solubilize the drug, since even in the case of aqueous solutions, precipitation of a drug that is poorly soluble could occur during injection, especially in those cases where the drug was solubilized using co-solvents or surfactants. A well-formulated parenteral solution is that which, besides having the required drug dose completely dissolved in the vehicle, allows mixing the drug into circulation, without risk of precipitation. This represents another reason to always perform injections in a slow manner.

On the other hand, there are vehicles capable of producing undesired effects in particular populations: phenobarbital, for example, is usually prepared in solutions with variable proportions of propylene glycol (PG), a solvent that can produce hyper osmolarity in children. Additionally, since the metabolic route responsible of metabolizing PG is not completely developed in children under the age of 4 years, repeated IV injections containing PG could generate toxicity in the paediatric population (Lim et al. 2014).

Regarding the formulation pH, it is desirable to formulate medications at physiological-like pH values. If there exist stability and/or solubility reasons not to do so, and a more extreme pH value is needed, the injected volume and the speed of injection should be carefully monitored, since injection may be accompanied by local pain and/or irritation. Injectable phenytoin is formulated at a pH of 10–12 because of the low solubility of the drug at lower pH values. Therefore, precautions should be taken during IV injection of phenytoin to avoid extravasation of the solution and consequent tissue damage, ranging from a simple local irritation to tissue necrosis and ulceration (Douglass 2018).

In the case of solutions that have a higher osmolarity than physiological fluids (anaesthetics, diuretics, parental nutrition, etc.), it is recommended to administer them in large calibre veins or even through a central IV route, to achieve a rapid access to the heart and the consequent dilution of the solution in a larger volume.

Prolonged time infusions as well as the use of highly irritating products may injure the vascular wall and produce venous thrombosis. Consequently, this route is reserved for cases of necessity, and for its use, maximum precautions of asepsis and a rigorous control of the injection technique are imposed.

6.2.2 Intramuscular (IM)

It consists on the injection of the medication into the muscle tissue, which can be done in different areas (Boylan and Nail 2002):

- Upper part of the arm: deltoid muscle admits approximately 2 mL. While it may result painful for the patient, this area generates the higher rate of absorption.
- Glutes: dorsogluteal muscle is the zone that admits higher volumes (7–8 mL approximately) although with lower rate of absorption due to the higher amount of adipose tissue.
- External thigh face: vastus lateralis muscle admits around 5 mL. It is the recommended zone for babies and children, since due to its minor muscle development, gluteal zone carries a high risk of nerve damage.

After IM injection, drug must be absorbed to reach the bloodstream, so there is a delay until the beginning of the therapeutic effect. Figure 6.2 comparatively shows plasma concentration profiles vs. time obtained after the administration of a drug through IV, IM and oral route. As previously stated, after an IV injection, the drug reaches maximum plasma concentration almost instantly, so the peak is often not observed, and instead a direct exponential decay is seen.

IM and oral routes present a defined phase of absorption, in which drug concentration raises slowly to a maximum and then decreases according to its elimination half-life. The time corresponding to the maximum concentration (time of maximum effect, t_{max}) is limited by the speed at which the release and absorption processes take place, so in general $t_{max\ oral} > t_{max\ IM}$, due to the greater complexity of the absorption process from the gastrointestinal tract.

Since the systemic elimination of a drug is essentially similar for the different routes of administration, only the speed and magnitude of absorption can be modified by the formulation. In the example of Fig. 6.2, the area under the curve (AUC) values is approximately equal in all three cases, indicating that the oral and IM preparations of the example are 100% bioavailable. In a real scenario, however, it

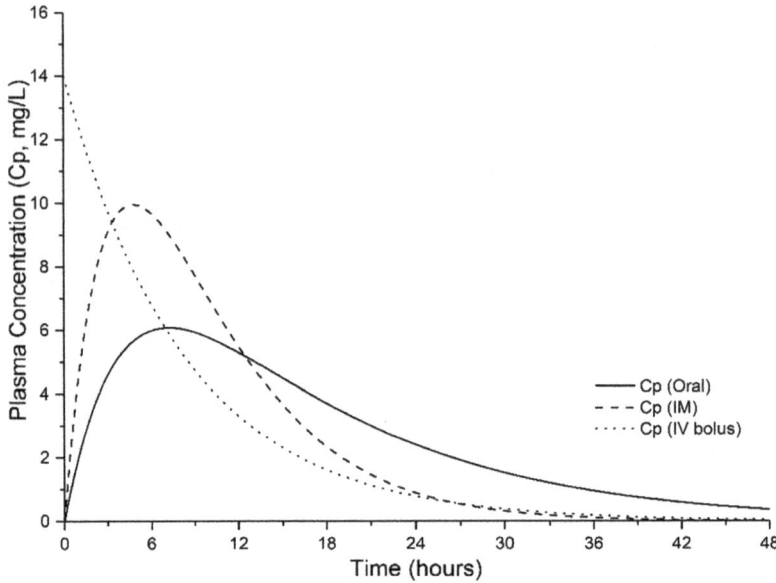

Fig. 6.2 Typical plasma concentration vs. time profiles for the same drug by three different routes: IV bolus (long dashes), IM (dotted line) and oral (continuous line)

is more common that $AUC_{oral} < AUC_{IM} < AUC_{IV}$, due to incomplete absorption or pre-systemic metabolism.

It is important to highlight that, even though the administration of drugs by IM route is usually safer than IV route, errors during IM injection can lead to blood clots, abscesses, scars and even nerve damage and paralysis (e.g. sciatic nerve) (Mishra and Stringer 2010).

6.2.2.1 Physiological Factors that Modify the IM Absorption of Drugs

After the IM administration of a pharmaceutical product, drug absorption occurs by diffusion from the muscle to the surrounding fluid and from there to the blood. Different muscular tissues have different blood supply: e.g. the blood flow to the deltoid muscle is higher than the gluteus muscle. Muscle blood flow also increases during exercise or with fever. In contrast, very low blood pressure is accompanied by poor muscle flow and capillary closure, compromising the drug absorption process. The presence of adipose tissue also contributes to slow down the absorption process, so it is common to observe unusual absorption profiles after IM injection to obese patients. The pharmacokinetic behaviour of the IM route may also be altered in neonates and preterm infants, as well as in pregnant women and elderly people.

6.2.2.2 Technological Factors that Modify the IM Absorption of Drugs

By changing the preparation vehicle, IM injections can be formulated to release the drug in a slow or rapid manner. Aqueous solutions are the base of IM immediate release preparations, since the drug is rapidly absorbed from the injection site. An example is lyophilized solids which are dissolved in an aqueous vehicle immediately prior to administration. The onset of the effect usually takes around 30 min, and its duration depends on the drug's half-life.

In the case of ionizable and poorly soluble compounds, it is always convenient to formulate them in buffer solutions of a pH close to the physiological; otherwise, they could precipitate at the injection site. The injectable solution of phenytoin, for example, is usually administered intravenously not only because its formulation pH (close to 12) is extremely irritating to the muscle but also because the low pH value in muscles and the lack of dilution (in contrast to the IV route) may cause the precipitation of the drug, which is then slowly redissolved, generating a slow and erratic absorption process (Fontes Ribeiro 2005).

On the other hand, drugs with very low aqueous solubility must be dissolved in other solvents like PG or mineral oils, whose viscosity delays the drug diffusion to the bloodstream. Additionally, the drug must be partitioned between the carrier and the physiologic aqueous environment for its absorption, which contributes to the relatively slow and sustained release profiles typically obtained with these vehicles.

All the aforementioned characteristics have been exploited for the design of prolonged action IM preparations that allow time intervals of several hours, days or even weeks between doses. Such preparations comprise both in oily solutions (e. g. oestradiol, testosterone) and aqueous suspensions forms (e.g. penicillin G procaine, methylprednisolone). Moreover, some drugs, including peptides and proteins, have been formulated as emulsions, suspensions, liposomes and even nanoparticles for IM injection, in order to achieve adequate pharmacokinetic profiles for each active ingredient (Hwang et al. 2016; Xie et al. 2018).

In general, hypertonic solutions are contraindicated by this route (as by subcutaneous administration) and should be administered by IV route with the already mentioned precautions.

6.2.3   Subcutaneous (SC)

A scheme of the different sites for parenteral drug administration is presented in Fig. 6.3. SC administration of drugs consists in injecting them under the skin into the adipose layer beneath the dermis, which is why it has also been called hypodermic administration. It is usually performed on the external side of the arm or thigh, or on the anterior face of the abdomen, and generally admits smaller injection volumes than IM route.

6.2.3.1 The Subcutaneous Absorption Process

In the case of small drug molecules (<1 kDa), the absorption occurs through the vascular endothelium of blood capillaries owing to their high permeability as well as the high fluid filtration-reabsorption rate at the capillaries (around 20–40 L/day, in

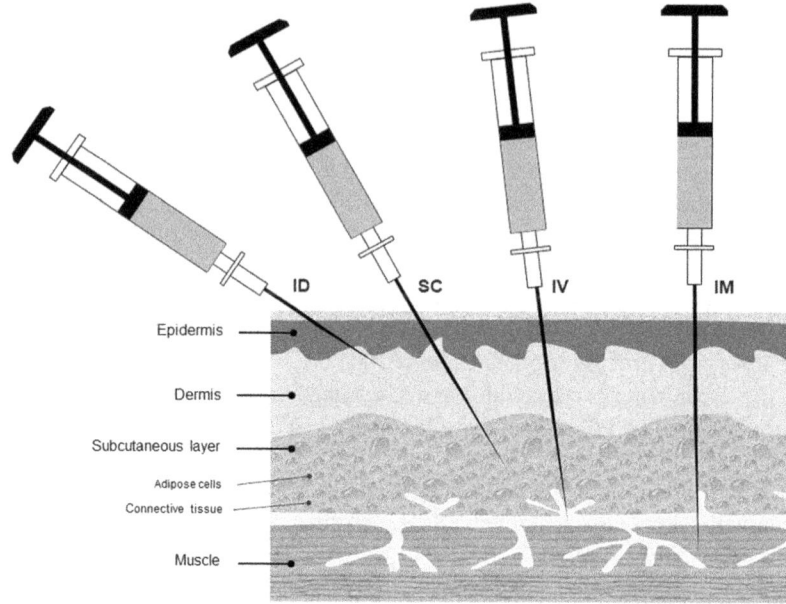

Fig. 6.3 Scheme of the different sites of parenteral administration of drugs. ID, intradermic; SC, subcutaneous; IV, intravenous; IM, intramuscular

comparison with approximately 2–4 L/day drained by the lymph). However, in the case of macromolecules and small particles (<100 nm approximately) for which the capillary endothelium results impermeable, the absorption is mainly attributed to lymphatic vessels, which provide an alternative route of absorption from the interstitial space (McLennan et al. 2005).

Another factor that must be taken into account is the blood flow at the administration site, which often limits the absorption rate. Since the SC tissue is less irrigated than the muscle, generally slower absorption kinetics are obtained by SC route in comparison with the IM route, even though the rate is usually higher than orally. The absorption rate of a drug can be reduced by vasoconstriction, by means of the local application of cold or the administration of a vasoconstrictor agent and may be accelerated by vasodilation and increased blood flow, as in the case of heat application, massage or exercise. For example, a small amount of adrenaline is sometimes combined with the anaesthetic drug lidocaine to restrict its distribution. Adrenaline acts as a local vasoconstrictor and avoids the migration of the anaesthetic from its SC site of administration (Gacto et al. 2009).

In normal blood flow conditions, drug absorption is produced mainly due to passive diffusion processes, since active transport mechanisms are no relevant for the SC route. The limiting factors of the absorption will be, therefore, the factors that influence diffusion: the area of the absorbent capillary membrane, the drug diffusion coefficient, the concentration gradient and the diffusion distance (capillary membrane thickness). Other factors that may limit the extent of drug absorption from the interstitial space include susceptibility to enzymatic degradation at the site of injection, the cell uptake by endocytosis, phagocytic mechanism or potential precipitation and/or aggregation of the drug.

6.2.3.2 Formulations for Subcutaneous Administration

As for the IM route, the injected solutions must be neutral or isotonic, otherwise they result irritating and may cause pain and necrosis. Unlike the IM route, though, it is not advisable to inject oily solutions by SC route, as they can become clogged and cause and sterile abscess.

One of the most interesting features of this route of administration is the forms of SC depot and continuous SC infusion, by which the drug is released slowly, maintaining stable blood levels for a prolonged time.

Among the best-known SC implants are contraceptives devices containing hormones such as levonorgestrel or etonogestrel, which consist in a tiny bar of around 40 mm long and 2 mm wide of a non-biodegradable plastic material that acts as a controlled release membrane. These devices are implanted surgically (prior local anaesthesia) in the inner part of the arm, from where they release the drug throughout a given period (3–5 years depending on the device), after which they must be removed (FDA 2016).

On the other hand, systems for constant SC infusion, such as insulin pumps or morphine infusers, allow small volumes of solution to be introduced at a very slow rate in the subcutaneous tissue.

There are several types of insulin pumps, but in general they consist in mechanic, small and portable devices electrically controlled (the patient carries them), which contain an insulin reservoir, connected to a cannula inserted subcutaneously, usually in the abdomen. These systems allow insulin release during 3 or 4 days, thus reducing the variability associated to injections. Over the years, more and better models have been designed, and nowadays there are programmable devices that allow a basal infusion adjustable to the patient schedule (working days, weekends) and insulin bolus on demand, to cover food ingestion (Partridge et al. 2016).

Finally, elastomeric infusion systems or infusers are lightweight devices, consisting of a cylindrical plastic container, within which there is a balloon or an elastomeric reservoir. The formulation to be infused is introduced in this balloon causing it to swell. The distended balloon exerts a constant pressure and expels the contents through a tube (at typical speed of 0.5–2 mL/h) connected to a catheter previously implanted in the patient's SC tissue. These systems are mainly used for the administration, both ambulatory and during hospital stays, of analgesic drugs in cancer patients under palliative care or in pain therapy (Lucendo Villarín and Noci Belda 2004; Mohseni and Ebneshahidi 2014).

6.2.4   Specialized Parenteral Routes

There are situations in which direct administration on a given region, tissue or organ may be needed, in order to achieve high drug concentrations in the site of action, almost immediately (Desai et al. 2007). Table 6.1 presents a summary of these specialized parenteral routes.

Table 6.1 Specialized parenteral routes of administration of medications

Route	Definition/characteristics
Intraspinal	Epidural: administration on or over the dura mater. The drug must be filtered through fat and veins to reach the nervous roots, thus delaying the beginning of the effect. It supports permanent catheter collation Intrathecal: direct administration in the cerebrospinal fluid of the subarachnoid space (between pia mater and arachnoid membranes). Immediate effect over nervous roots. It does not support permanent catheter These routes are used primarily for anaesthesia (e.g. lidocaine, bupivacaine) and pain management (e.g. opioid analgesic)
Intracerebroventricular	Direct administration to the cerebral ventricles. This route, like the previous one, is an option to the IV route in the case of CNS active drugs that are unable to cross the blood-brain barrier, when high, localized concentrations are required (e.g. antibiotics, chemotherapy of gliomas)
Intraarterial	Direct administration to an artery, generally for local effects over irrigated organs or tissues. For example, antineoplastic injected in the surroundings of the tumour, with a decrease of systemic adverse effects. It is also useful for the administration of vasodilators in arterial embolisms or contrast media for arteriography
Intracardiac	Direct administration into the heart, used only as emergency route during a cardiac arrest (adrenaline injection into cardiac chambers) due to the serious injuries that may be caused by the needle
Intradermic	Administration within the skin, at the dermis level, generally in the ventral zone of the forearm. Due to the extremely low blood supply to the dermis, intradermic administration implies almost null systemic absorption of the drug. It is usually used for vaccines and for local anaesthesia, as well as for diagnostic purposes in hypersensitivity tests
Intraarticular	Direct administration on a joint, generally for local effects. For example, anti-inflammatory corticosteroids for arthritis
Intralymphatic	Administration into a lymph node or a vessel. It is used, for example, for the administration of stem cells during the treatment of autoimmune diseases, antitumoural therapy and/or for diagnosis purposes (contrast reagents injection)
Intraosseous	Direct administration within a long bone medulla by injection through it. It is used to achieve systemic effects since the vascular bed of these bones quickly transports the drug to the rest of the body and hence is an option when the IV route is not possible, as in paediatric emergencies to administer antibiotics, antiepileptics, etc.
Others	Intravitreal (in the vitreous humour of the eye), intracavitary (cavities, e.g. lungs), intra-abdominal or intraperitoneal, intrapleural, etc.

6.3   Enteral Routes of Administration

6.3.1   Oral Administration of Drugs

When possible, the oral route is the first choice for the administration of drugs, since it is both convenient and economical. Neither special knowledge nor special supplies (syringes, needles) is required for its use: a simple and basic series of instructions allow patients to take their medication safely, reducing visits to health centres. Such advantages benefit both physicians and patients, leading to an increased compliance and, ultimately, to a higher probability of a successful drug therapy.

Although some drugs are specially targeted to gastrointestinal sites of action, such as bismuth subsalicylate for heartburn and ezetimibe for the reduction of cholesterol absorption, most pharmaceutical active ingredients exert their therapeutic effect outside the gastrointestinal (GI) tract. Therefore, they must be absorbed from the GI system to gain access to the systemic circulation and reach their site of action.

The absorption of drugs administered by the enteral route is determined, in part, by the physiological state of the GI tract, which is in turn affected by diet, hormones, autonomic nervous system, pathological states and other drugs.

Additionally, the physicochemical properties of the drug will also be determinant for its absorption kinetics. This, together with the possibility of losses due to pre-systemic metabolism, hinders the systemic bioavailability of drugs administered by the oral route.

6.3.1.1 Gastrointestinal Tract: Anatomical and Physiological Considerations

The main physiological processes that occur at the GI level are secretion, digestion and absorption. *Secretion* involves the transport of fluids, electrolytes, peptides and proteins to the lumen of the digestive tract (e.g. saliva, gastric acid, pancreatic secretions, etc.). *Digestion* is the breakdown of food into smaller structures, as a preparation for *absorption* of nutrient molecules from the GI lumen to the rest of the organism.

Drugs administered by the oral route move through the various parts of the digestive tract (oral cavity, pharynx, oesophagus, small intestine and large intestine; see Fig. 6.4), and unabsorbed compounds leave the body through the anal sphincter. The GI total transit time can vary from 0.4 to 5 days. As for the components of the diet, the most important site for the absorption of drugs is the small intestine (although there may be absorption in the colon or even in the stomach), and if absorptions is not completed in that region, it can become erratic or incomplete (Rowland and Tozer 1995).

Oral Cavity, Pharynx and Oesophagus

Once ingested, food materials are reduced to smaller and softer particles by chewing and insalivation. This step forms the bolus that advances into the pharynx through the epiglottis during the swallowing process. Some medications follow the same process, although most of them are swallowed without prior chewing.

Fig. 6.4 Anatomical scheme
of the human gastrointestinal
tract

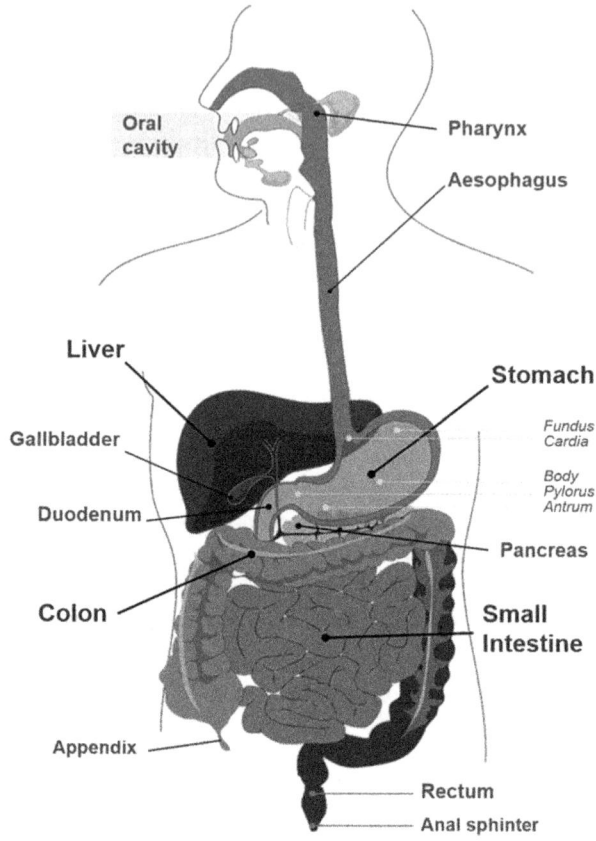

Saliva is the main oral secretion: it has a pH close to 7 and is secreted in volumes around 500–1500 mL per day, at a rate between 0 and 6 mL/min (Aps and Martens 2005). Its main components include ptyalin or salivary amylase and mucin, a glycoprotein that lubricates food, and can also interact with drugs. The oral cavity is followed by the pharynx, which is a part of both the respiratory and digestive systems. Next section is the oesophagus, which leads to a weak sphincter called cardia, in the proximal part of the stomach. Very little (or none) drug dissolution occurs in the oesophagus due to the mucous nature of its secretions that possess a lubricating function to facilitate swallowing (Brodin et al. 2010).

Stomach

The stomach is innervated by the vagus nerve. A local nervous plexus, hormones, mechanoreceptors sensitive to GI wall tension and chemoreceptors control the regulation of gastric secretions and gastric emptying. The digestive juices of the stomach come from the gastric glands, which cover almost the entire wall of the gastric body. These glands contain three main types of cells: mucosal cells of the neck, responsible for generating the viscous and alkaline mucus that protects the stomach surface; parietal or oxyntic cells, responsible for secreting hydrochloric acid; and gastric chief cells (aka peptic cells), producing pepsinogen, the precursor of pepsin. Gastrin, on the other hand, is released from the G cells, more abundant in the antral mucosa and in the duodenum (Hall 2015).

Briefly, the gastric wall distension caused by the arrival of food (as well as the presence of certain peptides and proteins) stimulates the release of the gastrin, which in turn stimulates the production of both acid and pepsinogen while potentiating the motor functions of the stomach. Pepsinogen is secreted as an inactive precursor but is converted to pepsin after contact with hydrochloric acid, which is an active protease at pHs 1.8–3.5.

Once the food has been mixed with the gastric secretions, the resulting product is called chyme, a muddy semi-solid paste that must pass through the pyloric sphincter to access the small intestine, in a process called gastric emptying (GE).

Small Intestine

In humans, the small intestine measures about 6 meters and possesses three anatomical and continuous areas: *duodenum, jejunum* and *ileum*. The most relevant feature of the small intestine is its absorptive function, for which it has an enormous superficial area (greater than 200 m²) available for nutrients. This is possible because the intestinal mucosa is folded into the lumen and covered with finger-like projections of tissue (villi) of about 1 mm in length, each with an arteriole and a venule that converge forming a profuse network of capillaries. Each villi also possess a blind lymphatic vessel. Villi walls are composed of columnar cells (enterocytes), whose luminal surface is constituted by the so-called intestinal microvilli, of about 1 μm long, which conforms the well-known intestinal brush border (Fig. 6.5) (Plá Delfina and Martín Villodre 1998).

The duodenum is the initial part of the small intestine (20–30 cm), to which the gastric content access and where the pancreas and the gallbladder drain their contents. The duodenal pH is approximately 6 to 6.5 due to the presence of bicarbonate that neutralizes the acid chyme drained by the stomach, pH that is optimal for the digestion of food by the pancreatic enzymes (proteases, amylases, lipases). The surfactant components present in the bile are responsible for the dispersion of the fatty components of the diet for their subsequent degradation by lipase enzymes; they may as well favour the dissolution of lipophilic drugs in this zone of the GI tract.

The duodenum is also the place where several prodrugs are enzymatically hydrolysed to their active forms. For the same reason, pharmaceutical active ingredients that may act as substrates of these enzymes (e.g. peptides) cannot be administered by oral route.

The remaining portions, jejunum and ileum, present increasing pH values (ca. 8 in the most distal area), decreasing internal diameter and decreasing number of folds, but the mucous membrane retains its absorptive capacity.

As in the stomach, peristaltic movements of contraction of the muscular walls provide the force that drives the contents along the small intestine.

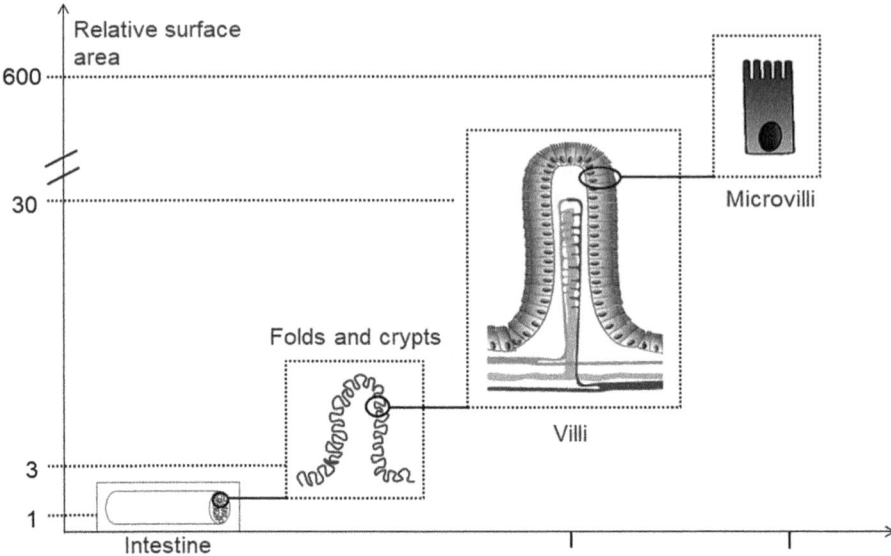

Fig. 6.5 Scheme of the morphological adaptations of the small intestine, responsible for the increase of the relative surface area available for absorption

Large Intestine

Separated from the small intestine through the ileocecal valve, the large intestine is subdivided into four main regions: the cecum, the colon, the rectum and the anus. There are no circular folds or villi in the large intestine walls, so drug and nutrient absorption is very limited. Nevertheless, some essential vitamins produced by intestinal bacteria as well as some drugs such as theophylline and metoprolol can be absorbed in this region. The lack of the solubilizing effect of chyme and digestive fluid (the colic content is solid or semi-solid due to the constant reabsorption of water) also undermines the dissolution and absorption of drugs. Drugs that may be absorbed at the colon are good candidates for sustained release dosage forms.

The main function of the large intestine is to absorb water from the liquid chyme to form semi-solid faeces that are excreted through the anus. It is covered with mucin, which acts as a lubricant and protects the mucosa. Aerobic and anaerobic microorganisms in the colon can metabolize some drugs: levodopa and lactulose, for example, are metabolized by enteric bacteria (Casabó Alós et al. 1998a).

6.3.1.2 Drugs Absorption in the Gastrointestinal Tract

As seen in Chap. 2, drug absorption can occur through different mechanisms (trans- or paracellular passive diffusion, active transport, endocytosis) depending on the physicochemical characteristics of the drug, mainly its hydrophilic/lipophilic balance and its molecular mass or volume. But irrespective of the way that molecules move across the GI membrane, the result is the drug access to blood capillaries (and to a lesser extent to the lymphatic vessels) in contact with the basement membrane of epithelial cells.

With the exception of vesicular transport or endocytosis, which admits the entry of undissolved particulate matter (macromolecules, nanoparticles), transport mechanisms require the drug to be previously dissolved in GI fluids to be absorbed. Therefore, absorption will depend on the solubility of the drugs, as well as their dissolution rate if they are administered as solids dosage forms. For drugs in solutions, the rate-limiting step of oral absorption is the drug permeability across the intestinal barrier. If both the solubility and permeability of a drug are improved, the result is an increase in the rate and degree of oral absorption, i.e. its bioavailability.

Therefore, oral bioavailability of pharmaceutical active ingredients may be influenced by numerous additional factors, both physiological and technological.

6.3.1.3 Primary Factors Influencing Oral Drug Absorption

Solubility

▶ **Definition** Solubility is the maximum dissolved concentration of a compound under given solution conditions (solvent, temperature, pH, etc.). It is one of the most relevant physicochemical properties for drug candidates, along with the degree of ionization and the oil/water partition coefficient, as indirect measures of the permeability of the drug across biological membranes. Indeed, one of the key factors for the success of a drug discovery project is the ability to find the adequate solubility/permeability balance, since although lipophilic drugs have low aqueous solubility, they generally have a high membrane permeability and vice versa (Kerns and Di 2008).

Another example of this relationship is found in the interplay between pH and degree of ionization: ionized forms are generally more soluble in water than nonionized ones, but the latter are more easily absorbed into the GI tract by passive diffusion. Therefore, since the small intestine pH values are in the range 5–8, drugs that are weak bases will exhibit a favoured absorption by predominance of the nonionized form, whereas weak acids will be mostly in ionized form (Gleeson 2008). Nevertheless, the large absorption surface of the small intestine is usually sufficient to compensate for this disadvantage and allow the complete absorption of many weakly acidic drugs.

Permeability

The permeability (P, cm/s) reflects the physiological and physicochemical properties of the GI membrane and their relationship with the compounds being absorbed, and it is a measure of the ease of the drug absorption process across the GI cell membranes. It is a property directly related to the transport mechanism, so its expression is modified according to whether it is by passive diffusion (Eq. 6.1), active transport (Eq. 6.2) or a combination of both (Eq. 6.3).

$$P_{\text{passive}} = \frac{J}{C} = \frac{D \cdot K}{h} \tag{6.1}$$

$$P_{\text{active}} = \frac{J}{C} = \frac{J_{\max} \, x \, C}{K_{\mathrm{m}} + C} \cdot \frac{1}{C} = \frac{J_{\max}}{K_{\mathrm{m}} + C} \tag{6.2}$$

$$P_{\text{total}} = P_{\text{passive}} + P_{\text{active}} \tag{6.3}$$

In the previous expressions, C is the drug concentration (mg/mL); K is the partition coefficient; D is the diffusion coefficient (cm^2/s); h is the cell membrane thickness (cm); J is the drug flow (mg/s.cm^2), $J_{\max}$ is the maximum drug flow, and K_{m} is the Michaelis-Menten constant which is inversely related to the drug affinity for a transporter (mg/mL) (Cao et al. 2008).

Therefore, passive permeability is related to membrane and drug properties, and thus, for a given drug/membrane system, it is a fixed value independent of the drug concentration. Conversely, the total effective permeability of drugs absorbed through both passive diffusion and active transport depends on drug concentration. Only at very low concentrations, when $C \ll K_{\mathrm{m}}$, P_{total} may be regarded as concentration-independent.

Dissolution Rate

▶ **Definition** Dissolution is defined as the mass transfer from a solid phase to the surrounded dissolution media or solvent. It is a dynamic property that changes over time, and from a physicochemical point of view, it can be represented as the inverse process to crystallization: disintegration of a crystal structure under the action of the solvent that surrounds it.

Solid oral dosage forms (capsules, tablets, suspensions, etc.) require the active ingredient to be dissolved in the GI fluids before absorption takes place. For poorly soluble drugs, this could be a slow process, particularly if high doses are required. In such cases, the dissolution rate becomes the rate-limiting step of drug absorption (Amidon et al. 1995).

The dissolution rate of a drug can be represented by Eq. 6.4, derived by Nernst and Brunner in 1904 from the theory of Noyes and Whitney by application of the Fick's diffusion law:

$$\frac{dC}{dt} = \frac{D \cdot A}{V \cdot h}(C_{\mathrm{s}} - C) \tag{6.4}$$

where C is the drug concentration (mg/mL), A is the surface area of the solid to be dissolved (cm^2), D the diffusion coefficient (cm^2/s), V the volume of solvent (mL) and h the thickness of the stagnant layer (cm) of the saturated solution of concentration C_{s} (Dokoumetzidis and Macheras 2006). Since the nature of the fluid and the agitation conditions in the GI tract are non-modifiable factors, the main variables that can be optimized to increase the dissolution rate are the surface area offered to the solvent (also related to the size and porosity of the particles) and the solubility. Therefore, dissolution kinetics may be improved by size-reduction techniques (micronization, nanotization) as well as by ionization, use of surfactants or disintegrants, among other strategies.

As a last remark, it is worth highlighting that the dissolution of the active ingredient from a solid dosage form is a complex process, very hard to predict in a mechanistic manner, particularly with expressions as simple as Eq. 6.4. The Noyes-Whitney and Nernst-Brunner models, originally derived for pure solids of defined geometries, are very useful for the study of the factors that affect solids dissolution, but if what is wanted is a mathematical expression of the dissolution process, mathematical modelling techniques are often used, adjusting known equations to the experimental data (Adams et al. 2002).

6.3.1.4 Secondary Factors Influencing GI Drug Absorption

Gastric Emptying (GE)

After oral administration, dosage forms quickly reach the stomach, where drug release and dissolution may begin. However, it is only after the GE process that the active ingredient and/or the remaining solid dosage form will be able to reach the initial portions of the small intestine. Since the duodenum is the primary site of drug absorption, a delay in GE time may translate into a lower rate (and generally also extent) of drug absorption.

The GE occurs both in fasting and fed conditions. However, the motility patterns are different in both cases. In fasting, or at the end of the postprandial period, the motor activity is under the control of the migrating motor complex (MMC), the electric and motor pattern responsible for the emptying of indigestible solid foods. The MMC has an approximate duration of 100 min and consists of four phases (Shargel et al. 2016):

1. Phase I (or relative quiescence) constitutes 50–60% of the cycle and is characterized by its inactivity, only with occasional contractile waves that do not generate propulsive movements.
2. Phase II (or intermittent contractions) constitutes 20–30% of the cycle and is when the frequency of contractions increases, although they remain irregular and do not generate propulsive phenomena.
3. Phase III (or intense contractions) lasts approximately 10 min and is when regular and frequent propulsive contractile waves are generated. During this phase the pylorus remains totally relaxed, allowing the emptying of the gastric contents that were not able to be transformed into chyme (non-digestible solids).
4. Stage IV (or deceleration) is a very short period, about 5 min, where the stomach returns to phase I.

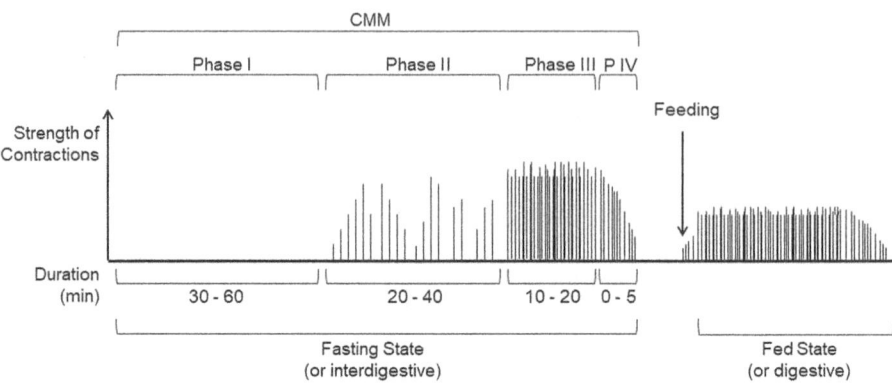

Fig. 6.6 Typical GI motility patterns in both fasting and digestive state (Gerk et al. 2016)

If at the end of the cycle food ingestion has not occurred, it starts again, so the MMC occurs cyclically during all inter-digestive periods. In the postprandial or fed state, i.e. after the ingestion of any drink or digestible substance, the MMC is replaced by regular contractions of less amplitude than in phase III that mix and propel the GI content. To pass through the pylorus, a particle must have less than a few millimetres (Rowland and Tozer 1995), and larger particles are emptied slower and with much less predictable kinetics (during phase III).

Figure 6.6 schematically represents the typical fasting and fed motility patterns.

Therefore, the arrival of the swallowed dosage form to the stomach will be accompanied by the interruption of the MMC, since drugs are usually taken concomitantly with some liquid or food. While food requires longer time for emptying, water is rapidly emptied due to the pressure gradient between the fundus and the duodenum (without motor involvement of the antrum), which is why the usual indication is to administer drugs under fasting state, with water. This ensures, in most cases, the higher rate of absorption of the active ingredient and a relatively rapid pharmacological effect.

Table 6.2 presents a summary of the main factors that affect GE (Casabó Alós et al. 1998b).

In the case of drugs with high solubility and high permeability, GE rate will control the absorption rate and the initiation of the therapeutic effect (Benet et al. 2011). On the other hand, the GE rate will be directly proportional to the maximum plasma concentration and inversely proportional to the time required to reach that concentration.

Intestinal Motility and GI Transit Time

Intestinal motility consists mainly of two types of movements: propulsion and mixing. Propulsion movements or peristalsis, responsible for the progress of the GI content, determines the transit time, while the mixing movements increase the dissolution rate of the drug while favouring its contact with the endothelial surface for absorption. As stated before, both movements are stimulated by the presence of

Table 6.2 Brief description of the main factors that affect gastric emptying

Patient-related factors	
Sex	In general, women present higher GE times than men
Emotional state	Stress increases the rate of GE; depression and fatigue, on the other hand, decrease it
Body position	If the patient is lying down, the GE is slowed if it is in left lateral decubitus and accelerates in the right lateral decubitus position
Diseases	Diabetes, hypothyroidism, and some GI lesions (duodenal or pyloric ulcer, pyloric stenosis) increase GE time. With hyperthyroidism, on the contrary, the GE is produced more rapidly
Exercise	Intense exercise may increase GE time
Food-related factor	
Type of food	In general, inhibitory activity of the different types of food over GE follows the following order: lipids > proteins > carbohydrates, in a concentration-dependent manner
Physical state/viscosity	GE rate of food usually exhibits the following order: liquids > semi-solids > solids
pH	At high concentrations, acids and bases are inhibitors of GE. This effect is only relevant in hyperchlorhydria state or after the administration of high doses of antacids
Volume	The GE rate responds to the pressure and the distension of the stomach wall. In that way, larger volumes accelerate the initiation of the GE. However, after this initial period, the process slows down, and the overall result is a delay of the GE
Temperature	Cold foods are considered digestive by accelerating the GE
Osmotic pressure	High concentrations of some ions, sugars and electrolytes tend to delay the GE
Others	
Drugs	Drugs with action on the autonomic nervous system usually affect GE: cholinergic and adrenergic blockers activate GE, while anticholinergics and adrenergics delay it

food in the intestine (Hall 2015). Therefore, administering drugs in fasting state and with a volume of water enhances their absorption properties not only by the higher rate of GE but also by the lower intestinal peristalsis.

GI transit time or residence time (RT) is a very important factor for drug absorption. If the drug progresses too rapidly through the GI tract, its absorption may be incomplete, which is especially critical in the case of poorly permeable compounds, as well as for prolonged and/or delayed-release dosage forms. Increasing the residence time in the GI tract (or decreasing motility) leads to a greater drug absorption potential. Excessive times, however, could lead to dose losses due to enzymatic or bacterial degradation.

Since the GI RT are shorter in the fasting state than in the fed state (e.g. small intestine RT is around 3–4 h in the fasting state and up to 12 h in the fed state (Gerk et al. 2016), the RT of a drug in the GI tract may be substantially modified by indicating its administration with food or in the fasting state.

Components, Volume and Properties of GI Fluids

As for GI fluids components, there are many endogenous substances capable of affecting the drug absorption. Gastric mucin can be combined with certain drugs (e. g. tetracyclines) and decrease its bioavailability (Braybrooks et al. 1975). Bile salts, on the other hand, can decrease the absorption of cationic drugs by forming insoluble complexes, while at the same time, by their surfactant action, they can favour the absorption of liposoluble drugs.

Other components present in GI fluids capable of affecting the bioavailability of pharmaceutical active ingredients are digestive enzymes (proteases, lipases, esterases, decarboxylases, among others), and their activity can be exploited for the design of prodrugs (esters, most commonly), inactive in their original form but active after a step of enzymatic biotransformation.

The intestinal bacterial presence is also a factor to consider, due to its high metabolic potential. More predominant in the colon area, human GI flora is characterized by hydrolysis and reduction reactions. Hydrolysis of the glucuronide conjugates excreted into the bile releases free parental drug molecules in the small intestine, which then may be reabsorbed and become systemically available, in a process called the enterohepatic circulation (see Chap. 5 for details).

The pH of the GI fluids is a key factor to assess its impact on drug absorption, since it is capable of modifying the degree of ionization and thus the solubility, dissolution rate and permeability of a drug. The GI pH depends on the individual's general health status, possible disease conditions, age, type of feeding and prandial state, and it can be modified by certain drugs. For example, anticholinergics and H_2 blockers increase gastric pH, so they can significantly decrease the bioavailability of some weakly basic drugs with pH-dependent solubility (Puttick and Phillips 1994).

Food

The study of the interactions between foods and drugs is a complex task, extremely dependent on which food and drug is considered and especially relevant in the case of drugs of narrow therapeutic index. These are very diverse and frequent interactions, although often difficult to detect, not only because of the diversity of food that a patient may ingest but also because they exhibit high inter- and intra-individual variability.

Above there is a list of some typical examples of food/drug interactions. Interested readers may refer to Marasanapalle et al. (2011), Melander (1978) and Sørensen (2002).

- High-fat foods stimulate the secretion of bile salts, which in turn may increase the solubility and/or bioavailability of certain liposoluble drugs such as spironolactone or griseofulvin.
- A high-protein content may increase gastric pH and, therefore, decrease the dissolution of weakly basic drugs.
- High-calorie foods slow the GE rate, delaying the absorption rate and the beginning of the therapeutic effect. For certain drugs, however, this may be

beneficial as a greater stomach residence time favours their complete dissolution (e.g. nitrofurantoin).

- Certain food components can interact with drugs (by complexation, adsorption, etc.) and reduce their bioavailability. Typical examples are tetracyclines with Ca (II) and Mg(II) ions in milk and the adsorption of digoxin to dietary fibre.
- Food may also influence the bioavailability of a drug at pre-systemic level. The juices of some fruits, like grapefruit, irreversibly inhibit various enzymes of cytochrome P450 (particularly intestinal CYP3A4) as well as certain efflux transporters (P-gp, MRPs), causing an increase in the bioavailability of the substrates of those enzymes and/or transporters. Ethanol in acute form is an enzymatic inhibitor, while its chronic consumption produces metabolic induction (phases I and II).

Case Study

A 65-year-old patient with documented paroxysmal atrial flutter for 10 years has been following an anticoagulation therapy for approximately 7 years. In addition, he has informed systolic congestive heart failure, hyperlipidemia, hypertension (not controlled) and a history of alcoholism.

On a routine control, the patient showed a subtherapeutic international normalized ratio (INR) level of 1.7 (normal values of INR: 2–3). The warfarin dose was then increased from 65 mg/week to 70 mg/week. Two weeks later, his INR was 4.77. The patient reported adherence to his warfarin treatment since his last appointment. The only change he reported was the incorporation of grapefruit to his diet (one grapefruit each day for 3 days during the previous week to INR draw). Due to the increase of the INR value, his warfarin dose was decreased to 8 mg daily. The patient was instructed to continue monitoring for signs and symptoms of bleeding or thromboembolism and to stop eating grapefruit. Five days later, the patient returned to the clinic, and his INR was 2.1. After this result, he was instructed to restart his dose to the initial 65 mg/week and has been within therapeutic range since then.

The (R) enantiomer of warfarin is metabolized by CYP1A2 and CYP3A4, both part of the cytochrome P-450 (CYP450) superfamily. On the other hand, grapefruit is able to affect a variety of isoenzymes of the CYP450, in particular, the 3A4, 1A2 and 2A6, which could contribute to the interaction with warfarin. Two compounds have been cited as potential responsible: the flavonoid naringin, which metabolite (naringenin) has demonstrated significant inhibition of CYP3A4, and dihydroxybergamottin (DHB) that has the same effects as naringin metabolite.

In this case, the inhibition of CYP3A4 isoenzyme, as a result of the ingestion of grapefruit, produces a decreased warfarin metabolism. This interaction increases the effects of warfarin, leading to an increase of INR values. This effect ceases when grapefruit is removed from patient's diet.

Case study adapted from Bodiford et al. (2013).

Age

Drug absorption in newborns is significantly reduced because they possess lower GE rate, lower acidity and volume of GI fluids and lower surface area and intestinal blood flow (O'Hara et al. 2015). On the other hand, gastric acid secretion and intestinal transit time are reduced/slowed in the elderly, which could negatively or positively affect a drug absorption rate, depending on the drug characteristics (Gidal 2006).

Formulation Characteristics

Both the excipients and the pharmaceutical form can affect the absorption of the active ingredient. In tablets, for example, the addition of disintegrating agents may improve the dissolution rate of a drug. Surfactants, such as Tween-80, can increase the solubility of poorly soluble drugs while improving drug permeability (Constantinides et al. 2016). Another example is enteric coatings, which resist the action of gastric fluids and disintegrate when the pH is increased at the intestinal level, thus preventing the acid degradation of certain drugs. This type of coatings is also used to protect the gastric mucosa, since many drugs orally administered are irritating to the stomach and may cause nausea, pain or a sensation of heartburn. However, since enteric-coated tablets maintain their integrity along the stomach and, on the other hand, they have a size larger than 2–3 mm, they will leave the stomach in phase III of MMC, which may lead to an erratic bioavailability of this type of pharmaceutical forms.

The formulation type, vehicle or pharmaceutical form is directly related to the absorption characteristics of the active ingredient. For example, dosage forms that carry the drug already dissolved (solutions, syrups, soft capsules) will generally imply faster absorption processes than those formulated with the drug in the solid state (tablets, suspensions), since the release process is very fast or not necessary.

It is also possible to modulate the bioavailability of a drug by modifying the release kinetic from the dosage form. The classic forms of prolonged and delayed intestinal release are the best-known example, although they will not be discussed here in detail (see, e.g. Qiu (2009); Qiu and Zhang (2009)).

Another strategy is to delay the GE of the dosage form, especially useful for drugs acting locally in the stomach (misoprostol, antacids, certain antibiotics used in the treatment of gastric ulcers associated with *Helicobacter pylori*), drugs with narrow absorption window (furosemide, atenolol, L-dopa), drugs unstable at intestinal pH (captopril) and/or drugs insoluble at the intestinal pH (e.g. weak bases, diazepam, verapamil). Examples of this type of vehicle are:

- Low-density systems: GE rate of the dosage form is delayed by maintaining long 'floating' times in the stomach (some for up to 12 h), which is achieved by incorporating air into the formulation (effervescent gas production, air chambers) or using low-density materials (fatty substances, oils, foams).
- Muco-adhesive systems: constituted mainly by hydrophilic polymers. In contact with the gastric fluid, they form a gel layer capable of interacting with the gastric mucosa. These systems are most commonly used for other routes of

administration (buccal, vaginal, transdermal), since the chara
tract (motility, pH, gastric mucin, inability to exert pressure, r
mucosa cells) make it difficult to achieve adhesion.
• Expandable systems: small-sized solid dosage forms that s
GI fluids, being able to increase their volume between 2
as a result of the osmotic absorption of water. Anim
these products, based on superporous hydrogels, showe
of 2–3 h (in fasting) and up to 24 h with food (Lopes et al. 2

6.3.2   Buccal and Sublingual Routes of Administration

Two sites within the oral cavity can mainly be used for systemic absorption of drugs: the sublingual region (SL) and the buccal region, the latter located between the oral mucosa and the mandibular arch. These two areas, unlike other buccal regions (such as the hard palate, gingiva or dorsal surface of the tongue), do not have keratinized epithelia, so they are more favourable for drug absorption.

The main characteristic of these routes of administration is the rapid onset of action, which is due to the high blood flow and lymphatic flow of the oral cavity. In addition, substances absorbed at this level reach the general circulation without loss due to a first-pass effect, as the buccal venous drainage flows directly into the superior vena cava.

Some advantages of administering drugs through the oral mucosa are:

• A rapid onset of the therapeutic effect is achieved, not only by the high blood supply of the area but also by the lack of GI factors that delay absorption (gastric emptying, presence of food, gastric disease, etc.).
• Portal circulation is avoided, which allows to improve the bioavailability of a drug (regarding the oral route) avoiding intestinal and first-pass hepatic metabolism.
• The active ingredient is not exposed to the aggressive GI medium, reason why it is possible the buccal administration of some drugs (e.g. peptides) that would otherwise be degraded by the GI pH or enzymes.
• It can be done in patients with swallowing difficulties, nausea or malabsorption syndrome and even in unconscious patients.

On the other hand, the greatest limitation of buccal administration is that, due to the small size of the oral cavity, only very potent drugs can be effectively delivered. The buccal mucosa offers about 200 cm^2 of area for drugs absorption, i.e. about 10,000 times less than duodenum (Madhav et al. 2009).

Another disadvantage is the difficulty in keeping the drug in the site, as well as the need for the patient to refrain from swallowing, talking or drinking during administration so as not to affect the residence time in which the medication is in direct contact with the mucous membrane. Moreover, this route is not applicable to bitter or bad-tasting drugs since, in addition to patient discomfort, this type of formulations generates excessive production of saliva, which increases the risk of swallowing (Rathbone et al. 2015).

While absorption may occur by any of paracellular and transcellular mechanisms studied, passive diffusion of drugs from the salivary aqueous phase through the membranes of oral mucosal cells predominates. Therefore, drugs of intermediate polarity are well absorbed, since excessive lipophilicity limits drug dissolution, in the same way that an increased polarity limits diffusion across cell membranes. The degree of ionization is also important: less ionized compounds at salivary pH are better absorbed.

6.3.2.1 Pharmaceutical Forms for Sublingual and Buccal Administration

Although there are several products on the market for these routes of administration, only a small fraction corresponds to systemic action products, while most are topical (not treated here).

Buccal administration products are usually prolonged-release medications, so they are formulated as patches or very adhesive tablets for higher comfort during prolonged contact with the oral mucosa (Montenegro-Nicolini and Morales 2017). SL tablets, on the other hand, are usually of rapid dissolution so that the active ingredient is rapidly absorbed preventing swallowing losses. In addition, they should be small, without angles or edges, and insipid, in order to not stimulate patients' salivation.

Classic examples of drugs administered by this route are the cardioactive nitroglycerine and isosorbide denitrate, indicated for the treatment of angina pectoris, congestive heart failure, acute myocardial infarction and other peripheral vascular diseases. They are potent drugs capable of achieving their effect (coronary vasodilatation and relief of left ventricular work by reducing venous return) with only a small absorbed drug amount. Administered by buccal or SL route, rapid onset action is achieved, with the additional advantages of:

- The effect can be stopped easily, which does not happen when the parenteral route is used.
- It avoids the extensive first-pass metabolism that these drugs suffer after oral administration.

Figure 6.7 shows the serum levels of nitroglycerine (NTG) obtained after sublingual, oral and transdermal administration (Blumenthal et al. 1977). It is observed that by SL route, the maximum concentration is reached before 10 min. However, the drug elimination rate is also rapid, so this route is only suitable for acute treatment. Chronic treatments are performed orally, despite the fact that relatively high doses are required to compensate the pre-systemic elimination of the drug.

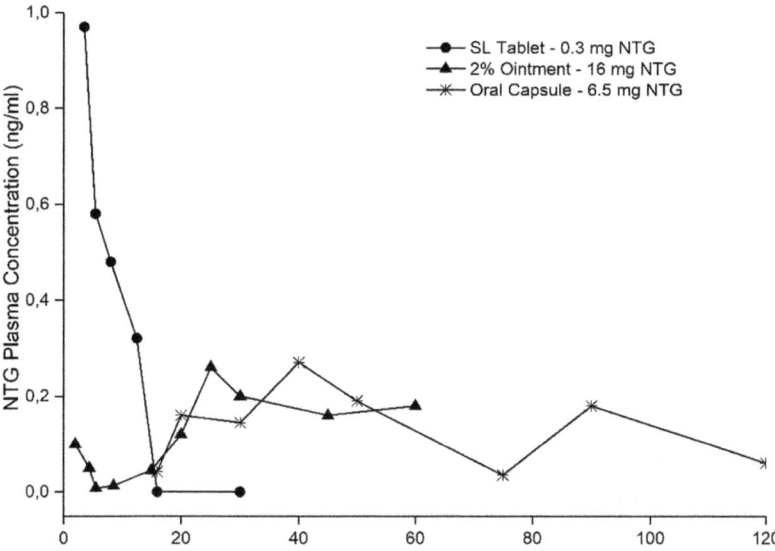

Fig. 6.7 Plasma concentrations of nitroglycerine (NTG) after the administration of a SL tablet of 0.3 mg NTG (black circles), 6.5 mg NTG oral capsule (asterisks) and a 2% ointment, equivalent to 16 mg of NTG (black triangles). (Adapted from Blumenthal et al. (1977))

6.3.3   Rectal Administration of Drugs

The rectum is the ending portion of the large intestine, approximately 15 cm long, from the colon to the anal sphincters. It can be used as a drug delivery route for both local and systemic effects, although this section focuses on the latter.

The rectal epithelium is composed of non-keratinized and villi-free cells, presenting a surface available for absorption of around 200–400 cm^2, significantly smaller than the duodenal mucosa. In general, the rectum is empty, occupied only by a few millilitres of fluid with pH ranging from 6 to 8, without buffer capacity, and very viscous due to the presence of mucin. This characteristic of the rectum hinders drug absorption, since it imposes serious limitations to the drug dissolution process (Garrigues Pelufo and del Val Bermejo Sanz 1998). The rectal area is perfused by the haemorrhoidal arteries, which in turn drain into the upper, middle and lower haemorrhoidal veins. The last two converge in the hypogastric vein and from there they access the inferior vena cava, which carries their contents to the heart. On the other side, the superior haemorrhoidal vein joins the mesenteric circulation, which feeds into the portal vein and from there to the liver. This is why the absorption of drugs through the rectal epithelium is often erratic and variable, since depending on the specific area where it occurs, part of the absorbed amount directly access the systemic circulation while another fraction suffer first-pass metabolism (Øie and Benet 2002).

There are several scenarios in which the rectal route of drug administration may become the route of choice due to one of the following characteristics:

- Applicable in cases of nausea, vomiting and inability to swallow (unconscious patients), as well as in the presence of diseases of the upper GI tract that affect oral drug absorption.
- Suitable for formulations with unpleasant taste (a particularly important factor in children).
- Allows achieving rapid systemic effects by giving a drug in a suitable solution (as an alternative to injection), with the additional advantage that such effect can be rapidly terminated in cases of toxicity or overdose.
- The absorption rate of the drug is not influenced by food or GE.
- Part of the metabolism of both enteric and first-pass hepatic elimination is avoided, which may result in a significant increase in bioavailability of extensively metabolized drugs (such as lidocaine; see Fig. 6.8) (de Boer et al. 1980).
- Avoids the contact of the drug with the aggressive conditions of the upper GI tract.

Figure 6.8 shows how, in the case of the anaesthetic and antiarrhythmic drug lidocaine, the administration of 200 mg by rectal route has a similar bioavailability, and even greater, than an identical dose of the drug administered orally.

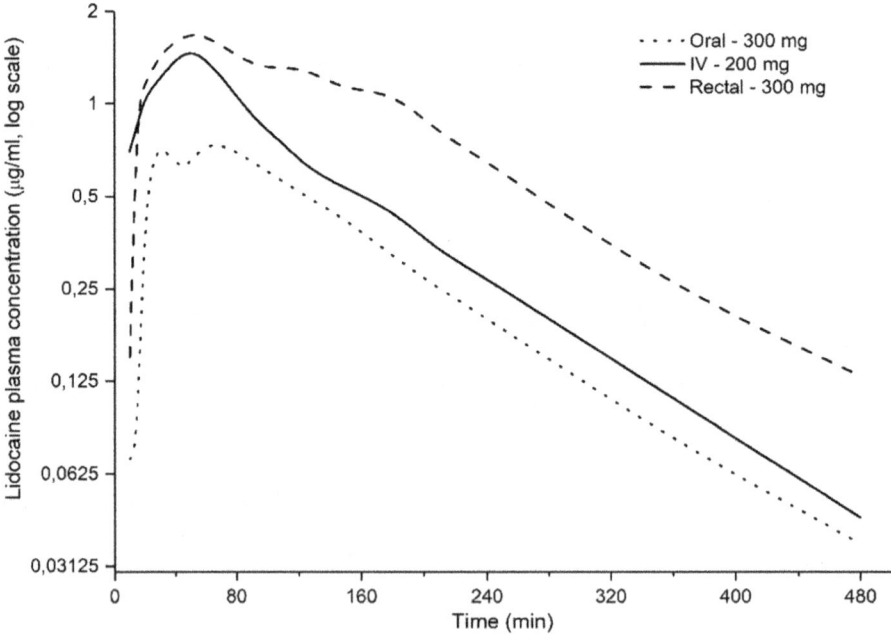

Fig. 6.8 Plasma concentrations of lidocaine after administration to a healthy volunteer by IV (line), oral (dots) and rectal (dash). (Adapted from de Boer et al. (1982))

There are, however, some drawbacks associated with rectal administration of pharmaceutical products, including:

- The interruption of the absorption process by defecation, which can occur especially when the drug is irritant.
- The reduced surface area may limit absorption, in the same way that the low volume of rectal fluids can lead to incomplete dissolution of the drug.
- It is possible the degradation of certain drugs by microorganisms in the rectum.
- Patient adherence may be a problem.

6.3.3.1 Absorption of Drugs by the Rectal Route

The predominant mechanism for the entry of drugs and other xenobiotics at the rectal level is the passive diffusion across epithelial cell membranes, other transport mechanisms not being considered relevant. Due to the reduced surface area and the shorter residence times in comparison to the oral administration, the physicochemical characteristics of the drug and the technological and bio-pharmaceutical aspects of the formulations are critical factors for the rectal absorption.

In general, lipophilic drugs are better absorbed than hydrophilic drugs, the latter usually presenting slow absorption kinetics and incomplete absorption. As mentioned in the previous sections, ionizable molecules reach their maximum absorption rate at pH values that minimize their ionization (Desai 2007).

With regard to the type of pharmaceutical dosage form, the rectal route allows for the administration of solid dosage forms, mainly suppository (emulsion or suspension based) and jelly capsules (of solutions or suspensions), as well as liquid dosage forms or enemas, classified according to their volume into macro-enemas (> 100 mL) or micro-enemas (1–20 mL) (Murdan 2013).

Absorption of drugs from aqueous and alcoholic solutions can occur very quickly, which has proven to be of great therapeutic value, for example, in the rapid suppression of acute seizures with diazepam (McTague et al. 2018). On the other hand, absorption from suppositories is generally slower and very dependent on the formulation's characteristics: suppository base, the use of surfactants or other additives, active ingredient particle size, etc. In general, suppositories with water-soluble base (polyethylene glycol, glycerine) release the drug by dissolution, while low melting point fat bases should melt at body temperature to release the drug. Some suppositories contain an emulsifying agent that keeps the fatty oil emulsified and the drug dissolved in it. Figure 6.9 shows the influence of the pharmaceutical form on the bioavailability of the drug promethazine.

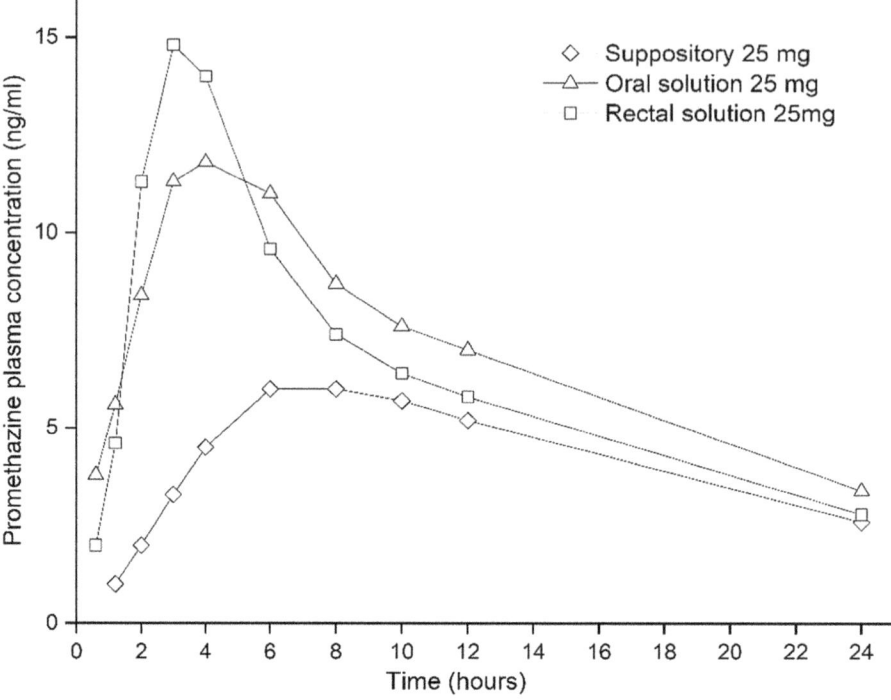

Fig. 6.9 Promethazine plasma concentrations after the administration of 25 mg to a human volunteer by oral solution (triangles), rectal solution (squares) and suppository (diamonds). (Adapted from de Boer et al. (1982))

6.4   Other Routes of Medication Administration

6.4.1   Pulmonary Administration of Inhaled Drugs

The pulmonary route of administration has been traditionally used for drug administration to the respiratory tract, in pathologies like chronic obstructive pulmonary disease or COPD, asthma, cystic fibrosis, etc. The main advantages of administering medications by inhalation include the rapid absorption and rapid onset of action (e.g. very important for bronchodilator and anti-inflammatory drugs) as well as the localization of drug activity in the lung with minimal systemic toxicity, which is particularly important in the case of anti-inflammatory corticosteroids such as beclomethasone, budesonide, fluticasone, etc.

But it would be a mistake to consider the lungs as an administration site only suitable for the local effect of drugs. Indeed, the respiratory tract may also be regarded as a systemic administration route, as in the case of inhalable insulin (Peters and Hulisz 2015). The lungs have a large surface area available for the systemic absorption of drugs: the alveolar-capillary barrier. This is a highly permeable and highly irrigated membrane, less than 0.5 μm thick, with a surface area of around 100 m^2 (Dagar 2007).

Human lungs, however, also have effective means for the elimination of deposited particles. In the upper airways, the ciliated epithelium contributes to mucociliary sweeping, by which the particles are drawn from the airways to the mouth. Already in the lungs, alveolar macrophages are able to phagocyte particles shortly after their deposition (Hickey 2006). Thus, effective inhalation therapy, and especially when a prolonged action of the drug is desired, requires bypassing the lungs' clearance mechanisms for the drug to be completely absorbed.

As for the pre-systemic losses of the drug, it is generally accepted that they are smaller by respiratory route than orally, but this is an aspect that must be addressed on a case-by-case basis. Although to a lesser extent, most metabolizing enzyme systems of the liver are also present in the lung. Bearing in mind that blood flow normalized to tissue weight is almost ten times greater in the lung, this organ may play a significant role in the overall systemic clearance of a drug (Farr et al. 1990).

To be able to reach the lungs, through the bronchial tree, a drug must be in aerosol form, generated by an appropriate device. Aerosols are relatively stable two-phase systems consisting of condensed and finely divided matter suspended into a continuous gaseous phase. Due to the size restrictions imposed by this route (see Sect. 6.4.1.2), dispersion must be colloidal, and the dispersed phase may be a liquid (mist), solid (suspension) or a combination of both.

There are three main devices for administering medications by inhalation (Garcia-Contreras et al. 2015):

- Inhalers in pressurized containers, which deliver fixed volumes by the actuation of a valve
- Nebulizers, capable of converting aqueous solutions or suspensions (of micronized drugs) into an aerosol, either by a high-velocity air stream or by ultrasonic energy
- Dry powder inhalers, both in single and multi-dose containers

6.4.1.1 Pulmonary Absorption of Drugs

Depending on their size, insoluble particles, such as powder and microorganisms, may become trapped in mucous secretions or be rapidly phagocytosed by alveolar macrophages, mechanisms that ensure the sterility of alveoli despite the continuous inhalation of the microbes in the air. In contrast, compounds which are able to dissolve in the pulmonary surfactant are capable of being absorbed by active transport and/or passive diffusion through both the aqueous pores and the epithelial membranes.

Volatile and non-volatile lipophilic compounds are rapidly absorbed through the lipid membranes, and the substantial blood supply to the area ensures that the compounds are then rapidly transferred to the systemic circulation. The absorption of hydrophilic compounds is generally slower and tends to decrease as the molecular weight increases, being carried out through the aqueous pores present in the pulmonary epithelium. Certain hydrophilic compounds that do not absorb well by GI route, do it by the respiratory route, such as sodium cromoglycate and gentamicin (Farr et al. 1990).

In addition, there are at least two pulmonary active transport systems, one for amino acids and one for organic anions. Sodium cromoglycate is transported by the latter, so it is considered that its absorption occurs both passively and actively (Nakamura et al. 2010).

It should be taken into account that evaluating the pulmonary absorption of non-volatile compounds is a complex process, since usually only 5 to 20% of the administered dose reach the lungs by the use of aerosols, while other fraction may be swallowed and access the systemic circulation through the gastrointestinal tract, complicating the interpretation of the results.

6.4.1.2 Factors Influencing Pulmonary Drug Deposition

Due to the anatomical characteristics of the respiratory tract, an effective drug deposition is only achieved by carefully tuning certain formulation characteristics, like density, porosity, hygroscopicity and, mainly, particle size. The aerosol particle size is critical for determining the degree of penetration into the bronchioles. To maximize the drug bioavailability by the pulmonary route, the particles/drops containing the active ingredient must be of such a size as to ensure their deposition in the deeper regions of the lung, i.e. 2–5 μm in diameter. In the case of larger particles (>5 μm), the inertia makes them travel a certain distance without modifying their path, until they impact and deposit, not advancing generally beyond the oropharyngeal cavity. Only smaller particles (<5 μm) are able to advance further into the bronchioles and reach the alveolar membrane by sedimentation or diffusion mechanisms (the latter only valid for particles of size ≪0.5 μm) (Hickey 2006).

Figure 6.10 shows the deposition of inert particles in the different regions of the respiratory tract (oropharyngeal, bronchial and alveolar), as a function of particle size (modelled data). It can be seen that the minimum of total deposition corresponds

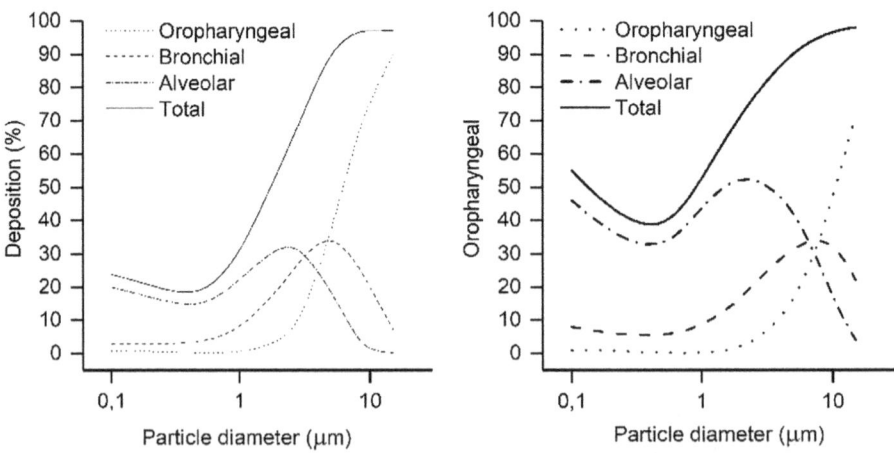

Fig. 6.10 Modelled data of the regional deposition of monodisperse inert particles as a function of particle size, for a healthy lung. Inhalation volume: 1,5 L. Left: flow 200 ml/s. Right: flow of 1000 ml/s. Adapted from Scheuch et al. (2006)

to a particle size of around 0.5 μm, when all the deposition mechanisms are inefficient: the particles are too large to be transported by diffusion and too small to be deposited by sedimentation and/or impaction. From an aerodynamic diameter >1 μm, deposition is increased by sedimentation, achieving the maximum alveolar deposition at diameters ca. 3 μm. For larger particles bronchial and oropharyngeal deposition prevails (Scheuch et al. 2006).

Finally, there are also physiological factors capable of influencing the bioavailability of inhalable drugs, such as mucus in the respiratory tract (highly variable among individuals and increased in some pathologies) or the respiratory pattern, among others. It can be seen in Fig. 6.10 that the deposition of particles is modified by the inspiratory airflow: at a higher flow rate (graph on the right), deposition mechanisms are less effective, both extra thoracic and alveolar.

6.4.2 Nasal Route of Drug Administration

Nasal administration refers to the absorption of drugs across the nasal mucosa, i.e. not accessing to the respiratory tract. It is a form of administration that can be used for both local and systemic therapies and is presented as an alternative, non-invasive route, especially useful in the case of extensively metabolized or labile drugs in the GI medium. It is limited, however, to very small volumes (25–200 μL), and thus only applicable to potent drugs with high water solubility. In addition, the active ingredient must have a molecular weight <1 kDa to be absorbed and should not be irritating or injurious to the nasal mucosa (Arora et al. 2002).

Table 6.3 presents some examples of drugs administered by this route, from local vasoconstrictors for rhinitis to biological products, including vaccines.

The nasal cavity begins in the nostrils, which gather and direct the entrance of air. Behind them is the region containing the turbinates and nasal epithelium and then the nasopharynx, where ends the septum that divides the nasal cavity into two halves and so the cavity becomes one.

The nasal mucosa is a very vascularized and easily accessible area, with a non-keratinized epithelium composed by ciliated cells, as well as mucous glands and goblet cells, responsible for producing and storing nasal mucus (Fig. 6.11). Compounds that reach the nasal epithelium rarely achieve prolonged contact times with the cells surface due to the presence of the mucus that cover the mucosa, which is constantly propelled by the cilia of epithelial cells. When this periciliary fluid interacts with the air flow, foreign compounds in the air are forced to move into the nasopharynx until they are swallowed, in a process named mucociliary clearance (Beule 2010).

Another problem associated with nasal drug delivery is the potential injury or irritation of the nasal mucosa caused by the active ingredient and/or other components of the formulation, such as preservatives. Although sporadic dosage is unlikely to damage epithelium and/or cilia, chronic applications may lead to more serious toxicity issues and may ultimately damage the cilia and compromise body's defences (Illum 2003).

Table 6.3 Examples of drugs administered by nasal route

Phenylephrine, naphazoline Local vasoconstrictors administered as drops
Ipratropium bromide Both for local and systemic action. In patients with perennial rhinitis, approximately 10% of the drug is absorbed intranasally
Triamcinolone, fluticasone, beclomethasone, dexamethasone Local and systemic anti-inflammatory corticosteroids for allergic rhinitis and administered by spraying
Levocabastine Antagonist of H_1 histamine receptor, with local and systemic action. It is used as an aerosol
Sodium cromoglycate As a spray or nasal solution, this drug stabilizes mast cells preventing the release of chemical mediators from inflammation. It is used for rhinitis prophylaxis and treatment
Live (attenuated) influenza viruses Influenza vaccine in the form of nasal spray
Butorphanol tartrate Opioid analgesic. It is given as nasal spray as preoperative or preanaesthetic medication, as well as for pain relief during parturition and for migraine or headache
Organic products Insulin, calcitonin, buserelin, desmopressin, nafarelin, oxytocin (peptide hormones), recombinant interferon-alpha D (protein)

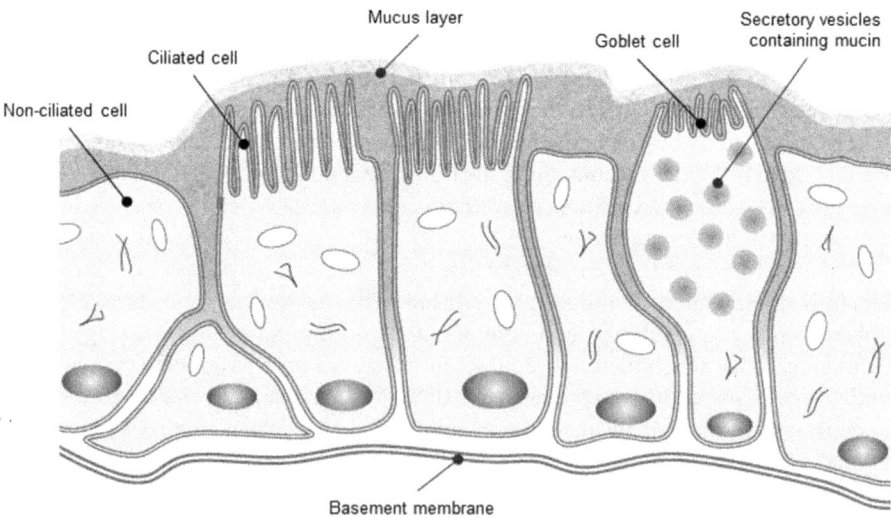

Fig. 6.11 Schematic representation of the nasal (columnar, non-keratinized) epithelium, where the ciliated cells responsible for the mucociliary clearance mechanism are observed

6.4.2.1 Physiological Factors Influencing the Nasal Absorption of Drugs

As in the case of pulmonary drug administration, nasal absorption occurs mainly by passive diffusion, both by the transcellular path (for medium- or low-polar active ingredients) and the paracellular path (very polar molecules with molecular weight <1 kDa, such as sodium cromoglycate). Nevertheless, the major obstacle to drug absorption by nasal route is the previously described mucociliary clearance of substances. This defence mechanism advances at a speed close to 5–6 mm/min; therefore, generally no particle remains to stay in contact with the nasal mucosa for more than 30 min (Merkus et al. 1998).

It is a route of administration with great interindividual variability mainly due to the several physiological factors relevant for the absorption process: nasal cavity dimensions, which determine the surface available for absorption; blood supply; metabolizing capacity of the mucosa; and the composition and volume of nasal secretions. Particularly important are certain pathological conditions, especially allergic and/or infectious diseases, which may greatly modify the level of nasal mucus secretion, as well as the area vascularization.

6.4.2.2 Technological Factors that Influence the Nasal Absorption of Drugs

In most nasal formulations, deposition site is controlled by adjusting variables such as particle diameter and size distribution and the velocity of the aerosol particles.

Inspired particles are prone to downward pulling of gravity, and thus large, dense particles fail to deep penetrate into nasal cavity. Moreover, the impaction mechanism is most likely to occur when the airstream carrying an aerosol particle changes direction; due to the shape of the nasal cavity, deposition of particles predominates within the anterior nasal region, where there is very little absorption.

On the other hand, although the posterior nasal mucosa has greater permeability, depositing a drug in this area will involve a faster ciliary elimination. Therefore, it is accepted that drugs with slow absorption should be deposited in the anterior part of the nose, while those that are quickly absorbed should be deposited in the back of the nose (Rogerson and Parr 1990).

The most commonly used dosage forms are solutions (nasal drops) and aerosols (nasal spray, both solutions and suspensions), although there are others, such as powder formulations, gel, creams or, more sophisticated, microspheres or liposomes. Drops are easy to formulate and apply, although it is difficult to control the applied dose. Conversely, aerosol formulations are packaged in suitable devices capable of delivering a well-defined dose. A key factor for the choice of the formulation type is the site of absorption preferred; aerosols are able to a deeper deposition of drugs than nasal drops.

Other factors that have to be considered are the pH, viscosity and osmolarity of the pharmaceutical formulation, as well as the presence of certain specialized excipients. With regard to the pH, it should be between 4.5 and 6.5 to avoid mucosa irritation. However, in the case of ionizable active ingredients, it may be considered to formulate at a pH outside that range to optimize their absorption by maximizing the proportion of the nonionized form (Arora et al. 2002).

Drug expulsion by the cilia may be reduced (at least to some extent) by the use of formulations with muco-adhesive materials (e.g. polyacrylic acid, cellulose polymers). They attach to the mucus layer prolonging the drug contact time with the epithelial surface.

Mucosal permeability modifying substances (absorption enhancers), such as bile salts, fatty acids, surfactants, chelants (such as EDTA), organic solvents (DMSO, ethanol) and others, may also be employed. There are many mechanisms by which nasal absorption of a drug can be promoted, the most common being the reversible modification of the epithelial barrier structure. However, toxicological aspect is fundamental, since most of these substances are irritating, and there is no safety data regarding to their long-term use (Cai et al. 2018).

References

Adams E, Coomans D, Smeyers-Verbeke J et al (2002) Non-linear mixed effects models for the evaluation of dissolution profiles. Int J Pharm 240:37–53

Amidon GL, Lennernäs H, Shah VP et al (1995) A theoretical basis for a biopharmaceutic drug classification: the correlation of in vitro drug product dissolution and in vivo bioavailability. Pharm Res 12:413–420

Aps JK, Martens LC (2005) Review: the physiology of saliva and transfer of drugs into saliva. Forensic Sci Int 150:119–131. https://doi.org/10.1016/j.forsciint.2004.10.026

Arora P, Sharma S, Garg S (2002) Permeability issues in nasal drug delivery. Drug Discov Today 7:967–975

Benet L, Broccatelli F, Oprea T (2011) BDDCS applied to over 900 drugs. AAPS J 13:519–547. https://doi.org/10.1208/s12248-011-9290-9

Beule AG (2010) Physiology and pathophysiology of respiratory mucosa of the nose and the paranasal sinuses. GMS Curr Top Otorhinolaryngol Head Neck Surg 9:Doc07. https://doi.org/10.3205/cto000071

Blumenthal HP, Fung L, McNiff EF et al (1977) Plasma nitroglycerin levels after sublingual, oral and topical administration. Br J Clin Pharmacol 4:241–242. https://doi.org/10.1111/j.1365-2125.1977.tb00703.x

Bodiford AB, Kessler FO, Fermo D, et al. (2013). Elevated international normalized ratio with the consumption of grapefruit and use of warfarin. SAGE Open Med Case Rep. 1 (2050313X13511602). https://doi.org/10.1177/2050313X13511602

Boylan J, Nail (2002) Parental products. In: Banker GS, Rhodes CT (eds) Modern pharmaceutics. Marcel Dekker, New York, pp 576–625

Braybrooks M, Barry B, Abbs ET (1975) The effect of mucin on the bioavailability of tetracycline from the gastrointestinal tract; in vivo, in vitro correlations. J Pharm Pharmacol 27:508–515

Brodin B, Steffansen B, Nielsen C (2010) Structure and function of absorption barriers. In: Steffansen B, Brodin B, Nielsen C (eds) Molecular biopharmaceutics. Pharmaceutical Press, London, pp 115–133

Cai Q, Feng L, Yap KZ (2018) Systematic review and meta-analysis of reported adverse events of long-term intranasal oxytocin treatment for autism spectrum disorder. Psychiatry Clin Neurosci 72:140–151. https://doi.org/10.1111/pcn.12627

Cao X, Yu LX, Sun D (2008) Drug absorption principles. In: Krishna R, Yu L (eds) Biopharmaceutics applications in drug development. Springer, Boston, pp 75–100. https://doi.org/10.1007/978-0-387-72379-2_4

Casabó Alós VG, Merino Sanjuán M, Jiménez Torres VN (1998a) Factores Fisiológicos en la Absorción Gastrointestinal (II). In: Berrozpe JD, Lanao JM, Guitart CP (eds) Biofarmacia y Farmacocinética. Síntesis S.A., Madrid, pp 163–187

Casabó Alós VG, Merino Sanjuán M, Jiménez Torres VN (1998b) Factores Fisiológicos en la Absorción Gastrointestinal (I). In: Berrozpe JD, Lanao JM, Guitart CP (eds) Biofarmacia y Farmacocinéitca. Síntesis S.A., Madrid, pp 145–162

Constantinides PP, Chakraborty S, Shukla D (2016) Considerations and recommendations on traditional and non-traditional uses of excipients in oral drug products. AAPS Open 2:3. https://doi.org/10.1186/s41120-016-0004-3

Dagar S (2007) Pulmonary delivery. In: Gibaldi M, Lee M, Desai A (eds) Gibaldi's drug delivery systems in pharmaceutical care. American Society of Health-System Pharmacists, Bethesda, p 80

de Boer AG, Breimer DD, Pronk J et al (1980) Rectal bioavailability of lidocaine in rats: absence of significant first-pass elimination. J Pharm Sci 69:804–807. https://doi.org/10.1002/jps.2600690716

de Boer AG, Moolenaar F, de Leede LG, Breimer DD (1982) Rectal drug administration: clinical pharmacokinetic considerations. Clin Pharmacokinet 7:285–311

Desai A (2007) Rectal, vaginal and urethral delivery. In: Gibaldi M, Lee M, Desai A (eds) Gibaldi's drug delivery systems in pharmaceutical care. American Society of Health-System Pharmacists, Bethesda, pp 95–102

Desai U, Lee EC, Chung K et al (2007) Lipid-lowering effects of anti-angiopoietin-like 4 antibody recapitulate the lipid phenotype found in angiopoietin-like 4 knockout mice. Proc Natl Acad Sci 104:11766–11771. https://doi.org/10.1073/pnas.0705041104

Dokoumetzidis A, Machera P (2006) A century of dissolution research: from Noyes and Whitney to the biopharmaceutics classification system. Int J Pharm 321:1–11

Douglass C (2018) Avoiding errors when administering injectable phenytoin to a child in status epilepticus. Nurs Child Young People 30:35–38. https://doi.org/10.7748/ncyp.2018.e965

Farr S, Kellaway I, Taylor G (1990) Drug delivery to the respiratory tract. In: Florence A, Salole E (eds) Routes of drug administration. Butterworth & Co., London, pp 48–77

FDA (2016). JADELLE (levonorgestrel implants) for subdermal use [WWW Document]

FDA and CDER (2014). Parenteral Dilantin (Phenytoin Sodium Injection, USP)

Florez J (1998) Absorción, distribución y eliminación de los fármacos. In: Florez J, Armijo JA, Mediavilla A (eds) Farmacología Humana. Masson S.A., Barcelona, pp 47–75

Fontes Ribeiro CA (2005) Influence of food and drugs on the bioavailability of antiepileptic drugs. In: Majkowski J, Bourgeois BFD, Patsalos PN et al (eds) Antiepileptic drugs: combination therapy and interactions – Google books. Cambridge Press, Cambridge, pp 93–110

Gacto P, Miralles F, Pereyra JJ et al (2009) Haemostatic effects of adrenaline–lidocaine subcutaneous infiltration at donor sites. Burns 35:343–347. https://doi.org/10.1016/j.burns.2008.06.019

Garcia-Contreras L, Ibrahim M, Verma R (2015) Inhalation drug delivery devices: technology update. Med Devices Evid Res 8:131. https://doi.org/10.2147/MDER.S48888

Garrigues Pelufo TM, del Val Bermejo Sanz M (1998) Absorción por Vía Perlingual, Bucal y Rectal. In: Berrozpe JD, Lanao JM, Guitart CP (eds) Biofarmacia y Farmacocinética. Síntesis S.A., Madrid, pp 349–366

Gerk PM, Yun ABC, Shargel L (2016) Phisyologic factors related to drug absorption. In: Shargel L, Yu A (eds) Applied biopharmaceutics & pharmacokinetics, 7th edn. The McGraw-Hill Companies, New York, pp 373–414

Gidal BE (2006) Drug absorption in the elderly: biopharmaceutical considerations for the antiepileptic drugs. Epilepsy Res 68:65–69. https://doi.org/10.1016/J.EPLEPSYRES.2005.07.018

Gleeson MP (2008) Generation of a set of simple, interpretable ADMET rules of thumb. J Med Chem 51:817–834. https://doi.org/10.1021/jm701122q

Hall JE (2015) Secretory functions of the alimentary tract. In: Hall JE, Guyton AC (eds) Guyton and hall textbook of medical physiology. Elsevier, Philadelphia, pp 817–832

Hickey AJ (2006) Delivery of drugs by the pulmonary route. In: Banker GS, Rhodes CT (eds) Modern pharmaceutics. Marcel Dekker, New York, pp 718–746

Hwang H, Kim H-S, Jeong H-S et al (2016) Liposomal angiogenic peptides for ischemic limb perfusion: comparative study between different administration methods. Drug Deliv 23:3619–3628. https://doi.org/10.1080/10717544.2016.1212951

Illum L (2003) Nasal drug delivery--possibilities, problems and solutions. J Control Release 87:187–198

Kerns EH, Di L (2008) Solubility. In: Kerns EH, Di L (eds) Drug-like properties: concepts, structure design and methods: from ADME to toxicity optimization. Elsevier, Burlington, pp 56–85

Lim TY, Poole RL, Pageler NM (2014) Propylene glycol toxicity in children. J Pediatr Pharmacol Ther 19:277–282. https://doi.org/10.5863/1551-6776-19.4.277

Lopes CM, Bettencourt C, Rossi A et al (2016) Overview on gastroretentive drug delivery systems for improving drug bioavailability. Int J Pharm 510:144–158. https://doi.org/10.1016/j.ijpharm.2016.05.016

Lucendo Villarín AJ, Noci Belda J (2004) Prevention and treatment of extravasation of intravenous chemotherapy. Enferm Clin 14:122–126. https://doi.org/10.1016/S1130-8621(04)73869-3

Madhav NV, Shakya AK, Shakya P et al (2009) Orotransmucosal drug delivery systems: a review. J Control Release 140:2–11. https://doi.org/10.1016/j.jconrel.2009.07.016

Marasanapalle VP, Li X, Jasti BR (2011) Effects of food on drug absorption. In: Hu M, Li X (eds) Oral bioavailability. John Wiley & Sons, Inc., Hoboken, pp 221–232

McLennan DN, Porter CJH, Charman SA (2005) Subcutaneous drug delivery and the role of the lymphatics. Drug Discov Today Technol 2:89–96. https://doi.org/10.1016/j.ddtec.2005.05.006

McTague A, Martland T, Appleton R (2018) Drug management for acute tonic-clonic convulsions including convulsive status epilepticus in children. Cochrane Database Syst Rev 1:CD001905. https://doi.org/10.1002/14651858.CD001905.pub3

Melander A (1978) Influence of food on the bioavailability of drugs. Clin Pharmacokinet 3:337–351. https://doi.org/10.2165/00003088-197803050-00001

Merkus FW, Verhoef JC, Schipper NG et al (1998) Nasal mucociliary clearance as a factor in nasal drug delivery. Adv Drug Deliv Rev 29:13–38

Mishra P, Stringer MD (2010) Sciatic nerve injury from intramuscular injection: a persistent and global problem. Int J Clin Pract 64:1573–1579. https://doi.org/10.1111/j.1742-1241.2009.02177.x

Mohseni M, Ebneshahidi A (2014) The flow rate accuracy of elastomeric infusion pumps after repeated filling. Anesth Pain Med 4:e14989. https://doi.org/10.5812/aapm.14989

Montenegro-Nicolini M, Morales JO (2017) Overview and future potential of buccal mucoadhesive films as drug delivery Systems for Biologics. AAPS PharmSciTech 18:3–14. https://doi.org/10.1208/s12249-016-0525-z

Murdan S (2013) Solutions. In: Aulton ME, Taylor KMG (eds) Aulton's pharmaceutics: the designs and manufacture of medicines. Churchill Livingstone/Elsevier, London, p 400

Nakamura T, Nakanishi T, Haruta T et al (2010) Transport of ipratropium, an anti-chronic obstructive pulmonary disease drug, is mediated by organic cation/carnitine transporters in human bronchial epithelial cells: implications for carrier-mediated pulmonary absorption. Mol Pharm 7:187–195. https://doi.org/10.1021/mp900206j

Øie S, Benet L (2002) The effect of route administration and distribution on drug action. In: Banker GS, Rhodes CT (eds) Modern pharmaceutics. Marcel Dekker, New York, pp 187–214

O'Hara K, Whright IM, Schneider JJ et al (2015) Pharmacokinetics in neonatal prescribing: evidence base, paradigms and the future. British Journal of Clinical Pharmacology 80 (6):1281–1288

Partridge H, Perkins B, Mathieu S et al (2016) Clinical recommendations in the management of the patient with type 1 diabetes on insulin pump therapy in the perioperative period: a primer for the anaesthetist. Br J Anaesth 116:18–26. https://doi.org/10.1093/bja/aev347

Peters DM, Hulisz DT (2015) The return of inhaled insulin. Pharm Choice Newsl. 17 (7), pp 334

Plá Delfina JM, Martín Villodre A (1998) Absorción gastrointestinal. In: Berrozpe JD, Lanao JM, Guitart CP (eds) Biofarmacia y Farmacocinética. Síntesis S.A., Madrid, pp 129–144

Puttick MP, Phillips P (1994) Itraconazole: precautions regarding drug interactions and bioavailability. Can J Infect Dis 5:179–183

Qiu Y (2009) Rational design of oral modefied-release drug delivery systems. In: Qiu Y, Chen Y, Zhang GGZ et al (eds) Developing solid oral dosage forms: pharmaceutical theory and practice. Elsevier, Burlington, pp 461–500

Qiu Y, Zhang G (2009) Development of modified-release solid oral dosage forms. In: Qiu Y, Chen Y, Zhang GGZ et al (eds) Developing solid oral dosage forms: pharmaceutical theory and practice. Elsevier, Burlington, pp 501–518

Rathbone MJ, Pather I, Şenel S (2015) Oral mucosal drug delivery and therapy. Advances in delivery science and technology. In: Rathbone MJ, Senel S, Pather I (eds) Oral mucosal drug delivery and therapy. Springer US, Boston, pp 17–30

Rogerson A, Parr G (1990) Nasal drug delivery. In: Florence A, Salole E (eds) Routes of drug administration. Butterworth & Co., London, pp 1–29

Rowland M, Tozer TN (1995) Absorption. In: Rowland M, Tozer TN (eds) Clinical pharmacokinetics and pharmacodynamics: concepts and applications. Lippincott Williams & Wilkins, Philadelphia, pp 119–136

Saxen MA (2016) Pharmacologic management of patient behavior. In: McDonald and Avery's dentistry for the child and adolescent. Elsevier, pp 303–327. https://doi.org/10.1016/B978-0-323-28745-6.00017-X

Scheuch G, Kohlhaeufl MJ, Brand P et al (2006) Clinical perspectives on pulmonary systemic and macromolecular delivery. Adv Drug Deliv Rev 58:996–1008. https://doi.org/10.1016/J.ADDR.2006.07.009

Shargel L, Wu-Pong S, Yu A (2016) Pharmacokinetics of oral absorption. In: Shargel L, Yu A (eds) Applied biopharmaceutics and pharmacokinetics, 5th edn. The McGraw-Hill Companies, New York, pp 177–204

Sørensen JM (2002) Herb–drug, food–drug, nutrient–drug, and drug–drug interactions: mechanisms involved and their medical implications. J Altern Complement Med 8:293–308. https://doi.org/10.1089/10755530260127989

Xie B, Liu Y, Guo Y et al (2018) Progesterone PLGA/mPEG-PLGA hybrid nanoparticle sustained-release system by intramuscular injection. Pharm Res 35:62. https://doi.org/10.1007/s11095-018-2357-x

Nanotechnology and Drug Delivery

7

Germán Abel Islan, Sergio Martin-Saldaña, Merari Tumin Chevalier, Vera Alejandra Alvarez, and Guillermo Raúl Castro

7.1 Introduction

In 1959, Prof. Richard Feynman (Nobel Prize awardee, 1965) delivered a talk in the American Physical Society called "There´s plenty of Room at the Bottom." The theoretical talk focused on the use of atoms to solve material problems of the societies. The term nanotechnology was coined 15 years later, in 1974, by Professor Norio Taniguchi at the University of Tokyo (Japan). Prof. Taniguchi was working in the concepts of ion and laser beams to make building blocks at nanoscale.

Now, nanotechnology could be defined as the manipulation of atoms and molecules to design and develop novel structures and devices with special purposes and superior biophysical properties. The origin of the prefix "nano" comes from the Greek word νᾶνος (i.e., *nanus* in Latin), *meaning* "dwarf" or minute, very small. From a scientific point of view, the nanoscale is in the range of 1×10^{-8} to 10^{-9} m, which can be correlated with atom and molecule volumes. A nanometer is very small, one billionth of a meter. For example, a human hair diameter is in the range of 70–90 μm (equivalent to $7\text{–}9 \times 10^4$ nm), close to human eye resolution limit (Table 7.1). A nanofiber is about 1000 times smaller than a human hair, and electronic microscopes are required to see it. For comparative purposes, 1 nm is the size of one molecule of glucose and the linear size of 10 hydrogen atoms properly

G. A. Islan (✉) · G. R. Castro
Laboratorio de Nanobiomateriales, CINDEFI, Departamento de Química, Facultad de Ciencias Exactas, Universidad Nacional de La Plata-CONICET (CCT La Plata), La Plata, Argentina
e-mail: germanislan@biol.unlp.edu.ar

S. Martin-Saldaña · M. T. Chevalier
Gihon Laboratorios Químicos SRL, Mar del Plata, Argentina

V. A. Alvarez
UNMdP, CONICET, Instituto Investigación de Ciencia & Tecnología de Materiales INTEMA, Grupo Materiales Compuestos Termoplásticos (CoMP), Mar Del Plata, Argentina

© Springer Nature Switzerland AG 2018
A. Talevi, P. A. M. Quiroga (eds.), *ADME Processes in Pharmaceutical Sciences*,
https://doi.org/10.1007/978-3-319-99593-9_7

Table 7.1 Equivalence of nanometers to other units

1 nanometer
10 angstroms
1×10^{-3} μm
1×10^{-9} meter
3.281×10^{-9} feet
39.37×10^{-9} inches

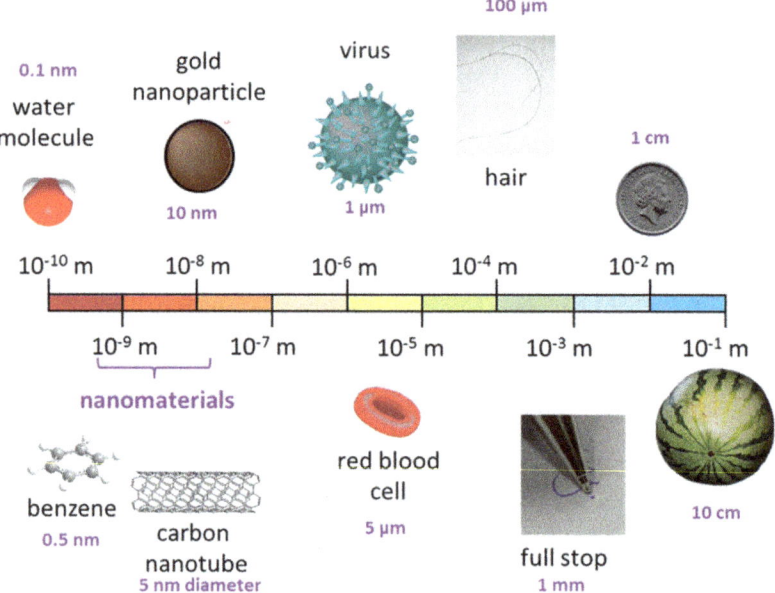

Fig. 7.1 Physical dimensions of materials in nanotechnology. (This work is licensed under a Creative Commons Attribution 4.0 International License. https://chembam.com/definitions/nano technology/)

aligned. Also, the double DNA is approximately 2–12 nm in cross section depending on the topology.

▶ **Definition** The most common definition of nanomaterial speaks of any physical object with at least having 1–100 nm in one inner or outer dimension. However, there are some controversies related to this definition since "nanotechnological effects" can be reached in sizes larger than 100 nm for devices used in biological systems. In fact, the increased drug solubilities, reduced toxicity, improved bioavailability, and physiological response in pharmaceuticals, cosmeceuticals, and foods are not strictly limited to 100 nm ranges (Fig. 7.1).

Two main strategies are currently used to produce different types of nanomaterials from bulk materials: top-down and bottom-up processing. Top-down nanofabrication implies sculpting (or milling) materials making self-structures

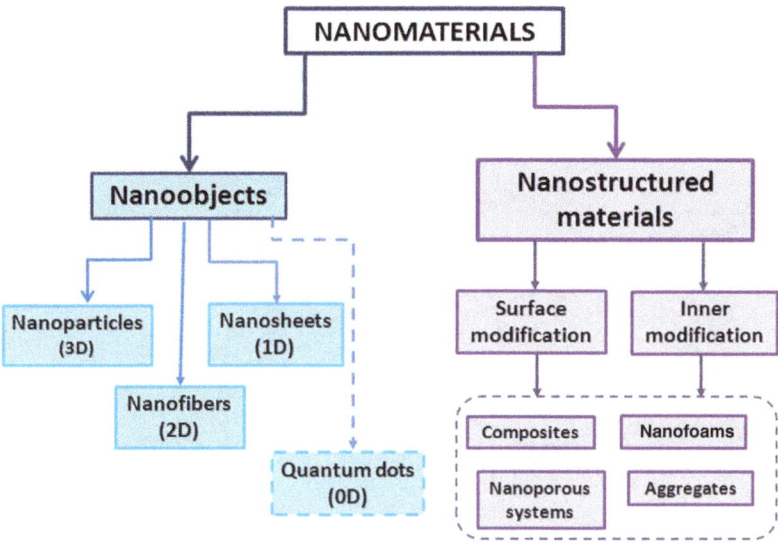

Fig. 7.2 General classification of nanomaterials

with defined topology. Typical examples are semiconductors currently used in microelectronics. On the contrary, bottom-up is a methodology in which atoms and/or molecules are stacked onto each other to create a nanostructure. There are a large diversity of nanomaterials considering the structure and the chemical composition of the reagents. The nanomaterials can be classified based on dimensions, chemical compositions, or structural characteristics. The International Organization for Standardization established the first ISO norm for nanoobjects in 2008 and later revised it in 2015: ISO/TS 80004-2:2015. The ISO norm established 3 types of nanoobjects, based on a 1–100 nm size in at least 1 dimension of the object. The ISO classification defined NPs as 3D objects (three external dimensions in the nanoscale), nanofibers as 2D objects (two external dimensions in the nanoscale), and nanoplates as 1D object (one external dimension in the nanoscale). Besides, quantum dots (i.e., crystal semiconductors) are considered zero-dimensional (point-like) nanoobjects. Also, complex material structures containing surface or inner nanoscale structures must be considered as complex nanoobjects (Fig. 7.2).

Paradigm shifts in the materials science arena are encouraged by the lack or scarcity of natural resources, the intensive use of raw materials, the high value-added products, and the social consensus at global scale for the development of green processes, all of which motivated a huge impulse of nanotechnology over the world. In fact, more than 60 countries developed or are in the trend to develop national nanotechnology research programs (Liu et al. 2009). The results of nanotechnology initiatives in several countries can be observed in the increase of publication numbers from 2001 to 2010, from 173 to 1730 papers in the nanotechnology field reported in the Science Citation Index, implying an 84-fold growth in 9 years with a total amount of 11,991 publications (Biglu et al. 2011). Moreover, a good correlation

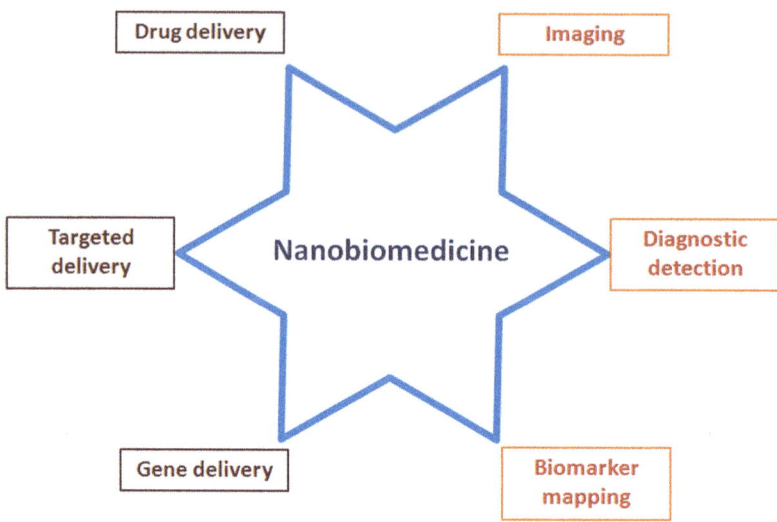

Fig. 7.3 Potential uses of nanoobjects in biomedicine

between publications and patents is observed in the nanotechnology field. The economic relevance of nanotechnology can be followed by the increased number and type of patents published around the world. The evolution of "nano-patents" with non-overlapping applications obtained from 15 repositories during 1991–2008 went from 224 to 10,067. Interestingly, the top patent applicants came from big companies but also top universities and research centers from the USA, Taiwan, the Russian Federation, and China (Dang et al. 2010). In 2017, the US Patent Office (USPTO) published 20,187 patents; among them 9145 were granted patents and 11,042 were applications. In the same year, the European Patent Office (EPO) published 4019 nanotech patents, 2386 of which were granted and 1633 were applications (Anonymous 2018). Despite the multiplicity of application fields of nanotechnological products, 79% of the overall patents come from the biomedicine area, including diagnostic and surgical applications (Bonta et al. 2017). The main applications of nanoobjects in the biomedical area can be categorized in diagnostics and therapeutics. For diagnosis different nanoobjects can be used for imaging and biomarker detection, whereas therapeutics include gene and molecular delivery and targeted therapies (Fig. 7.3).

Nanotechnology is providing powerful tools for the pharmaceutical industry such as polymeric, solid lipid, metallic, and magnetic NPs, liposomes, dendrimers, and other nanostructures. This field is now called nanopharmaceutics and allows the understanding of some of the major mechanisms of current diseases on a molecular basis. Drug delivery mediated by nanosystems is particularly relevant since it requires a small amount of the nanodevice to deliver molecules under physiological conditions, in the therapeutic window, with tailored kinetics, minimizing undesirable side effects. However, some pertinent considerations must be taken into account for the design and synthesis of drug delivery nanosystems.

7.2 Physicochemical Principles of Nanosized Drug Delivery Systems (DDS)

7.2.1 Surface-to-Volume Ratio

The effect of size on the physical properties of nanoobjects of the same chemical composition but with different diameters is dramatic. In fact, the surface-to-volume ratio changes when the dimension of an object changes. A lower volume implies high relative surface area. For example, let us consider the sphere:

The equations of surface and volume of the sphere are as follows:

$$\text{Surface} : S = 4\pi r^2 \tag{7.1}$$

$$\text{Volume} : V = \frac{4}{3}\pi r^3 \tag{7.2}$$

$$\text{Aspect ratio} : \frac{S}{V} = \frac{3}{r} \tag{7.3}$$

By decreasing the radius of the sphere, the relative surface area per volume unit (i.e., aspect ratio) of the object increases. In a NP, more electrons will be exposed to the surface of the nanoobject when the diameter of the NP decreases. Accordingly, the NP would be more reactive to the environmental conditions. This effect associated with the resonant oscillation of electrons conducted at the interface between negative and positive permittivity materials stimulated by incident light is known as surface plasmon resonance (SPR). This phenomenon can be observed in Fig. 7.4 where the increase of gold colloidal nanoparticle diameter implies a color shift from red to violet (Amendola et al. 2017).

7.2.2 Effect of Particle Shape and Aspect Ratio

The design of nanoobjects for biomedical applications requires meeting the characteristics to perform the desired tasks avoiding undesirable side effects in the organism. In general, ratio (ratio aspect) is considered the main descriptor for nanoobjects, but not all nanoobjects are spheres or spheroids (Fig. 7.2). The curves of aspect ratio of different objects showing the correlation between surface and volume are shown in Fig. 7.5.

In nanotechnology, topological changes of nanoobjects with the same chemical composition can induce different physical and biological responses, including toxic ones (Hassellöv et al. 2008). For example, spiked structures with the same surface area-to-volume ratio and chemical composition have higher reactivity than cubic NPs. Many factors can affect the synthesis of nanoobjects with different aspect ratios. Particularly, the NPs' shapes depend on an equilibrium process based on

Fig. 7.4 Effect of gold NPs' sizes on light absorption. (Source: Colloidal gold – Wikipedia. (By Aleksandar Kondinski (kondinski.webs.com) [GFDL (http://www.gnu.org/copyleft/fdl.html) or CC BY-SA 4.0 (https://creativecommons.org/licenses/by-sa/4.0)], via Wikimedia Commons)

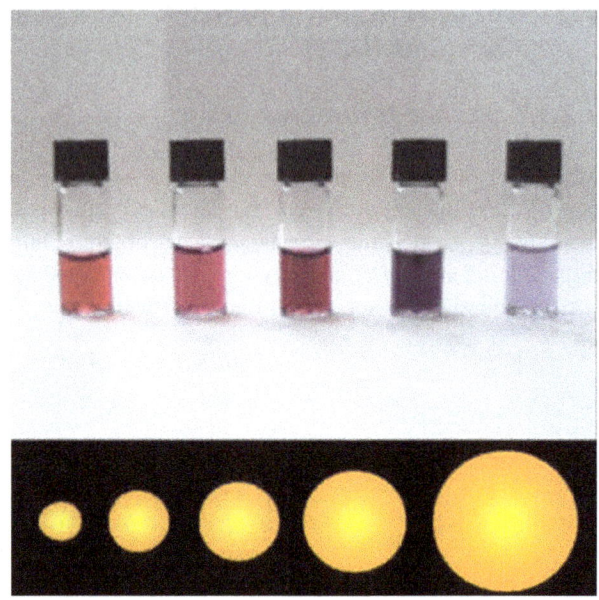

energy minimization and entropy, which is described by the equation of energy distribution postulated by Boltzmann:

$$F(E) = A.e^{-\left(\frac{E_a}{K_b t}\right)} \tag{7.4}$$

where $F(E)$ is the function of free energy distribution, A is constant, E_a represents free energy, k_b denotes the Boltzmann constant, and T is temperature expressed in kelvin degrees.

Experimentally, the aspect ratio of NPs depends on many intrinsic and extrinsic parameters during their preparation. Intrinsic parameters are the concentrations and molar ratios of reactants (substrates, precursors, and catalysts if required); among extrinsic parameters, we can mention environmental factors such as temperature, stirring rate, co-templates, pressure, media, etc.

NPs with different aspect ratios show distinctive effects on cell function and viability. For instance, mesoporous silica NPs with larger aspect ratios display more pronounced effects on cell metabolism including changes in cell proliferation, adhesion, cytoskeleton formation, and apoptosis, compared with the same mesoporous silica NPs with smaller aspect ratio (Huang et al. 2010).

7.2.3 Surface Chemistry of Nanoobjects

The effects of contact between nanoobjects and the cell surface depend on many biophysical parameters, and one of the most relevant are the chemical surface

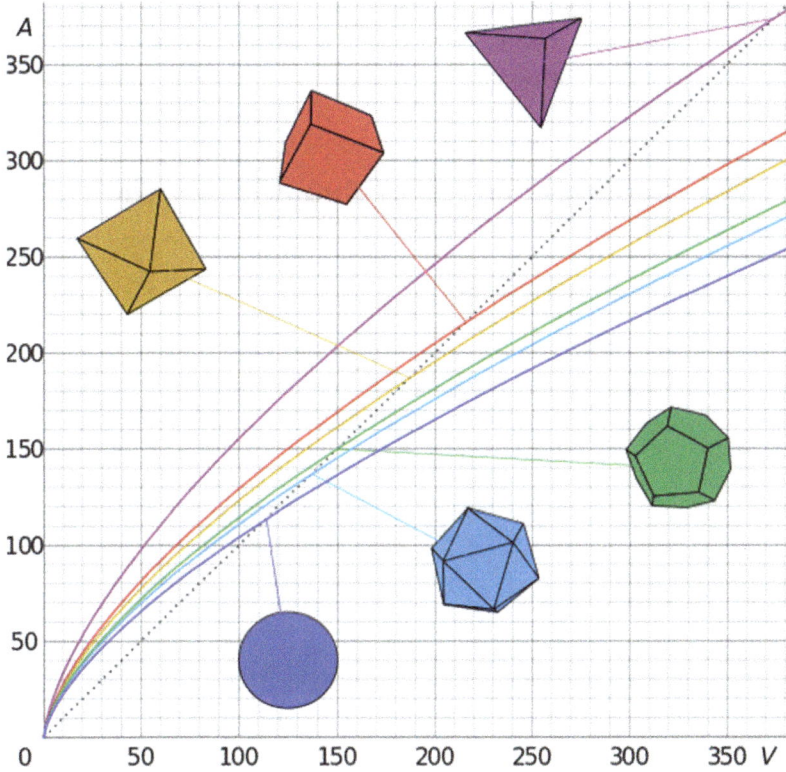

Fig. 7.5 Correlation between surface area and volume for different geometries. (Source: Wikipedia. By derivative work: Cmglee [CC BY-SA 3.0 (https://creativecommons.org/licenses/by-sa/3.0)], via Wikimedia Commons)

characteristics of the NP. Mammalian cell membranes are negatively charged which implies that nanodevices that are positively charged are able to interact with the cell by electrostatic attraction. On the other side, nanodevices with negative surface charge can be easily opsonized by macrophages in vivo.

Surface modifications of nanodevices are relevant for many reasons such as specific targeted delivery to cells that show specific markers or receptors in the cell membrane surface or the escape from the immune systems with a consequent increase of circulation time in the body (Kamaly et al. 2012; Saneja et al. 2017; Unger et al. 2008).

The main parameter to determine the characteristic charge and stability of the NPs is the zeta (ζ) potential, defined as the electric potential at the interface double layer between the liquid media and the nanodevice. NPs with ζ potential of -30 mV are considered strongly negative; between -10 and 10 mV, neutral; and higher than $+30$ mV, strongly positive.

Accordingly, polycationic molecules (i.e., polylysine, polyethyleneimine, chitosan) are sometimes used for NP capping, being able to permeate membranes and disrupt cells wall of microorganisms for gene and drug delivery (Barratt 2000).

7.3 Drug Encapsulation and In Vitro Release from Nanoparticulate Systems

7.3.1 Drug Loading Methods

Drug encapsulation (DE) is the process in which a certain drug is incorporated into a selected nanosystem. Two parameters are commonly associated with this concept: drug loading (DL) and entrapment/encapsulation efficiency (EE). They are defined as

$$\text{DL} = \frac{W_t}{M_t} \tag{7.5}$$

$$\text{EE} = \frac{W_t}{W_o} \times 100 \tag{7.6}$$

where W_t is the total mass of drug incorporated into the NPs, W_o is initial mass of the drug used to make the formulation, and M_t is the mass of the nanocarrier.

The DL considers the amount of drug incorporated into the NPs in relationship with the total mass of the carrier. This value is interesting to estimate what amount of NPs needs to be administered to reach the therapeutic concentrations of the drug in the respective target tissues/organs. On the other hand, the EE is an interesting parameter to evaluate the efficacy of the process to incorporate the drug into designed nanosystems after preparation. This is very important for scaling up processes and industrial applications (Jyothi et al. 2010).

Different techniques have been reported to optimize the drug encapsulation in a wide range of nanosystems. The basis of drug encapsulation is the presence of drug-matrix interactions due to chemical interactions such as hydrogen bonding, ionic or dipole interactions, hydrophobic affinity or covalent bonding, or just physical interactions such as entrapment, precipitation, or adsorption on the surface. However, in most cases more than one loading mechanism is present (Marco et al. 2010). It is also important to mention that the *DE* strictly depends on the nature of the drug and the type of NPs (e.g., solid lipid NPs, dendrimers, biopolymeric NPs, liposomes, polymeric micelles) and the method selected to prepare the NPs (Ochubiojo et al. 2012). For example, the incorporation of the drug into NPs can be modified by changing the pH of synthesis to displace the equilibrium of the drug to species with more affinity to the NP, setting a temperature in which the drug solubility in the matrix is higher or simply changing the sequence of adding the excipients.

The drug can be incorporated into NPs by two main methods:

- Addition of the drug during preparation
- Addition to the formed NPs

During the first one, the drug is solubilized in the NP matrix and then NPs are synthesized. At this point, and to obtain desirable EE, it is important to test the solubility of the drug in the components that will constitute the NPs. For example, during preparation of lipid NPs, liquid oils are added to the melted matrix to increase the solubility of the drug and therefore increase the drug loading values (Islan et al. 2016a, b). In the case of polymeric matrices, a solution of the drug and the polymer is simultaneously prepared and then NPs are formed by the addition of a crosslinking or stabilizer agent (Islan et al. 2015).

In the second case, the NPs are prepared and the drug is incorporated in a subsequent step. The drug can be loaded by absorption/adsorption mechanisms promoted by a drug concentration gradient (Singh and Lillard 2009). Also, the drug can be chemically bonded to the surface of the NPs by covalent reactions (Popat et al. 2012). The presence of weak forces like hydrogen bonds or electrostatic interactions can also play an important role to capture the drug inside the NPs (Pakulska et al. 2016; Zhang et al. 2010).

7.3.2 Determination of Drug Loading: Direct/Indirect Techniques

Once NPs are loaded with the desired drug, the next step is the determination of how much drug was effectively incorporated into NPs.

In the direct method, the NPs need to be purified in order to remove the non-encapsulated drug and the excess of surfactant, stabilizer, and reducing or crosslinking agents previously used during NP preparation (Robertson et al. 2016). The methods include filtration, size exclusion chromatography, centrifugation, density gradient separation, or dialysis (Gaspar et al. 2016). Purification of samples not only improves formulation quality but also separates NPs into discrete sizes and shapes. After purification, many methods include the use of organic solvents like dichloromethane and acetonitrile, or pH-based extraction to disintegrate the NP matrix and release all the drug content (Amini et al. 2017). In a final step, the drug is analytically determined by UV-Vis spectroscopy, HPLC, fluorescence, electrochemical detection, or other spectroscopic methods (de Oliveira et al. 2017).

In the indirect method, the non-encapsulated drug is analytically measured after separation of NPs (de Oliveira et al. 2017; Martins et al. 2017). The separation process can be carried out by physical methods like dialysis through membranes, centrifugal filters, or ultracentrifugation. The encapsulation efficiency in this case is determined as follows:

$$EE = \frac{W_0 - (C_i \times V_t)}{W_0} \times 100 \tag{7.7}$$

where C_i is the concentration of the non-encapsulated drug after NPs separation, W_o is the initial mass of the drug used to make the formulation, and V_t is the total volume of the formulation. In all cases, a high sensitivity method is desirable to drug determination since some matrix components can interfere with the detection of the drug.

7.3.3 Mechanisms of Drug Release from NPs

Drug release from NPs depends on various environmental factors and is governed by two main processes: diffusion of the drug and matrix erosion/degradation as shown in Fig. 7.6a (Fu and Kao 2010). The physicochemical parameters which are involved in these mechanisms are described below.

7.3.4 Drug Solubility

It is expected that a decrease in particle size results in a higher surface free energy, enhancing the dissolution rate of the drug. This fact becomes an interesting issue for hydrophobic drugs with low bioavailability. NPs have proved to be useful since their high specific surface areas contribute to increasing the dissolution rate and solubility of this kind of drugs (Kwok and Chan 2014). However, some works have reported that in the case of nanosized crystallites the dissolution could be inhibited due to their small size, which could be attributed to a kinetic phenomenon (Tang et al. 2004). In this regard, every time that a nanosystem containing a drug is developed, the dissolution profile needs to be tested. It is commonly observed that the dissolution of a drug from NPs occurs in two steps. During the first one, the NP is solvated with a consequent solvation of the encapsulated drug at the time that the solvent takes away the molecules associated to the NP's surface. In a second stage, the diffusion of drug molecules from the inside of the matrix begins to release the cargo to the bulk dissolution media. This process can be mathematically modeled to predict the behavior of a nanoparticulate formulation considering Fick's diffusion laws. The presence of additional barriers (commonly known as "coatings") on the NP surface generates a diffusion boundary layer, which is responsible for the delay in the release delivery of the drug (Amoozgar et al. 2014).

7.3.5 Diffusion of the Drug Through the NP Matrix

One example of the process is shown in Fig. 7.6b, in which the SLN matrix was modified by the addition of liquid oils that not only modify the crystalline degree of

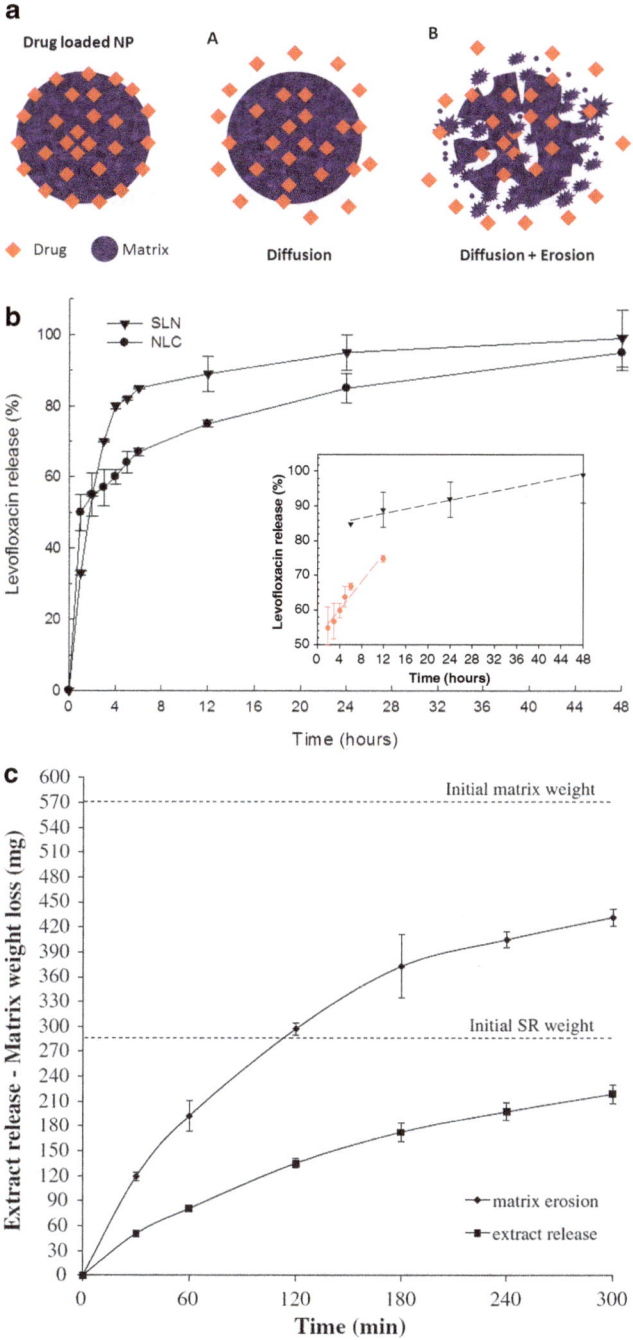

Fig. 7.6 Processes involved in drug release from NPs: (**a**) Scheme of diffusion and erosion mechanisms. (**b**) A release profile of levofloxacin in which the release can be changed by increasing the diffusion of the drug across the NP matrix. (With permission from Islan et al. 2016a). (**c**) The release of the drug as function of matrix erosion. (Reproduced with permission from Gallo et al. 2013)

the matrix but also the diffusion of the drug through the lipid matrix. As a result, a faster drug release is obtained since the mobility of the drug is higher in a liquid/oil phase than in a crystalline lipid (Islan et al. 2016a, b).

7.3.6   Degradation or Dissolution of the NP Matrix

Figure 7.6c describes the process in which a component is being released as the matrix is eroded (Singh and Lillard 2009).

It is possible to find different suitable mathematical models that can predict the release profile of a drug from NPs. The use of mathematical programs could become a routine tool for optimization and design of novel nanoformulations in terms of composition, geometry, dimensions, and preparation procedures to select the desired administration route, drug dose, and release profile (Siepmann and Siepmann 2008). These models consider the diffusion of the drug from the NP core and surface in addition to erosion processes of the matrix. One of the best-known models is Higuchi's model which contemplates that the release rate is proportional to the square root of time:

$$Q = [\,D\,(2 \times C_{\mathrm{T}} - C_{\mathrm{s}}) \times C_{\mathrm{s}} \times t\,]^{1/2} \qquad\qquad (7.8)$$

where Q is the amount of drug released at time t and per unit area of exposed NP surface, D is the coefficient of diffusion of the drug into the NP matrix, C_{T} is the total concentration, and C_{S} is the solubility of the drug in the polymer matrix (Higuchi 1967).

This model is valid when the drug is uniformly dispersed in a polymer matrix and no degradation of the components is observed. Other modifications of the Higuchi equation have been implemented to take into account parameters or phenomena like porosity, swelling, and erosion of the matrix (Siepmann and Siepmann 2008).

7.3.7   Methods to Measure Drug Release from NPs

Different methods have been proposed for the assessment of release from NPs. According to the NP characteristics, the method must ensure the effective separation of the NPs from the release media in order to measure the released drug (Zhou et al. 2016). Among these techniques, the membrane diffusion method shows to be the most reproducible; it is based on the utilization of dialysis bag/tube, reverse dialysis bag, or diffusion cells (Souza 2014). Other available techniques to force the separation of the NPs at each release time include ultracentrifugation, filtration, and centrifugal ultrafiltration. Usually all the release studies are assessed under controlled agitation and well-defined temperatures, related to the simulated physiological conditions (Modi and Anderson 2013). Some methods are shown in Fig. 7.7.

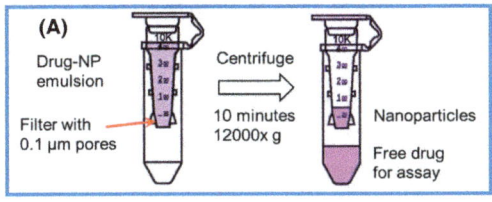

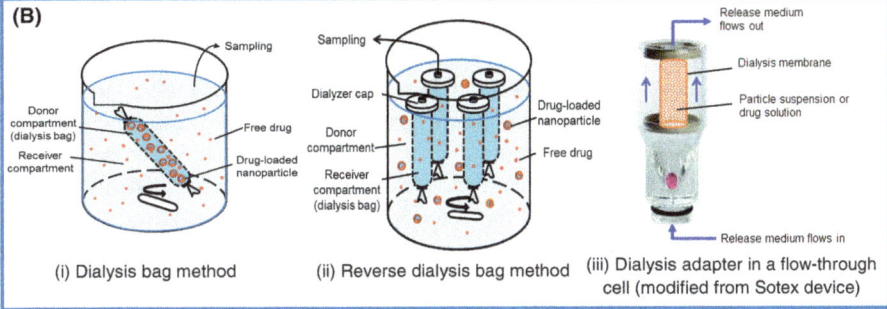

Fig. 7.7 Methods to assess the release of drug from NPs: (**a**) centrifugal ultrafiltration devices; (**b**) membrane diffusion devices. (With permission from Zhou et al. 2016)

7.4   Administration Routes of NPs

The selected administration route is determinant to define the NP transport kinetics and delivery efficiency. The main routes of administration of NPs include the following: oral, intravenous, subcutaneous, intradermal, intramuscular, nasal, and pulmonary; each administration pathway corresponds to a specific applications and distinctive nanosystem features (Anselmo and Mitragotri 2014).

7.4.1   Oral Delivery

The uptake of non-soluble, particulate matter through the digestive tract is in close relationship with particle size. Microparticles in the gut lumen are able to release drugs that then enter the body by persorption. On the other hand, the absorption of NPs has attracted the attention of researchers aiming to develop effective carriers that enable the oral absorption of drugs which may be poorly absorbed or are susceptible to gastrointestinal degradation. It has been observed that NPs from 50 nm to 20 μm are absorbed mainly through the Peyer's patch regions of the small intestine, with little translocation occurring through non-lymphoid gut tissue, with hydrophobic particles being associated to higher uptake and anionic particles being linked to reduced absorption (Bellmann et al. 2015).

NPs, then, can be absorbed following ingestion and in this manner gain access to the blood and consequently to other organs. This process involves local NP interaction with the gastrointestinal tract mucosa. An important parameter to take into account when designing NPs for oral administration is pH: in the human stomach, pH ranges from 1.2 to 2.0 in a fasted state to approximately 5.0 after ingestion (which is followed by gradual re-acidification); meanwhile, the pH of the human duodenum is between 6 and 7. It must be noticed that the intestinal compartment is highly complex from a chemical viewpoint (Bergin and Witzmann 2013).

Example

Xiao et al. co-loaded CD98 siRNA (siCD98) and camptothecin (CPT) into CD98 Fab'-functionalized NPs for oral administration. In order to protect NPs during their transit along the gastrointestinal tract, these were embedded in a hydrogel (chitosan/alginate) that can be degraded in the colon. An improved therapeutic efficacy was observed when the proposed formulation was compared to non-embedded drug-loaded NPs or non-functionalized siCD98/CPT-NPs. Specific and selective release of NPs in the colonic lumen and drug internalization into target cells was confirmed, evidencing a potential for clinical applications in colon cancer-targeted combination therapy (Xiao et al. 2014).

Wang et al. have attempted to overcome the difficulties presented by oral administration of water-soluble chemotherapeutics using topotecan (TPT) as a model drug to be carried in a novel lipid formulation containing core-shell lipid NPs (CLN). An increase in both plasma area under the concentration-time profile (AUC) and peak plasma concentration (Cmax) of TPT was observed after the oral administration of TPT-CLN in rats, which confirmed a significant enhancement of oral absorption resulting in an improved in vivo antitumor efficacy compared to the free drug (Wang et al. 2017).

7.4.2 Pulmonary Delivery

The pulmonary route offers a noninvasive method of drug administration, as well as a high surface area with rapid absorption due to high vascularization and circumvention of the first-pass effect (Sung et al. 2007). According to the lung anatomy, the deposition of particles in the different regions of the lungs depends on the particle size. Impaction, sedimentation, and diffusion have been recognized as the three different mechanisms of DDS.

▶ **Definition** *Impaction*: The aerosol particles pass through the oropharynx and upper respiratory passages at a high velocity and finally are deposited in the oropharynx regions. This mechanism is generally observed for particles bigger than 5 μm.

Sedimentation: Mainly due to gravitational forces. Particles sizing from 1 to 5 μm are slowly deposited in the smaller airways and bronchioles, provided a sufficiently long time span.

Brownian motion: Plays an important role in the deeper alveolar areas of the lungs. Particles from 0.5 to 1 μm are deposited in the alveolar region. Meanwhile, smaller particles are exhaled (Yang et al. 2008).

Sedimentation is the most convenient method of particle deposition when considering nanoparticulate systems. After being released from an aerosol, these nanoscale DDS tend to form aggregates in the micrometer size range having enough mass to sediment and extend their residence in the bronchiolar region. It is in this way that NPs are able to achieve the desired therapeutic effects when drugs are administered through the pulmonary route (Paranjpe and Muller-Goymann 2014).

Example

Lendon et al. have reported effective and low toxicity lung delivery of a miRNA-145 inhibitor using functionalized cationic lipopolyamine NPs to repair pulmonary arteriopathy and improve cardiac function in rats with severe pulmonary arterial hypertension (Mclendon et al. 2016).

It has been studied that chemotherapeutics delivered to the lungs can considerably enhance drug effectiveness in lung-resident cancers and may represent an improvement when compared to intravenous administration. Kaminskas et al. have explored the efficacy of a PEGylated polylysine dendrimer, conjugated to doxorubicin over lung-resident cancer. Twice-weekly intratracheal instillation of the proposed nanosystem led to a >95% reduction in lung tumor burden against 30–50% using intravenous administration of doxorubicin solution showing how PEGylated dendrimers may be valuable inhalable DDS to improve anti-lung-resident cancer activity (Kaminskas et al. 2014).

7.4.3 Parenteral Delivery

The parenteral administration route is a common form of delivery for active pharmaceutical ingredients with poor oral bioavailability and a narrow therapeutic index (Gulati and Gupta 2011). Some common parenteral administration routes include subcutaneous, intramuscular, transdermal, and intravenous administration.

Intravenous administration remains the most frequent route to deliver nanoscaled systems. After intravenous administration, nanocarriers circulate throughout the body to finally abandon the circulatory system and migrate to other tissues by endocytosis, shear forces, or passive diffusion through fenestrations in the capillary network (Hickey et al. 2015).

Taking this into consideration, increasing attention has been paid to the development of NP drug carriers possessing adequate size for intravenous administration and displaying an extended bloodstream half-life in order to enable drug release into the vascular compartment in a continuous and controlled manner (Saneja et al. 2017).

Kennedy et al. performed a study showing that intradermally delivered NPs using dissolving microneedles could successfully enter into the lymphatic system. The proposed system represents a minimally invasive alternative to subcutaneous delivery of nanoparticulate carriers. However, it has also been observed that a specific NP modification must be done to prevent subsequent biodistribution via the systemic circulation (Kennedy et al. 2017).

Other new approach proved the versatility of recombinant protein polymers as drug nanocarrier skeleton for subcutaneous administration. This kind of studies may enable chronic self-administration in patients (Dhandhukia et al. 2017).

7.4.4   Topical Delivery

Drug-loaded NPs for topical administration are intended for a local effect, avoiding the need for systemically administered drug therapies; minimizing the dosage used to reach therapeutic levels in the targeted site, in this case the skin; and reducing off-target adverse effects (Goyal et al. 2016).

The skin is composed of three different layers that play different key roles: the epidermis, dermis, and hypodermis. The epidermis contains four histologically distinct layers, from the innermost stratum basale via the stratum spinosum and stratum granulosum (SG) to the superficial stratum corneum (SC), a layer rich in keratin that limits drug penetration through the skin (Wu and Guy 2009).

NPs' potential ability to overcome this dermal barrier remains a challenge and an attractive but controversial field of study. Uncertainty in this particular approach may be overcome by controlling several physicochemical parameters such as size, surface area, hydrophilicity/lipophilicity grade, and NP concentration, as well as formulations paying specific attention to the selected vehicle (Labouta et al. 2011).

Santini et al. synthesized magnetic (iron oxide) NPs coated with an amphiphilic polymer for topical administration, comparing the skin penetration with that of the non-coated NPs. The authors confirmed, by transmission and scanning electron microscopies, the efficient penetration of the nanosystem through all the skin layers with controllable kinetics. Results obtained in this research suggest that amphiphilic polymer/magnetic NP combination improves the penetration of NPs, while non-coated NPs resulted in preferential capture by the mononuclear phagocyte system (Santini et al. 2015).

Fluorouracil (5-FU) (which is used in the treatment of various malignancies) presents severe side effects related to its systemic administration. Its topical delivery to treat skin cancer could circumvent these shortcomings, but it is limited by the drug's poor permeability through the skin. Safwat et al. loaded 5-FU into gold NP (GNP)-based topical delivery system incorporated into gel and cream bases in order to enhance its efficacy against skin cancer and reduce its systemic side effects. 5-FU GNP gel and cream had around twofold higher permeability through mice skin compared with free 5-FU gel and cream formulations. GNP gel and cream achieved 6.8- and 18.4-fold lower tumor volume compared with the untreated mice in a mouse model having A431 skin cancer cells implanted in the subcutaneous space (Safwat et al. 2018).

Antifungals are associated with off-target effects and limited tissue penetration when systemically administered, and antimicrobial resistance is a growing problem. To address this, Mordorski et al. recently investigated topical nitric oxide-releasing NPs (NO-NP), which inhibited *Trichophyton rubrum* in a murine model of dermal dermatophytosis after 3 days, with complete clearance after seven. Additionally, NO-NP decreased tissue cytokines such as IL-2, IL-6, and IL-10 and TNF-α, indicating attenuation of the host inflammatory response (Mordorski et al. 2017).

7.4.5 Toxicological Considerations for the Administration of NPs

Nanotoxicology is emerging as an important branch of nanotechnology; it refers to the study of interactions of nanostructures with biological systems in order to determine the influence of physical and chemical properties such as size, shape, surface, chemistry, composition, and aggregation of nanostructured materials in potential toxic responses. It has already been reported how some drug nanocarriers can cause serious harmful effects; thus, this new area of research deserves special attention (Sharma et al. 2012).

NPs have demonstrated their capability to reach clinical applications, and their unique set of properties is thought to underlie their biomedically exploitable features. Taking this into consideration, it is crucial to study the mechanisms of potential toxicity of these nanosystems and to be able to attend the problem of extrapolating knowledge from in vitro to in vivo and finally to human scenarios (Yildirimer et al. 2011).

Delivering the drug in a safe and effective manner without causing systemic toxicity should be the researcher's priority when a new drug delivery nanosystem is proposed. In this section we have described several administration routes to deliberately deliver NPs into the body; it is important to notice that NPs can also reach several organs in both unwanted and, so far, unpredictable ways. NPs, therefore, can induce adverse health effects (Dhawan and Sharma 2010).

Particle size, particle composition and charge, and particle surface area as well as agglomeration and dispersibility behavior are the main key physicochemical properties that can influence NP toxicity (Sharma et al. 2012).

7.5   Nanodispositives Designed for the Treatment of Illnesses

The selection of a suitable nanocarrier to deliver a given drug is the most important aspect in the design of new (nano)therapeutic approaches (Zhang et al. 2017). In this regard, some considerations related with the drug/matrix need to be contemplated for the best selection:

– The hydrophilic/hydrophobic nature of the drug
– The presence of polarizable or pH-dependent functional groups of the matrix/ drug molecules
– The physicochemical properties of the NP surface
– The administration route of NPs
– The therapeutic needs in terms of the amount of drug release in time
– The presence of diffusion barriers to be overcome by the NP
– The stability of the formulation
– Biocompatibility of the components
– The pathology to treat

Among the different nanocarriers with special applications in the medical field, two of the most well-known systems are described in the present chapter: lipid-based nanosystems and polymeric NPs (Mozafari 2007).

7.5.1   Lipid-Based Systems

NPs based on lipid compositions have emerged as one of the most interesting vehicles for drug delivery, with low toxicity for clinical applications. They are suitable carriers for entrapment of hydrophobic drugs (e.g., antithumorals or antiparasitaries) and may provide controlled release of the drug. According to the spatial disposition and the structure of the lipids/surfactants/drugs, two main types of lipid carriers can be found: liposomes and solid lipid NPs.

7.5.1.1 Liposomes

These systems are formed by closed spherical vesicles with a lipidic bilayer membrane composed of natural or synthetic amphipathic lipids. The bilayers enclose an inner aqueous core, in which the hydrophilic drugs can be loaded, while lipophilic drugs can be encapsulated in the lipid layer. A scheme of liposomes is shown in Fig. 7.8a, b.

In the last years, liposomes have emerged as interesting oral and parenteral dosage forms to protect the entrapped drugs from degradation under physiological conditions (e.g., gastrointestinal media) and to increase the bioavailability of poorly absorbed drugs. Various liposomal formulations can be found in clinical uses, and lots are undergoing clinical trials (Allen and Cullis 2013). Doxorubicin hydrochloride (Dox-HCl) liposomal injection (Caelyx® in Europe, Doxil® in the USA) was inserted in the market in 1995 and represented the first antecedent of a nanosystem

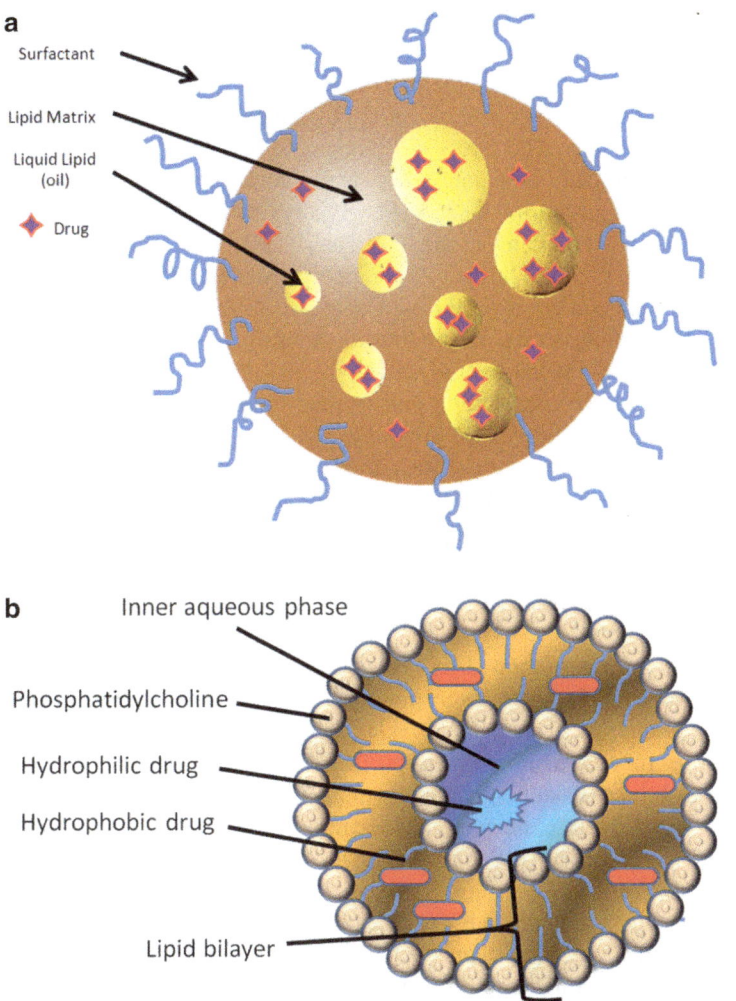

a

Surfactant

Lipid Matrix

Liquid Lipid (oil)

◆ Drug

b

Inner aqueous phase

Phosphatidylcholine

Hydrophilic drug

Hydrophobic drug

Lipid bilayer

Fig. 7.8 Scheme of: (**a**) liposome structure (Modified from Islan et al. 2016b) and (**b**) a modified solid lipid NPs: a NLC. (Modified from Islan et al. 2017)

approved for clinical uses in cancer therapy. Other nanodevices were later developed such as DaunoXome® (liposomes loaded with daunorubicin citrate) for advanced AIDS-related Kaposi sarcoma and AmBisome® (liposomes loaded with amphotericin B) for fungal infections. The rapid development of new and feasible techniques for liposome preparation have strongly impacted in the increased number of formulations that entered clinical trials, particularly to improve the therapeutic action of traditional anticancer and antibiotic drugs (Bozzuto and Molinari 2015).

7.5.2   Solid Lipid NPs (SLN) and Nanostructured Lipid Carriers (NLC)

The SLN and NLC are nanovehicles composed of lipids (triglycerides, glyceride mixtures, or waxes) which solidify at room temperature and are stabilized with surfactants (Fig. 7.8b). The lipids are biocompatible with low toxicity. The most commonly used in SLN preparation are stearic acid, cetyl alcohol, cetyl palmitate, Dynasan® 116, Compritol® 888 ATO, glyceryl monostearate, Precirol® ATO5, Imwitor® 900, or myristyl myristate. Particularly, NLC are second-generation SLN, in which a small amount of liquid oils are incorporated in the solid lipid matrix (Islan et al. 2016a). This addition produces a decrease in the highly ordered crystalline structure of the solid matrix, improving its stability and achieving enhanced drug encapsulation (Scioli Montoto et al. 2018).

These kinds of NPs are able to encapsulate both hydrophobic and hydrophilic drugs by different techniques. Improvement in bioavailability, bypassing the first-pass metabolism and providing a sustainable release of the drug encapsulated into SLN/NLC, has been observed in some cases (Islan et al. 2017). These nanosystems exhibit a wide range of applications in cancer, infections, and neurological applications.

Example

Paclitaxel (an antitumoral drug) has been loaded into cationic SLN, and the surface was modified with a substrate of somatostatin to improve the anti-glioma efficacy (Banerjee et al. 2016). Also, NLC have demonstrated to be efficient carriers for monoterpenes with antiproliferative activity against HepG2 and A549 cancer cell lines (Rodenak-Kladniew et al. 2017). In another application, encapsulation of carbamazepine was carried out in SLN and NLC and showed interesting *pharmacokinetic properties* in comparison with the free drug (Scioli Montoto et al. 2018).

Other studies demonstrate the potential of SLN and NLC in the treatment of bacterial infections. SLN encapsulating the antituberculosis drugs rifampicin, isoniazid, and pyrazinamide have been tested in animal models, and results showed higher efficacy after being administered every 10 days (Pandey et al. 2005). In a similar approach, SLN of rifabutin exhibited great antibacterial activity after pulmonary delivery (Gaspar et al. 2016). Furthermore, SLN and NLC were good vehicles for the co-delivery of antibiotics and enzymes as potential treatment of lung infections Islan et al. 2016a).

7.5.3   Polymeric Nanoparticles (PNPs)

According to the spatial disposition of the polymeric chains, the PNP can be classified in two major types: nanocapsules and nanospheres (Fig. 7.9). Nanocapsules show a vesicular structure enclosed by a solid shell. They can act as drug reservoirs, transporting the drugs inside a hydrophilic or hydrophobic core.

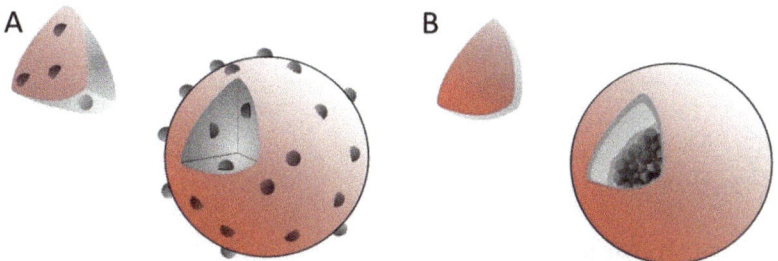

Fig. 7.9 Polymeric NPs: (**a**) nanospheres; (**b**) nanocapsules. (Reproduced with permission from El-Say and El-Sawy 2017)

The other type of PNP, nanospheres, are solid matrices composed of polymers, in which the drugs are embedded in the sphere center or adsorbed in the NP's surface (El-Say and El-Sawy 2017).

On the other hand, considering the nature of the polymers, PNPs can be classified in two subtypes: biodegradable and non-biodegradable NPs.

Biodegradable polymers can be metabolized by the organism following degradation kinetics that can be predicted and modeled. Most of them are approved by the US Food and Drug Administration (FDA) as safe for human uses. Some examples of biodegradable polymers are poly(lactide-co-glycolide) (PLGA), poly(lactide) (PL), and poly(glycolide) (PGA), which are the most widely studied materials for drug delivery. In fact, different controlled-release products composed of PLGA can be found in the market for the treatment of different pathologies, i.e., Lupron Depot® (prostate cancer), Somatuline® LA (acromegaly), and Risperdal®Consta™ (psychiatric disorders) (Kumari et al. 2010). They are useful carriers for different applications and can encapsulate small molecules, genes, proteins, and many others. However, a disadvantage of this kind of polymers is the production of acidic products after being degraded, which could denature proteins and DNA, or modify the pH in tissues (Liu et al. 2006).

Other attractive polymer is poly(caprolactone) (PCL) which is commonly used to obtain long-term controlled-release profiles since it exhibits a slow degradation rate. PCL NPs have been studied for the delivery of drugs for cancer and diabetes and for antifungal treatments (Chan et al. 2010).

In addition, some biopolymers obtained from natural sources can be included in this group, such as gelatin, pectins, chitosan, alginate, and dextrans (Landriscina et al. 2015). Particularly, chitosan NPs have been widely used in drug delivery of antihormonal and immunization agents. Another natural polymer, gelatin, has been used as raw material for the preparation of NPs for the delivery of proteins, DNA, or small molecules (Nitta and Numata 2013).

In the group of non-biodegradable polymers, some examples are polyethylenimine (PEI), poly(l-lysine) (PLL), and poly(methyl methacrylate) (PMMA). Some applications of this kind of NPs involve the development of dental resins and gene delivery (Imazato et al. 2017). However, their application in living organisms is limited due to their non-biodegradable nature which could be associated to systemic toxicity.

7.6   Active Molecular Targeting

Active targeting refers to the presence of an active moiety on the NP surface which is able to bind to receptors overexpressed at the targeted tissue enhancing the internalization of DDS. This approach would benefit from an increased specificity and reduced off-target adverse reactions (Friedman et al. 2013).

An efficient host-guest interaction is crucial to design DDS with a good clinical performance. However, biological systems complexity remains a challenge to the successful application of therapeutic NPs. Hence, the development of strategies to improve NP interaction with the host has become crucial, and different engineered strategies for surface modification were being developed over the last years.

Over the last 30 years, new therapeutic nanocarriers have been developed due to their ability to improve drug delivery to the target site and overcome biological barriers (Amin et al. 2015). The interaction between the DDS and biological systems presents challenging drawbacks including lack of proper interaction, unspecific targeting and inability to enter targeted cells, uncontrolled drug release, or rapid clearance by the mononuclear phagocyte system (Manzoor et al. 2012). In the last decade, several studies have found that NP surfaces have a major influence in the nanocarrier performance, because they are responsible for establishing contact with biological systems. In this regard, NP surface engineering plays a main role in DDS performance in vivo, including stability in biological milieu, and could help in overcoming the limitations of other DDS (Balasubramanian et al. 2018).

7.6.1   Surface Modification: Strategies

Drug delivery to selected tissues or cells needs that specific interactions between target cell and NPs are favored. The ability of nanomaterials to accommodate molecules used for this objective, which are called targeting ligands, allows them to specifically target desired tissues or cells. These targeting ligands include great variety of molecules such as oligosaccharides, peptides, proteins, antibodies, or nucleic acids, among others (Alibakhshi et al. 2017; Belfiore et al. 2018).

Surface properties play the most important role in the NPs' permanence in plasma. Surface modifications such as attachment of bioactive and penetrating molecules, surface coating, stimulus-responsive or stimulus-sensitive materials, and modification of surface charge, texture, and shape allow NPs to deliver drugs selectively to targeted cells. This kind of modifications plays a main role in colloidal stability in plasma as well as targeting capabilities. NPs' surface could be modified during their synthesis or after it (De Crozals et al. 2016).

In this sense there is a wide diversity of strategies to functionalize NPs depending on their materials' nature. Gold NPs are mainly functionalized by thiol groups which allows an easy functionalization by ligand exchange (Ruan et al. 2015). Inorganic NPs like iron oxide are easily oxidized under aerobic conditions, but surface oxidation could be controlled by an oxidant transfer agent resulting in NP stabilization (De Crozals et al. 2016). Silanization is also a usual technique used to functionalize iron oxide NPs' surface (Walter et al. 2015). Functionalization of silica NPs is mainly based on the condensation of silane ligands on the surface of silane groups, forming a Si-O-Si covalent bond as a result. By this method different functional moieties such as nucleic acids or polyesters, among others, could be introduced (De Crozals et al. 2016).

Liposomes are surface functionalized in similar ways to other carriers to selectively target unhealthy cells, but there are limitations such as accelerated clearance and nonspecific binding of serum proteins (Noble et al. 2014).

PEGylation is an extended alternative to inhibit NP aggregation and avoid nonspecific interactions by altering the surface properties of the NP used in the preparation of effective ligand-targeted liposomes (Papi et al. 2017). Indeed, this PEG coating allows in linking different active targeting molecules in order to reach specific tissue/cells, specially PEG-peptide conjugates (Hamley 2014).

The use of hydrophilic polymers is a well-stablished practice, PEG mainly, which could be grafted, conjugated, or absorbed to NPs' surface, allowing targeting functionality. NP surface charge could be used as a tool for electrostatic decoration with different moieties (Chevalier et al. 2017). Low affinity can be offset increasing avidity through the surface functionalization of multiple molecules in the case of weak binding ligands. Conjugation is one of the mainly used techniques which allows to functionalize NPs with targeting molecules. Polyamidoamine (PAMAM) dendrimers containing abundant amino surface groups provide different possibilities for dendrimer surface decoration by conjugation of various moieties (Dehshahri and Sadeghpour 2015).

Covalent binding and linker chemistry are two of the conjugation methods most used to functionalized different kinds of NPs in order to increase target specificity (Amin et al. 2015). "Click" chemistry is widely used due to its efficiency for coupling different synthetic or natural biomolecules such as peptides, DNA, RNA, aptamers, or antibodies between other ligands to achieve active targeted DDS (De Crozals et al. 2016).

7.6.2 Molecular Markers of Pathologies

Some diseases are characterized by abnormal production of molecular markers which differentiates diseased cells from normal healthy cells. NPs with appropriate moieties attached to their surface which allows DDS to specifically interact with diseased cells causing targeted release of the drug have been developed during the last years. Representative ligands for targeted DDS development are usually peptides, antibodies and their fragments, and peptide-like molecule such as nucleic acid ligands and sugars (Kamaly et al. 2012).

Due to its incidence, cancer is probably the most studied disease in terms of active targeted NP development. One of the most studied markers are matrix metalloproteinases (MMP) which is overexpressed by many cancer cells and has been used as a target for the development of mesoporous silica NPs, liposomes, and polymeric micelles (Zhang et al. 2016).

Example

A recent study describes a glucose-mediated GLUT2 targeting MMP2-responsive PEG detachment, triphenylphosphonium-mediated mitochondrial targeting, and a glutathione-sensitive intracellular paclitaxel release conjugate which could be used to enhanced paclitaxel delivery showing better tumor inhibition rate and lower body weight loss (Ma et al. 2018).

Hyaluronic acid is a polymer which is present in elevated concentration in tissues that display an increased cell division (such as cancer cells) and which can bind to two types of receptors playing a significant role in cancer initiation and progression (CD44 and RHAMM). CD44 is overexpressed in cancer cells, and on that basis, different DDS decorated by hyaluronic acid such as polymeric NPs, liposomes, and inorganic NPs have been developed in the last years (Ramzy et al. 2017).

One of the most frequent strategies to developed nanomedicines against brain cancer involves the use of ligands targeting anti-epidermal growth factor receptor (EGFR). EGFR is a transmembrane protein receptor of the HER/erbB family with tyrosine kinase activity along the inner cytoplasmic domain (Shonka et al. 2017). For example, cationic solid lipid NPs functionalized with an anti-EGFR moiety were recently developed to target malignant glioblastoma cells (Tapeinos et al. 2017).

In a recently published study, Palao-Suay et al. successfully synthesized active targeted polymeric NPs conjugating a bromide derivative of triphenylphosphonium lipophilic cation and peptide ligand LTVSPWY to amphiphilic PEG-b-poly-MTOS. The peptide ligand LTVSPWY strongly and preferentially interacts with human EGFR of cancer cells, without binding to nonmalignant mammary epithelial cells (Palao-Suay et al. 2017).

Another commonly used targeting moiety is folic acid since it is known that cancer cells overexpress folate receptors (among others). Exploiting this knowledge, PEGylated doxorubicin-loaded liposomes targeted with folic acid, transferrin, or both were evaluated by Sriraman et al. showing a marked improvement in cell association in human cervical carcinoma (HeLa) cell monolayers and a significantly higher cytotoxic effects in vivo (Sriraman et al. 2016).

Ontak (denileukindiftitox, Eisai, Inc.), approved in 2008, represents the first FDA-approved nanoparticulated system that combines targeting proteins with cytotoxic molecules. This medication includes an interleukin (IL)-2 receptor antagonist initially designed to target the cytocidal action of diphtheria toxin toward cells that overexpress the IL-2 receptor on non-Hodgkin's peripheral T-cell lymphomas. Ontak was not observed to be myelosuppressive, nor was it associated with significant organ toxicity in clinical trials (Ventola 2017).

Wu et al. synthesized an aptamer (S2.2)-guided Ag-Au nanostructure (aptamer-Ag-Au) that targets the surface of human breast cancer cells (MCF-7) via the specific interaction between S2.2 aptamer (a 25-base oligonucleotide) and MUC1 mucin expressed in MCF-7 cells with high affinity and specificity (Wu et al. 2012).

Combretastatin A-4 (CA-4), which is a tubulin-binding vascular disrupting agent that selectively targets tumor endothelium, presents side effects which limit its therapeutic dose, especially due to its cardiovascular toxicity. Liu et al. recently developed a tumor vasculature-targeted delivery vehicle based on peptide-modified cross-linked multilamellar liposomal vesicles to deliver CA-4 combined with doxorubicin. Authors used RIF7 as a tumor vasculature-targeting peptide, which shows enhanced antitumor efficacy in vitro and in vivo (Liu et al. 2018).

Cheng et al. recently developed a tumor-targeting, redox-responsive DDS with bioactive surface based on immobilizing peptide-based amphiphile C12-CGRKKRRQRRRPPQRGDS onto mesoporous silica NPs (Cheng et al. 2017).

Mao et al. developed a silk fibroin-based cyclic with pentapeptide cRGDfk and chlorine e6 conjugated in order to deliver 5-FU, showing good biocompatibility and security profile in vivo (Mao et al. 2018).

Another approach focused on developing DDS which target mitochondria or other specific organelles to treat neurodegenerative diseases (Pezzini et al. 2017). Neurodegenerative diseases such as amyloidosis are characterized by the extracellular deposition of insoluble fibrillar proteinaceous aggregates and could be treated and diagnosed by NPs like gadolinium-based NPs (Plissonneau et al. 2016).

Rittchen et al. used leukemia inhibitory factor NPs for myelin repair in mice, showing in vivo potential for the therapeutic targeted delivery of drugs or biologics to potently improve the function of oligodendrocyte precursor cells and enhance central nervous system remyelination in multiple sclerosis patients (Rittchen et al. 2015).

Certain receptors are overexpressed during inflammation process offering molecular targets for anchoring drug carriers through specific interactions. Tang et al. developed pH-sensitive NPs composed of siRNA chemically crosslinked with multi-armed PEG carriers with mannose targeting moieties which were able to deliver tumor necrosis factor alpha (TNF-α) siRNA into macrophages and selectively accumulate in mice liver protecting them from inflammation-induced liver damage (Tang et al. 2017). Other approaches such as mannose functionalized dendrimer NPs targeting macrophages have been proposed to modulate atherosclerosis among other inflammation-related diseases (He et al. 2018).

Clinical validation of active targeted NPs is still limited. However, some next-generation nanoparticulated drugs in clinical trials employ active targeting approaches, and many of them such as Neulasta, Pegfilgrastim (synthetic G-CSF), or Ontak are approved by FDA nowadays (Anselmo and Mitragotri 2016).

7.6.3   Effect of Targeting

NPs functionalized with targeting moieties usually present stronger interactions with cells than those non-functionalized. "Active" targeting involves a transport mechanism to guide a particle toward its intended target, but in most of the particle-based systems, the term active is referred to ligand receptor interactions. That kind of interactions requires close proximity between the NP and the target.

NP surfaces are covered by plasma protein corona immediately after their administration creating a specific biological barrier which determines NP in vivo performance, including its effect and toxicity (Chen et al. 2017). NPs usually present enhanced cellular adhesion and are affected by hemodynamic forces, endothelium interactions, and physicochemical properties of NPs (Dai et al. 2018). Another important variable which could be crucial to NPs' performance in vivo involves the varying flow rates in the bloodstream related to different shear stresses, showing a decrease in the cellular association of targeted particles during systemic circulation.

Effects from the biological environment, especially the formation of biomolecular corona, interaction of the NP with the extracellular matrix, and other biological parameters such as blood flow rate, play a main role in DDS performance in vivo (Dai et al. 2018; Di Ianni et al. 2017).

7.7   Chapter Conclusions

Nowadays nanotechnology represents an expanding field that is providing powerful tools for the pharmaceutical industry such as polymeric, solid lipid, metallic, and magnetic NPs, liposomes, dendrimers, and other nanostructures. The development of nanopharmaceutics allows the understanding of some of the major mechanisms of current diseases on a molecular basis. Particularly, drug delivery mediated by nanosystems is relevant since NPs are able to reach the target organ and deliver the drug with a desirable profile according to the pathology needs, minimizing undesirable side effects.

The main physicochemical properties of nanosystems are related with the increase in the surface-to-volume ratio of the particles, their permeability across biological barriers, and the surface modification of nanodevices. These parameters are relevant in drug loading and release and particularly for the targeted delivery to cells showing specific markers or receptors in the membrane surface.

The mechanisms of drug release and the selection of rational administration routes of NPs can improve the delivery of drugs to the specific sites for treatment of different pathologies such as cancer, infections, or metabolic diseases.

References

Alibakhshi A, Abarghooi Kahaki F, al AS (2017) Targeted cancer therapy through antibody fragments-decorated nanomedicines. J Control Release 268:323–334

Allen TM, Cullis PR (2013) Liposomal drug delivery systems: from concept to clinical applications. Adv Drug Deliv Rev 65:36–48

Amendola V, Pilot R, Iatì MA et al (2017) Surface plasmon resonance in gold nanoparticles: a review. J Phys Condens Matter 29:1–48

Amin ML, Joo JY, Yi DK et al (2015) Surface modification and local orientations of surface molecules in nanotherapeutics. J Control Release 207:131–142

Amini Y, Amel Jamehdar S, Sadri K et al (2017) Different methods to determine the encapsulation efficiency of protein in PLGA nanoparticles. Biomed Mater Eng 28:613–620

Amoozgar Z, Wang L, Brandstoetter T et al (2014) Dual-layer surface coating of PLGA-based nanoparticles provides slow-release drug delivery to achieve metronomic therapy in a paclitaxel-resistant murine ovarian cancer model. Biomacromolecules 15:4187–4194

Anonymous (2018) Top ten countries in nanotechnology patents in 2017. http://statnano.com/news/62082. Accessed on 12 June 2018

Anselmo AC, Mitragotri S (2016) Nanoparticles in the clinic. Bioeng Transl Med 1:10–29

Anselmo AC, Mitragotri S (2014) An overview of clinical and commercial impact of drug delivery systems. J Control Release 190:15–28

Balasubramanian V, Liu Z, Hirvonen J et al (2018) Bridging the knowledge of different worlds to understand the big picture of cancer nanomedicines. Adv Healthc Mater 7: 1700432

Banerjee I, De K, Mukherjee D et al (2016) Paclitaxel-loaded solid lipid nanoparticles modified with Tyr-3-octreotide for enhanced anti-angiogenic and anti-glioma therapy. Acta Biomater 38:69–81

Barratt GM (2000) Therapeutic applications of colloidal drug carriers. Pharm Sci Technolo Today 3:163

Belfiore L, Saunders DN, Ranson M et al (2018) Towards clinical translation of ligand-functionalized liposomes in targeted cancer therapy: challenges and opportunities. J Control Release 277:1–13

Bellmann S, Carlander D, Fasano A et al (2015) Mammalian gastrointestinal tract parameters modulating the integrity, surface properties, and absorption of food-relevant nanomaterials. Wiley Interdiscip Rev Nanomed Nanobiotechnol 7:609–622

Bergin IL, Witzmann FA (2013) Nanoparticle toxicity by the gastrointestinal route: evidence and knowledge gaps. Int J Biomed Nanosci Nanotechnol 3:163

Biglu MH, Eskandari F, Asgharzadeh A (2011) Scientometric analysis of nanotechnology in MEDLINE. Bioimpacts 1:193–198

Bonta R, Uppala S, Manchikanti P (2017) Patents in nanobiotechnology: a cross jurisdictional approach. Recent Pat Biotechnol 11:52–70

Bozzuto G, Molinari A (2015) Liposomes as nanomedical devices. Int J Nanomedicine 10:975–999

Chan JM, Valencia PM, Zhang L et al (2010) Polymeric nanoparticles for drug delivery. Methods Mol Biol 624:163–175

Chen D, Ganesh S, Wang W et al (2017) Plasma protein adsorption and biological identity of systemically administered nanoparticles. Nanomedicine 12:2113

Cheng Y-J, Zhang A-Q, Hu J-J et al (2017) Multifunctional peptide-amphiphile end-capped mesoporous silica nanoparticles for tumor targeting drug delivery. ACS Appl Mater Interfaces 9:2093–2103

Chevalier MT, Rescignano N, Martin-Saldaña S et al (2017) Non-covalently coated biopolymeric nanoparticles for improved tamoxifen delivery. Eur Polym J 95:348–357

Dai Q, Bertleff-Zieschang N, Braunger JA et al (2018) Particle targeting in complex biological media. Adv Healthc Mater 7:1700575

Dang Y, Zhang Y, Fan L et al (2010) Trends in worldwide nanotechnology patent applications: 1991 to 2008. J Nanopart Res 12:687

De Crozals G, Bonnet R, Farre C et al (2016) Nanoparticles with multiple properties for biomedical applications: a strategic guide. Nano Today 11:435–463

de Oliveira JK, Ronik DFV, Ascari J et al (2017) A stability-indicating high performance liquid chromatography method to determine apocynin in nanoparticles. J Pharm Anal 7:129–133

Dehshahri A, Sadeghpour H (2015) Surface decorations of poly(amidoamine) dendrimer by various pendant moieties for improved delivery of nucleic acid materials. Colloids Surf B Biointerfaces 132:85–102

Dhandhukia JP, Li Z, Peddi S et al (2017) Berunda polypeptides: multi-headed fusion proteins promote subcutaneous administration of rapamycin to breast cancer in vivo. Theranostics 7:3856

Dhawan A, Sharma V (2010) Toxicity assessment of nanomaterials: methods and challenges. Anal Bioanal Chem 398:589–605

Di Ianni ME, Islan GA, Chain CY et al (2017) Interaction of solid lipid nanoparticles and specific proteins of the corona studied by surface Plasmon resonance. J Nanomater 2017:1

Di Marco M, Shamsuddin S, Razak KA et al (2010) Overview of the main methods used to combine proteins with nanosystems: absorption, bioconjugation, and encapsulation. Int J Nanomedicine 5:37–49

El-Say KM, El-Sawy HS (2017) Polymeric nanoparticles: promising platform for drug delivery. Int J Pharm 528:675–691

Friedman A, Claypool S, Liu R (2013) The smart targeting of nanoparticles. Curr Pharm Des 19:6315–6329

Fu Y, Kao WJ (2010) Drug release kinetics and transport mechanisms of non-degradable and degradable polymeric delivery systems. Expert Opin Drug Deliv 7:429–444

Gallo L, Piña J, Bucalá V et al (2013) Development of a modified-release hydrophilic matrix system of a plant extract based on co-spray-dried powders. Powder Technol 241:252–262

Gaspar DP, Faria V, Gonçalves LMD et al (2016) Rifabutin-loaded solid lipid nanoparticles for inhaled antitubercular therapy: physicochemical and in vitro studies. Int J Pharm 497:199–209

Goyal R, Macri LK, Kaplan HM et al (2016) Nanoparticles and nanofibers for topical drug delivery. J Control Release 240:77–92

Gulati N, Gupta H (2011) Parenteral drug delivery: a review. Recent Pat Drug Deliv Formul 5:133–145

Hamley IW (2014) PEG – peptide conjugates. Biomacromolecules 15:1543–1559

Hassellöv M, Readman JW, Ranville JF et al (2008) Nanoparticle analysis and characterization methodologies in environmental risk assessment of engineered nanoparticles. Ecotoxicology 17:344–361

He H, Yuan Q, Bie J et al (2018) Development of mannose functionalized dendrimeric nanoparticles for targeted delivery to macrophages: use of this platform to modulate atherosclerosis. Transl Res 193:13–30

Hickey JW, Santos JL, Williford JM et al (2015) Control of polymeric nanoparticle size to improve therapeutic delivery. J Control Release 219:535–547

Higuchi WI (1967) Diffusional models useful in biopharmaceutics. Drug release rate processes. J Pharm Sci 56:315–324

Huang XL, Teng X, Chen D et al (2010) The effect of the shape of mesoporous silica nanoparticles on cellular uptake and cell function. Biomaterials 31:438–448

Imazato S, Kitagawa H, Tsuboi R et al (2017) Non-biodegradable polymer particles for drug delivery: a new technology for "bio-active" restorative materials. Dent Mater J 36:524–532

Islan GA, Tornello PC, Abraham GA et al (2016a) Smart lipid nanoparticles containing levofloxacin and DNase for lung delivery. Design and characterization. Colloids Surf B Biointerfaces 143:168–176

Islan GA, Cacicedo ML, Bosio VE et al (2016b) Advances in smart nanopreparations for oral drug delivery. In: Smart pharmaceutical Nanocarriers. Imperial College Press, London, pp 479–521

Islan GA, Cacicedo ML, Rodenak-Kladniew B et al (2017) Development and tailoring of hybrid lipid nanocarriers. Curr Pharm Des 23: 6643 - 6658

Islan GA, Mukherjee A, Castro GR (2015) Development of biopolymer nanocomposite for silver nanoparticles and Ciprofloxacin controlled release. Int J Biol Macromol 72:740–750

Jyothi NVN, Prasanna PM, Sakarkar SN et al (2010) Microencapsulation techniques, factors influencing encapsulation efficiency. J Microencapsul 27:187–197

Kamaly N, Xiao Z, Valencia PM et al (2012) Targeted polymeric therapeutic nanoparticles: design, development and clinical translation. Chem Soc Rev 41:2971–3010

Kaminskas LM, McLeod VM, Ryan GM et al (2014) Pulmonary administration of a doxorubicin-conjugated dendrimer enhances drug exposure to lung metastases and improves cancer therapy. J Control Release 183:18–26

Kennedy J, Larrañeta E, McCrudden MTC et al (2017) In vivo studies investigating biodistribution of nanoparticle-encapsulated rhodamine B delivered via dissolving microneedles. J Control Release 265:57–65

Kumari A, Yadav SK, Yadav SC (2010) Biodegradable polymeric nanoparticles based drug delivery systems. Colloids Surf B Biointerfaces 75:1–18

Kwok PCL, Chan H-K (2014) Nanotechnology versus other techniques in improving drug dissolution. Curr Pharm Des 20:474–482

Labouta HI, El-Khordagui LK, Kraus T et al (2011) Mechanism and determinants of nanoparticle penetration through human skin. Nanoscale 3:4989–4999

Landriscina A, Rosen J, Friedman AJ (2015) Biodegradable chitosan nanoparticles in drug delivery for infectious disease. Nanomedicine 10:1609–1619

Liu H, Slamovich EB, Webster TJ (2006) Less harmful acidic degradation of poly(lactic-co-glycolic acid) bone tissue engineering scaffolds through titania nanoparticle addition. Int J Nanomedicine 1:541–545

Liu X, Zhang P, Li X et al (2009) Trends for nanotechnology development in China, Russia, and India. J Nanopart Res 11:1845–1866

Liu Y, Kim YJ, Siriwon N et al (2018) Combination drug delivery via multilamellar vesicles enables targeting of tumor cells and tumor vasculature. Biotechnol Bioeng 115:1403–1415

Ma P, Chen J, Bi X et al (2018) Overcoming multidrug resistance through the GLUT1-mediated and enzyme-triggered mitochondrial targeting conjugate with redox-sensitive paclitaxel release. ACS Appl Mater Interfaces 10:12351–12363

Manzoor AA, Lindner LH, Landon CD et al (2012) Overcoming limitations in nanoparticle drug delivery: triggered, intravascular release to improve drug penetration into tumors. Cancer Res 72:5566–5575

Mao B, Liu C, Zheng W et al (2018) Cyclic cRGDfk peptide and Chlorin e6 functionalized silk fibroin nanoparticles for targeted drug delivery and photodynamic therapy. Biomaterials 161:306–320

Martins LG, khalil NM, Mainardes RM (2017) Application of a validated HPLC-PDA method for the determination of melatonin content and its release from poly(lactic acid) nanoparticles. J Pharm Anal 7:388–393

Mclendon JM, Joshi SR, Sparks J et al (2016) Lipid nanoparticle delivery of a microRNA-145 inhibitor improves experimental pulmonary hypertension. J Control Release 28:67–75

Modi S, Anderson BD (2013) Determination of drug release kinetics from nanoparticles: overcoming pitfalls of the dynamic dialysis method. Mol Pharm 10:3076–3089

Mordorski B, Costa-Orlandi CB, Baltazar LM et al (2017) Topical nitric oxide releasing nanoparticles are effective in a murine model of dermal Trichophyton rubrum dermatophytosis. Nanomed Nanotechnol Biol Med 13:2267–2270

Mozafari MR (2007) Nanomaterials and nanosystems for biomedical applications. Springer, Dordrecht

Nitta SK, Numata K (2013) Biopolymer-based nanoparticles for drug/gene delivery and tissue engineering. Int J Mol Sci 14:1629–1654

Noble GT, Stefanick JF, Ashley JD et al (2014) Ligand-targeted liposome design: challenges and fundamental considerations. Trends Biotechnol 32:32–45

Ochubiojo M, Chinwude I, Ibanga E et al (2012) Nanotechnology in drug delivery in: recent advances in novel drug carrier systems

Pakulska MM, Donaghue IE, Obermeyer JM et al (2016) Encapsulation-free controlled release: electrostatic adsorption eliminates the need for protein encapsulation in PLGA nanoparticles. Sci Adv 2:e1600519

Palao-Suay R, Rosa Aguilar M, Parra-Ruiz FJ et al (2017) Correction to: multifunctional decoration of alpha-tocopheryl succinate-based NP for cancer treatment: effect of TPP and LTVSPWY peptide. J Mater Sci Mater Med 28:182. https://doi.org/10.1007/s10856-017-5963-y

Pandey R, Sharma S, Khuller GK (2005) Oral solid lipid nanoparticle-based antitubercular chemotherapy. Tuberculosis 85:415–420

Papi M, Caputo D, Palmieri V et al (2017) Clinically approved PEGylated nanoparticles are covered by a protein corona that boosts the uptake by cancer cells. Nanoscale 9:10327–10334

Paranjpe M, Muller-Goymann CC (2014) Nanoparticle-mediated pulmonary drug delivery: a review. Int J Mol Sci 15:5852–5873

Pezzini I, Mattoli V, Ciofani G (2017) Mitochondria and neurodegenerative diseases: the promising role of nanotechnology in targeted drug delivery. Expert Opin Drug Deliv 14:513–523

Plissonneau M, Pansieri J, Heinrich-Balard L et al (2016) Gd-nanoparticles functionalization with specific peptides for ß-amyloid plaques targeting. J Nanobiotechnol 14:60

Popat A, Ross BP, Liu J et al (2012) Enzyme-responsive controlled release of covalently bound prodrug from functional mesoporous silica nanospheres. Angew Chem Int Ed 51:12486–12489

Ramzy L, Nasr M, Metwally AA et al (2017) Cancer nanotheranostics: a review of the role of conjugated ligands for overexpressed receptors. Eur J Pharm Sci 104:273–292

Rittchen S, Boyd A, Burns A et al (2015) Myelin repair in vivo is increased by targeting oligodendrocyte precursor cells with nanoparticles encapsulating leukaemia inhibitory factor (LIF). Biomaterials 56:78–85

Robertson JD, Rizzello L, Avila-Olias M et al (2016) Purification of nanoparticles by size and shape. Sci Rep 6: 27494

Rodenak-Kladniew B, Islan GA, de Bravo MG et al (2017) Design, characterization and in vitro evaluation of linalool-loaded solid lipid nanoparticles as potent tool in cancer therapy. Colloids Surf B Biointerfaces 154:123–132

Ruan S, Yuan M, Zhang L et al (2015) Tumor microenvironment sensitive doxorubicin delivery and release to glioma using angiopep-2 decorated gold nanoparticles. Biomaterials 37:425–435

Safwat MA, Soliman GM, Sayed D et al (2018) Fluorouracil-loaded gold nanoparticles for the treatment of skin Cancer: development, in vitro characterization and in vivo evaluation in a mouse skin Cancer xenograft model. Mol Pharm 15:2194–2205

Saneja A, Kumar R, Singh A et al (2017) Development and evaluation of long-circulating nanoparticles loaded with betulinic acid for improved anti-tumor efficacy. Int J Pharm 531:153–166

Santini B, Zanoni I, Marzi R et al (2015) Cream formulation impact on topical administration of engineered colloidal nanoparticles. PLoS One 10:e0126366

Scioli Montoto S, Sbaraglini ML, Talevi A et al (2018) Carbamazepine-loaded solid lipid nanoparticles and nanostructured lipid carriers: physicochemical characterization and in vitro/ in vivo evaluation. Colloids Surf B Biointerfaces 167:73–81

Sharma A, Madhunapantula SV, Robertson GP (2012) Toxicological considerations when creating nanoparticle-based drugs and drug delivery systems. Expert Opin Drug Metab Toxicol 8:47–69

Shonka N, Venur VA, Ahluwalia MS (2017) Targeted treatment of brain metastases. Curr Neurol Neurosci Rep 17:37

Siepmann J, Siepmann F (2008) Mathematical modeling of drug delivery. Int J Pharm 364 (2):328–343

Singh R, Lillard JW (2009) Nanoparticle-based targeted drug delivery. Exp Mol Pathol 86:215–223

Souza SD (2014) A review of in vitro drug release test methods for nano-sized dosage forms. Adv Pharm 2014:1–12

Sriraman SK, Salzano G, Sarisozen C et al (2016) Anti-cancer activity of doxorubicin-loaded liposomes co-modified with transferrin and folic acid. Eur J Pharm Biopharm 105:40–49

Sung JC, Pulliam BL, Edwards DA (2007) Nanoparticles for drug delivery to the lungs. Trends Biotechnol 25:563–570

Tang R, Wang L, Orme CA et al (2004) Dissolution at the nanoscale: self-preservation of biominerals. Angew Chemie Int Ed 43:2697–2701

Tang Y, Zeng Z, He X et al (2017) SiRNA crosslinked nanoparticles for the treatment of inflammation-induced liver injury. Adv Sci 4: 1600228

Tapeinos C, Battaglini M, Ciofani G (2017) Advances in the design of solid lipid nanoparticles and nanostructured lipid carriers for targeting brain diseases. J Control Release 264:306–332

Unger F, Wittmar M, Morell F et al (2008) Branched polyesters based on poly[vinyl-3-(dialkylamino)alkylcarbamate-co-vinyl acetate-co-vinyl alcohol]-graft-poly(d,l-lactide-co-glycolide): effects of polymer structure on in vitro degradation behaviour. Biomaterials 29:2007–2014

Ventola CL (2017) Progress in nanomedicine: approved and investigational nanodrugs. P T 42:742–755

Walter A, Garofalo A, Parat A et al (2015) Functionalization strategies and dendronization of iron oxide nanoparticles. Nanotechnol Rev 4:581–593

Wang T, Shen L, Zhang Z et al (2017) A novel core-shell lipid nanoparticle for improving oral administration of water soluble chemotherapeutic agents: inhibited intestinal hydrolysis and enhanced lymphatic absorption. Drug Deliv 24:1565–1573

Wu P, Gao Y, Zhang H et al (2012) Aptamer-guided silver-gold bimetallic nanostructures with highly active surface-enhanced Raman scattering for specific detection and near-infrared photothermal therapy of human breast cancer cells. Anal Chem 84:7692–7699

Wu X, Guy RH (2009) Applications of nanoparticles in topical drug delivery and in cosmetics. J Drug Deliv Sci Technol 19:371–384

Xiao B, Laroui H, Viennois E et al (2014) Nanoparticles with surface antibody against CD98 and carrying CD98 small interfering RNA reduce colitis in mice. Gastroenterology 146:1289–1300

Yang W, Peters JI, Williams RO (2008) Inhaled nanoparticles-a current review. Int J Pharm 356:239–247

Yildirimer L, Thanh NTK, Loizidou M et al (2011) Toxicological considerations of clinically applicable nanoparticles. Nano Today 6:585–607

Zhang RX, Ahmed T, Li LY et al (2017) Design of nanocarriers for nanoscale drug delivery to enhance cancer treatment using hybrid polymer and lipid building blocks. Nanoscale 9:1334–1355

Zhang X, Wang X, Zhong W et al (2016) Matrix metalloproteinases-2/9-sensitive peptide-conjugated polymer micelles for site-specific release of drugs and enhancing tumor accumulation: preparation and in vitro and in vivo evaluation. Int J Nanomedicine 11:1643–1661

Zhang Y, Zhi Z, Jiang T et al (2010) Spherical mesoporous silica nanoparticles for loading and release of the poorly water-soluble drug telmisartan. J Control Release 145:257–263

Zhou Y, He C, Chen K et al (2016) A new method for evaluating actual drug release kinetics of nanoparticles inside dialysis devices via numerical deconvolution. J Control Release 243:11–20

Part II

The Importance of ADME Properties in Burgeoning Pharmaceutical Topics

Relationship Between Pharmacokinetics and Pharmacogenomics and Its Impact on Drug Choice and Dose Regimens

8

Matías F. Martínez and Luis A. Quiñones

8.1 Introduction

8.1.1 Drug Metabolism: Pharmacokinetics and Pharmacodynamics

▶ **Definition** Pharmacokinetics can be defined as the relationship between the administered dose and the plasma concentration of a drug, which implies the study of the different processes of absorption, distribution, metabolism, and excretion, in short "what the organism does with the drug". Pharmacokinetics determines the concentration of drugs in the drug-exposed subject and therefore contributes to the intensity of the observed response. Modifications in pharmacokinetics help explaining different responses among different individuals, where different physiological situations may be found, including short or advanced age, organic failure (renal, hepatic), hypo-hypervolemia, etc. Pharmacokinetic parameters vary among subjects and also depending on the route of administration.

As shown in Fig. 8.1 (right superior box), the first part of the curve represents the absorption process. The kinetics of absorption depends on the interplay between passive and active absorption and the activity of efflux pumps. When the drug is absorbed, it could go to the systemic circulation without modifications or suffer a first-step metabolism. Absorption-related processes and first-pass metabolism could affect the maximum level that the drug can achieve (plasmatic peak). The second part of the curve is dominated by the elimination or excretion process. Here the metabolism of the drug and its urinary and fecal excretion are the most important processes.

M. F. Martínez · L. A. Quiñones (✉)
Laboratory of Chemical Carcinogenesis and Pharmacogenetic (CQF), Department of Basic and Clinical Oncology (DOBC), Faculty of Medicine, University of Chile & Latin American Society of Pharmacogenomics and Personalized Medicine (SOLFAGEM), Santiago, Chile
e-mail: matiasmartinez@ug.uchile.cl; lquinone@med.uchile.cl

© Springer Nature Switzerland AG 2018
A. Talevi, P. A. M. Quiroga (eds.), *ADME Processes in Pharmaceutical Sciences*,
https://doi.org/10.1007/978-3-319-99593-9_8

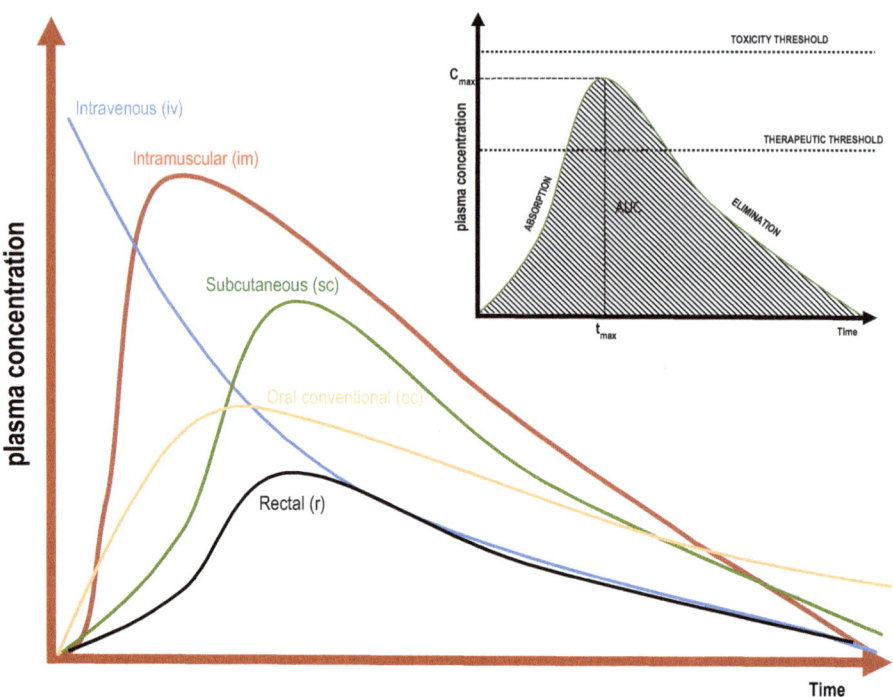

Fig. 8.1 Pharmacokinetics curves of several routes of administration. *In the right superior box, the main basic pharmacokinetics parameters are shown*

On the other hand, pharmacodynamics is defined as the relationship between the plasma concentration and its effect, in short "what the drug does in the body." Although the concentrations of the drugs are frequently measured in plasma, their site of action is the target tissue, and it would be ideal to evaluate the drug levels in this compartment, which is currently impossible due to the anatomical location and the limitations of the available analytical methods.

▶ **Important** There is a relationship between plasma concentration and concentration in the target tissue. The different pharmacokinetic-pharmacodynamic models that allow us to establish this relationship have proved to be very useful in explaining many clinical observations. By means of pharmacokinetic-pharmacodynamic simulations, different therapeutic ranges can be calculated, within which the therapeutic effect would be found without adverse reactions to the medication.

8.1.1.1 The ADME Process

From the moment a drug is administered until we finally excrete it from our organism, different processes occur. These encompass the processes of absorption, distribution, metabolism, and excretion (abbreviated: ADME). Although some drugs exert their effect without entering systemic circulation (creams that serve as a barrier, some laxatives and others), most of them must enter the organism via enteral or parenteral (absorption) to be transported by the blood to the target organ, where they will cross cell membranes and finally reach the molecular target. Therefore, the body is a nonhomogeneous compartment in which the drug is distributed. Subsequently, the drug molecules are processed by enzymes of phase I (cytochrome P450 (CYP450), epoxide hydrolase, and others) and phase II (glutathione S-transferases, UDP glucuronosyl transferases, sulfotransferases, and others) which have evolved to biotransform endogenous and exogenous compounds. Thus, depending on the phase I and II enzyme activity, the drug level varies increasing or decreasing the therapeutic response to the drug. The activity of metabolism enzymes may vary depending on genetics and also by interaction with other molecules. Finally, urinary or biliary excretion are the main (but not the only) excretion routes that allow the drugs to be removed (Fig. 8.2) (Brunton et al. 2017).

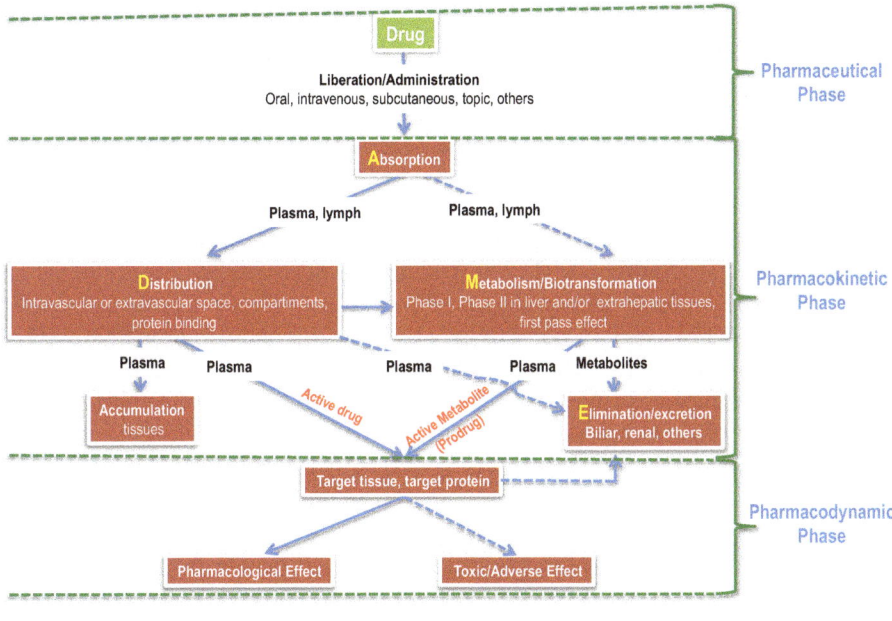

Fig. 8.2 The drug phases in the body

8.1.2   Efficacy and Safety of Drugs

▶ **Definition** Efficacy is the ability of a drug to produce an effect and is related to its affinity for the receptor and to the activation of the latter to produce a biological response. The degree of activation of a biological system by a drug is known as intrinsic activity or efficacy.

The efficacy of a drug is the maximum pharmacological effect that can be obtained; once the maximum pharmacological effect has been obtained, no increment in the effect would be observed even if the doses are increased. The measure of efficacy is the maximum effect, the lower this is, the less effective the drug is in producing an effect. In this sense the dose does not influence efficacy. For example, despite increasing the dose of codeine analgesia is not achieved in the preoperative period. Similarly, opioids are more effective in relieving high-intensity pain compared to maximum aspirin doses: the maximum analgesic effect of the latter is lower than opiates (Sear 2004).

On the other hand, the safety of drugs depends upon the ability of them to induce damage in the body, i.e., to elicit toxicity. Toxicity is the ability of a substance, artificial or natural, to produce a harmful effect when in contact with living systems. No chemical substance can be considered nontoxic since any substance can produce a toxic effect if a sufficient dose is administered. This is represented in the famous phrase of Paracelsus "only the dose makes poison" (Klaassen and Watkins 2013). In this sense, adverse drug reactions (ADRs) can be considered a form of toxicity (a response to a drug which is noxious and unintended and which occurs at doses normally used) (Edwards and Aronson 2000; WHO 2002). The term toxicity is most often applied to the effects of overdoses (accidental or intentional), i.e., to the presence of high blood concentrations or to the exacerbated pharmacological effects that appear during the correct use of the drug (e.g., when the metabolism of the drug is temporarily inhibited by a disease or the administration of another drug). It is important to highlight that a toxic effect is always dose-related and occurs as an exaggerated therapeutic effect and a side effect is not associated with therapeutic effect and it may be or may be not dose-related (Edwards and Aronson 2000). Since all drugs can cause adverse reactions, whenever a medication is prescribed, a risk-benefit analysis must be carried out to evaluate the probability of obtaining benefits against the risk of adverse reactions to the drug (Brunton et al. 2017).

8.1.3   Drug Variability, Pharmacogenetics, Pharmacogenomics, and the Role of Epigenetics

The success of modern medicine partially depends on effective pharmacological treatments. A well-known fact is that individuals respond differently to drug therapy and that no medication is 100% effective in all patients. Consequently, the margin of response to a pharmacological treatment varies, since some individuals can obtain the expected effects, while others do not obtain a therapeutic result or may even experience adverse effects at therapeutic doses (Zhou et al. 2008; Wilkinson 2005; Xie and Frueh 2005).

The existence of an inter-individual heterogeneity in the response to drugs, which affects both efficacy and safety, can be mediated by the alteration of the pharmacokinetics and pharmacodynamics of drugs. These variability mechanisms are shaped by the genetic-environmental interaction. The contribution of each factor varies with each drug (Wijnen et al. 2007; Evans and McLeod 2003).

▶ **Definition** In 1959 Friedrich Vogel defined pharmacogenetics as the genetic variation in the incidence of adverse effects in the patients (Vogel 1959). In 1962, this definition was changed to the study of the genetic variants that cause variability in response to drugs (Kalow 1962). Currently, the most accepted definition is "the discipline that allows identifying the genetic basis of the inter-individual differences in the drug response."

Pharmacogenomics studies the set of variations in genes related to pharmacokinetics and pharmacodynamics and their relationship with pharmacological response. In other words, pharmacogenomics studies the relationship between genetic information and the risk of incidence of a determined event, or the genetic contribution to its incidence, understanding an event either as an adverse effect or the therapeutic success (Meyer 2000, 2004).

▶   **Important** Pharmacogenetics and pharmacogenomics are tools that allow explaining and predicting variability in ADME process and the association with the clinical success or therapeutic failure due to lack of effectivity or security. Both disciplines are important tools for personalized medicine (Roses 2000).

On the other hand, pharmacoepigenetics and pharmacoepigenomics are new promising areas which are impacting on pharmacology. Both areas study the epigenetic basis for variation in drug response. Epigenetic changes include DNA methylation, histone modifications and RNA-mediated silencing. Altering any of these epigenetic mechanisms leads to inappropriate gene expression and the development of cancer and other epigenetic diseases (Ingelman-Sundberg and Gomez 2010). The most studied epigenetic mechanism that influences gene expression is DNA methylation. Nowadays, possible therapeutic applications represent a major component of pharmacoepigenetics, thus some epigenetics drugs which act as inhibitors of DNA methyltransferases and histone deacetylases have been studied, mainly in cancer treatment (Peedicayil 2008).

8.2 How We Can Explain Response Variability Due to Genetic

Genetic variability could affect every process in the normal physiology, therefore, pharmacokinetics and xenobiotics detoxification are not exceptions. A mutation could affect protein amount, due to a dysregulated expression or variation in the copy number of genes, or protein activity, which may be enhanced or decreased for genetic variants. Knowing the type of mutation, we can start to understand the potential effect in the protein and the consequence in the involved process.

8.2.1 Type of Mutations and Their Impact on Protein Function: Polymorphisms

When a genetic variant that leads to at least two non-rare phenotypes has a frequency of 1% or greater in a population is called polymorphism (Brookes 1999).

The most common type of mutation is a single base change (SNP, single nucleotide polymorphism), that is, the replacement of one nucleotide by the other in the same position. Depending of the DNA region where this change occurs, the effect on the phenotype may be different; depending of the change in the amino acid codification, protein structure and function could be affected. A synonymous substitution is when the base change encodes the same original amino acid; a missense substitution is a change in the encoded amino acid, so it changes the protein structure, and this could affect its functionality. Finally, a nonsense substitution changes the encoded amino acid for a stop codon, so the produced protein is shorter than the original, affecting its structure and function (Fig. 8.3).

Other types of mutation are deletions. In this case a variable number of bases are "eliminated" from the original sequence with consequences in the structure and function of the protein, depending on the deletion size the transcription process could originate a short protein or no protein.

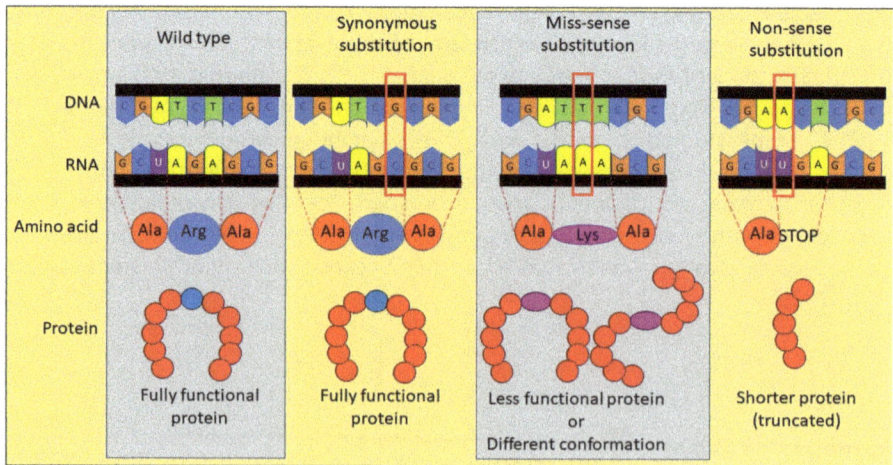

Fig. 8.3 Type of mutation in coding region

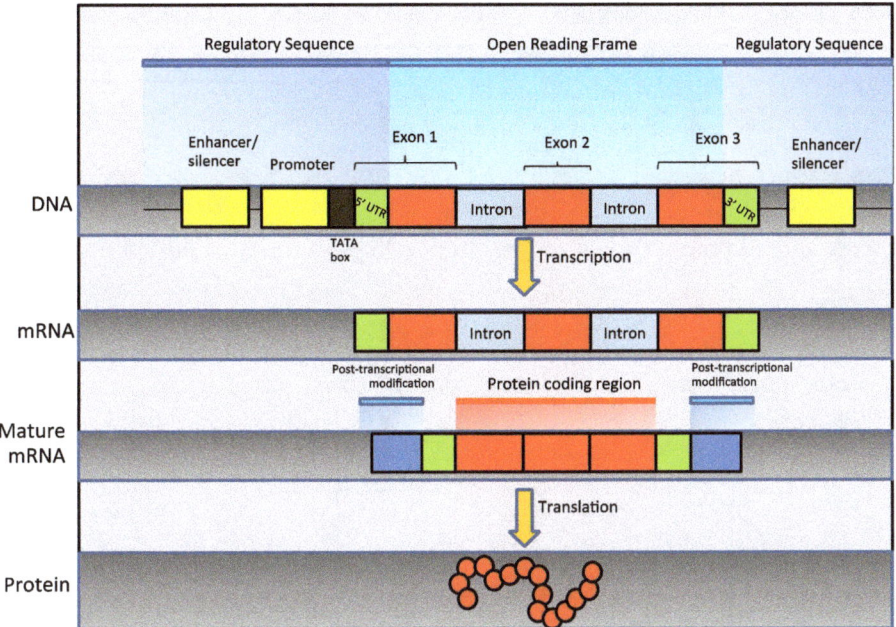

Fig. 8.4 Basic structure of a gene

Copy number variation (CNV) is another type of mutation that could affect the pharmacokinetics, an example is the duplication or multiplication of whole genes which allows to have a higher amount of protein. On the other hand, the loss of one of two or more copies of a whole gene would often result in a lower amount of protein and, sometimes, a lower effect.

Figure 8.4 shows the basic structure of a gene; the untranslated region comprises the promoter zone and enhancer/silencer binding sequence. This is called "regulatory sequence," and a mutation here affects the gene expression impacting on the amount of transcript and, as a consequence, on the amount of protein. For example, if a mutation occurs in an enhancer site resulting in a weak binding with the enhancer, therefore, there will be a lower expression of the genes, decreasing the amount of transcript. Conversely, if this mutation allows a tight binding of the enhancer, the amount of transcript will increase and, as a consequence, the amount of protein.

On the other hand, if there is a mutation in the open reading frame region, it could affect the structure and activity of the protein. When a non-synonymous substitution occurs in the protein coding region (i.e., exons), the amino acid change could modify the tertiary structure of the protein affecting, for example, its activity, the affinity for the substrate, or the protein stability. When the substitution is in the intron, primarily in the splice-donor or splice-acceptor sites, it could lead to an alternative (or novel) splicing or suppress splicing in that intron producing a larger protein than normal. A intronic mutation could also affect the regulatory system of splicing, affecting the normal functions of proteins due to problems in mRNA maturation.

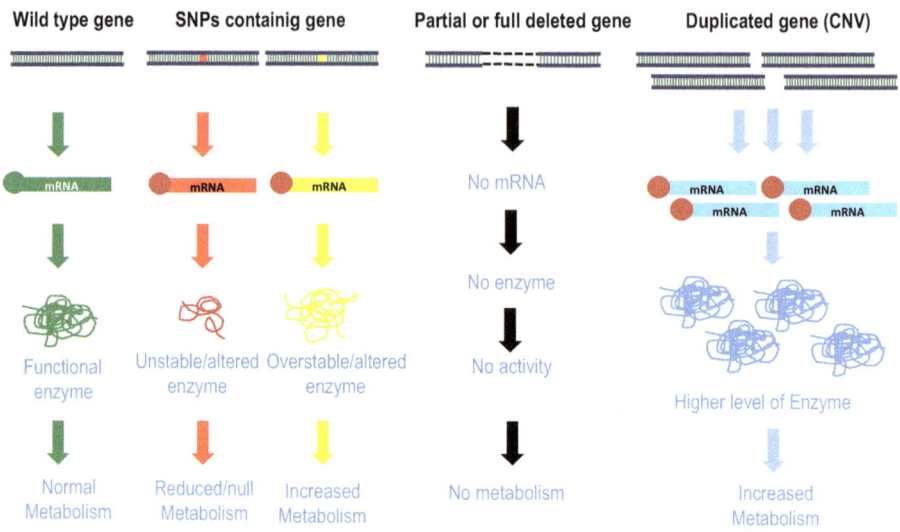

Fig. 8.5 Gene variants in drug-metabolizing enzymes

Pharmacogenetics has mainly focused on SNPs, pronounced snips, but every variant could be a factor explaining pharmacological variability. Several genetic variations are known. Figure 8.5 shows the different possibilities of genetic variations and their potential phenotypic expression in drug-metabolizing enzymes.

8.2.2 Drug-Metabolizing Enzymes

The most studied area in pharmacogenetics and pharmacogenomics is the drug metabolism. Depending on the degree of functionality of drug metabolism, the patients could be categorized in four main groups: poor metabolizer (PM), extensive metabolizers (EM), intermediate metabolizers (IM) and rapid/ultrarapid metabolizers (UM).

▶ **Definition** Those patients with a lower functionality are classified as "poor metabolizer." These patients could carry a variant allele (in heterozygote or homozygote form) that decrease the genetic expression of a metabolic enzyme, having a deletion of the gene or carrying a SNP that affects the activity of the enzyme (see Fig. 8.5). Those patients with a normal metabolism or without any variantallele affecting the metabolism are called "extensive metabolizers, normal metabolizer, or wild-type subject." Patients with an accelerated or augmented metabolism are called "rapid or ultrarapid metabolizers." These patients probably carry a variant that increases the genetic expression, by a mutation in the site of union of a silencer or by carrying an extra copy of the whole gene (Fig. 8.5). Patients carrying a variant allele in a heterozygote form such as their enzyme activity is between the extensive and poor metabolizer are called intermediate metabolizers.

In the xenobiotic metabolism, phase I is related to functionalization reactions and the principal family of enzymes which are cytochrome P450 (CYP450), phase II is related to conjugation reactions, and this conjugation is with more hydrophilic molecules to favor the elimination.

8.2.2.1 Phase I Enzymes

The main phase I enzymes, CYP450, is a super family of metabolic enzymes that catabolize phase I metabolism generally introducing or exposing a hydrophilic group in drug. CYP450 is the most important metabolizing system responsible for the oxidation of numerous endogenous (endobiotic) and exogenous (xenobiotic) compounds. It is expressed in many tissues, being more abundant in the liver. Based on similarities in amino acid sequences (Orellana and Guajardo 2004), it is classified into families, subfamilies, and isoforms. In humans, 18 CYP families and 43 subfamilies have been identified. Of these, 57 genes and 59 pseudogenes have been sequenced. The nomenclature of these enzymes and their variants is widely agreed, though it is constantly updated (http://drnelson.utmem.edu/CytochromeP450.html) (Nelson 2009).

These enzymes are very polymorphic, and the relationships with drug response have been described widely in scientific literature (Shalan et al. 2019; Annalora et al. 2017; Munro et al. 2018). Generally, one CYP enzyme metabolizes more than one drug, and a drug is metabolized by more than one CYP enzyme.

Besides drug detoxification, CYP enzymes participate in bio-activation of prodrugs and generation of toxic metabolites; thus, if these enzymes are less functional, the therapeutic effect will be minor due to less transformation to the active drug (Zhou et al. 2009; Walsh and Miwa 2011). Regarding safety, biotransformation could turn a drug into a toxic or reactive metabolite; thus, an enhanced process of activation in this case could be more harmful for the patient (Walsh and Miwa 2011; Bolleddula et al. 2014).

Among all CYP families, the most studied are 1, 2, and 3 families, all of them polymorphic. In this respect, polymorphisms are usually named with the number of the enzyme followed by a star and then the number of variant (e.g., CYP3A4*1B), but the best way to name them, when the variant corresponds to a SNP, is using the "rs number," a unique number assigned by NCBI dbSNP to each single nucleotide polymorphism.

In CYP450 superfamily there are a wide spectrum of relevant polymorphisms. There are SNPs related to a higher activity of the enzyme, as, for example, CYP1A1*2C, which is an amino acid change near to the active site and has been linked to a rising in the enzymatic capacity. Other SNPs are related to a higher level of genetic expression, for example, CYP1A1*2A, which is a substitution in the 3′ region of polyadenylation. Moreover, a polymorphism could be related to a suppression of the activity, as, for example, CYP3A4*26, which is a nonsense substitution; therefore, the protein has no activity (Preissner et al. 2013).

Several studies suggest that differences in basal levels of CYP constitute one of the main sources of inter-individual variability in the response to xenobiotics (Lin and Lu 2001; Fujikura et al. 2015; Quiñones et al. 2008).

8.2.2.2 Phase II Enzymes

This phase of the metabolism is associated to conjugation with hydrophilic molecules to favor the elimination of a drug. After this stage, drugs are generally inactive, and their final fate is urinary or fecal excretion. In a few cases, after phase II a chemical compound could acquire toxic activity, for example, nephrotoxicity of hydroquinone and bromobenzene is mediated via quinone-glutathione conjugates.

In drug metabolism glutathione S-transferases (GST) and uridine 5′-diphospho-glucuronosyltransferases (UDP – glucuronosyl transferase, UGT) metabolize a half of drugs, approximately. Other important phase II enzymes are sulfotransferases (SULTs), N-acetyltransferases (NATs), and thiopurine S-methyltransferase (TPMT).

Gene-related deficiency in phase II enzymes is related to an increased risk of toxicity, and clinical recommendations when a drug metabolized by these enzymes is administered to a patient with such deficiency are focused on avoiding adverse effects and improve safety in patients (Jancova et al. 2010).

Glutathione S-transferases (GSTs). Human glutathione S-transferase is a family of multigenes of soluble dimeric enzymes (Strange et al. 2001), which include alpha (α), mu (μ), pi (π), zeta (ζ), sigma (σ), kappa (κ), omega (ω), and theta (τ), with a broad subcellular distribution and partial superposition of specificities. Its main function is to detoxify pollutants, carcinogens, and mutagens, by conjugation with glutathione (GSH). They also play a role in the protection of tissues against reactive oxygen species (ROS) and lipid hydroperoxides during oxidative stress (Hayes and Strange 1995; Gallagher et al. 2006).

As an example, UGT is related to bilirubin metabolism, and its deficiency generates Gilbert's syndrome characterized by jaundice, so it is also associated with hepatic damage after administration of atazanavir, a drug used in HIV patients, due to a reduced conjugation and elimination. Besides, TPMT is related to metabolism of mercaptopurine and thioguanine, and its deficiency is associated with moderate or severe myelosuppression depending on metabolizer type, with poor and intermediate metabolizers being at higher risk than extensive metabolizer.

8.2.3 Drug Transporters

Drug transporters participate in many pharmacokinetic processes: absorption, distribution, and excretion; therefore, their dysfunction could affect the therapeutic goals related to sub-optimal or supraoptimal plasmatic levels (Lin et al. 2015).

There are two families of important drug transporters: ATP-binding cassette (ABC) and solute linking carriers (SLC), each of them having different functions. They could act as efflux or influx transporters; they appear in enterocytes affecting drug absorption, in renal tubules and hepatocytes impacting the urinary or biliary excretion of their substrates, and at the blood-brain barrier, affecting drug distribution to the brain (Vasiliou et al. 2009).

The ABC family comprises a widespread type of transporters in different species, and they possess a well-conserved nucleotide-binding domain where ATP is hydrolyzed. ABC transporters are expressed predominantly in the liver, intestine, blood-brain barrier, blood-testis barrier, placenta, and kidney. ABC proteins transport many endogenous substrates, including inorganic anions, metal ions, peptides, amino acids, sugars, and others, including drugs. A well-studied member of this family is ABCB1 (ATP-binding cassette subfamily B member 1) or P-glycoprotein (P-gp), which can contribute to drug elimination and affect drug absorption and distribution to the therapeutic target. It can be inhibited or upregulated by drug interactions (Fromm 2002).

SLC superfamily is a very diverse family including the SLC22 subfamily, the organic cation transporters (OCTs), organic zwitterion/cation transporters (OCTNs), organic anion transporters (OATs), and others such as the human SLC6 family members serotonin, norepinephrine, and dopamine transporters (SERT, NET, and DAT, respectively), among others (Colas et al. 2016).

▶   **Important** SLC21A6 or SLCO1B1 gene (solute carrier organic anion transporter family member 1) encodes the transporter OATP1B1 (organic anion–transporting polypeptide 1B1). This protein is found in liver cells; it transports compounds from the blood into the liver so that they can be cleared from the body. As an example, the rs4149056 C allele is related to a decreased activity of the transporter in vitro (Kameyama et al. 2005; Tirona et al. 2001) and reduced clearance of some drugs in vivo (Niemi et al. 2004; Pasanen et al. 2007), and it has been associated with an increased risk to develop myopathy after administration of simvastatin, a drug used for hypercholesterolemia, due to a decreased arrival to the liver (Group 2008).

8.2.4   Other Polymorphic Targets

Other pharmacogenomically relevant targets are related to specific process linked to certain drugs' pharmacodynamics. For example, VKORC1 (vitamin K epoxide reductase complex subunit 1) is the therapeutic target of warfarin. The polymorphism -1639 G>A has been described as a key factor in dosage modification, because of their effect on producing less amount of VKORC1 protein, where a patient will need a smaller dose of warfarin to reach the same therapeutic effect (Zhu et al. 2007).

DPYD gene, which encodes dihydropyrimidine dehydrogenase (DPD), is a limiting enzyme in fluoropyrimidines catabolism, and its reduced function is related to a high risk of toxicity due to fluorouracil, tegafur, and capecitabine driving to neutropenia, nausea, vomiting, severe diarrhea, stomatitis, mucositis, and hand-foot syndrome (Lee et al. 2014; Froehlich et al. 2015; Meulendijks et al. 2015).

Another case is glucose-6-phosphate dehydrogenase (G6PD), which is the enzyme that converts glucose-6-phosphate into 6-phosphogluconolactone, the first step of the pentose phosphate pathway. It is particularly important in erythrocytes because along with 6-phosphogluconate dehydrogenase, they are the only available source of NADPH required to protect erythrocytes from oxidative stress. A deficiency of this enzyme generates a high risk to develop hemolytic anemia due to a high oxidative stress sometimes generated by drugs (Mason et al. 2007; Cappellini and Fiorelli 2008; Nkhoma et al. 2009).

Finally, HLA is a member of the major histocompatibility complex (MHC) gene. HLA molecules are expressed in almost all cells and are responsible for presenting peptides to immune cells. Variations in HLA-B have been associated with several autoimmune conditions. Several variants in HLA-B have been associated with other adverse drug reaction phenotypes. Patients with the HLA-B*15:02 genotype have an increased risk of developing Stevens-Johnson syndrome from treatment with carbamazepine (Ferrell and McLeod 2008), whereas HLAB*58:01 is associated with an increased risk of severe cutaneous adverse reactions in response to allopurinol (Hung et al. 2005).

8.3 Relationship Between Pharmacogenetics and ADME Processes

Every pharmacokinetic process could be affected by pharmacogenetics, and the effect can be monitored through drug plasmatic levels. Figure 8.6 shows plasmatic concentration versus time curves for three patients who received the same dose of the

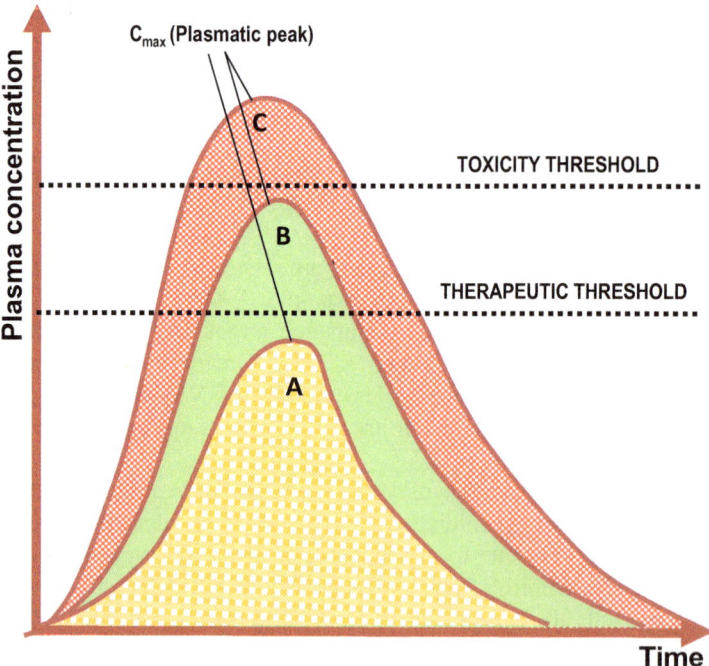

Fig. 8.6 Scheme of plasma concentration *versus* time curve in patients with different genotypes of P-gp, affecting absorption of drugs

same medication per os. The therapeutic threshold (the minimum level that must be reached to have a pharmacological effect) is not reached by the patient A; therefore, this treatment will fail. The toxicity threshold (the minimum level in which there is a probability to get adverse effects) is reached by the patient C; thus, the treatment could fail due to a toxic event. The range between therapeutic and toxicity thresholds is called the therapeutic range. Drug plasma concentrations should be within this range to reach clinical goals.

In the next subsections, we have assumed that we are treating subjects with polymorphisms affecting exclusively absorption-, distribution-, metabolism-, or excretion-related genes, and the correspondent impact on pharmacokinetic profiles is discussed.

8.3.1   Absorption

The activity of the transporter P-gp is very variable. It could vary due to inducers or inhibitors of this protein or due to genetic differences in the encoding gene (*MDR1*). A less functional P-gp will increase the absorbed fraction of the drug due to a decrease in the return of the drug to intestinal lumen. On the other hand, an overexpressed P-gp will decrease the absorbed drug (Hoffmeyer et al. 2000; Fromm 2002). Figure 8.6 represents pharmacokinetic curves for three patients differing only in P-gp polymorphism. Patient A carries the *MDR1* Ser893 polymorphism, a variation that promotes a higher activity of P-gp in enterocytes. Patient B has a "wild-type" gene, and finally patient C is homozygous for the SNP C3435T, a variant that negatively affects the expression of the gene, so there is a lower amount of P-gp in cell surface decreasing the efflux.

Note that patient A does not reach the therapeutic threshold and patient C is exposed to drug levels above the toxicity threshold. The absorption rate is different in each patient, being the highest in patient C. The slope of the ascendant phase is higher than in patient A or B, but the elimination rate is similar in every patient (parallel curves in the elimination phase).

Another important point to highlight is the plasmatic peak. In patient C, the peak (Cmax) is higher and occurs earlier than in the wild-type patient and the patient with under-expression of P-gp.

However, contrary to this hypothetical example, it is well known that several transporters are involved in the absorption phase of some drugs, with some transporters favoring drug uptake and others favoring the efflux. Therefore, it is necessary to know which specific transporters have an impact on the absorption of a given drug to predict the effect of a polymorphic variant on the drug pharmacokinetics.

8.3.2 Distribution

Polymorphisms in plasma proteins have not been extensively described yet. Let's assume a hypothetical situation were a polymorphism in albumin affecting drug-binding sites has been described, which is linked to a reduced bound fraction in plasma. Thus, if we study a drug with a high protein binding, as, for example, warfarin, and remembering that the free drug is the therapeutically active compound, we will have more therapeutically active warfarin, reaching a therapeutic threshold with a relatively low dose compared to wild-type carrier subjects. Furthermore, the toxicity threshold will be reached more easily Therefore, this patient should receive lower doses than a wild-type one, mainly to prevent hemorrhagic episodes.

Some transporters could affect the distribution of drugs thus affecting the pharmacological success. This is the case of the abovementioned P-gp protein, which is also expressed in the blood-brain barrier, and an overexpression of it could decrease the entrance of active drugs to the central nervous system. Other transporters as MRP3 and MRP4 (encoded by ABCC2 and ABCC4 genes, respectively) are expressed in the liver; polymorphisms rs12762549 in MRP2 and rs11568658 in MRP4 decrease the activity of transporters and could decrease the arrival to this organ and affect both distribution and metabolism.

8.3.3 Drug Metabolism (Biotransformation)

Drug metabolism is the most studied area in pharmacogenetics and pharmacogenomics. Variations in phase I and phase II enzymes have a proven relationship with therapeutic response, and the effect of a rapid or slow metabolism can be easily studied through the plasmatic level of a drug.

At this level, we have two different scenarios: the most common is an active drug in a rapid or ultrarapid metabolizer that will be eliminated faster than in an extensive metabolizer, making it difficult to reach efficacy. The other frequent scenario involves a patient with a poor metabolism, for whom the drug will stay longer in the organism, thus favoring the incidence of adverse reactions. Voriconazole is a good example of it. This is an antifungal drug metabolized by the CYP2C19 enzyme. In this gene we could find different genotypes associated with variations in metabolism. CYP2C19*17 is related to a higher metabolism, and CYP2C19*3 is related to a poor metabolism (Sim et al. 2006; Ikeda et al. 2004; Li et al. 2016). Plasma levels of voriconazole after an oral dose in our three hypothetical patients can be observed in Fig. 8.7. The genotype of patient A is CYP2C19*17/*17 (rapid metabolizer), patient B is CYP2C19*1/*1 (wild-type genotype), and patient C has the genotype CYP2C19*3/*3 (poor metabolizer).

Patient A has lower plasma levels due to an accelerated metabolism. The consequences of not reaching the therapeutic level are the progression to severe sepsis and septic shock and potential death of the patient. Patient B will likely reach therapeutic goals with this treatment. Patient C has the highest plasma concentrations, and this treatment will probably produce toxicity and should be

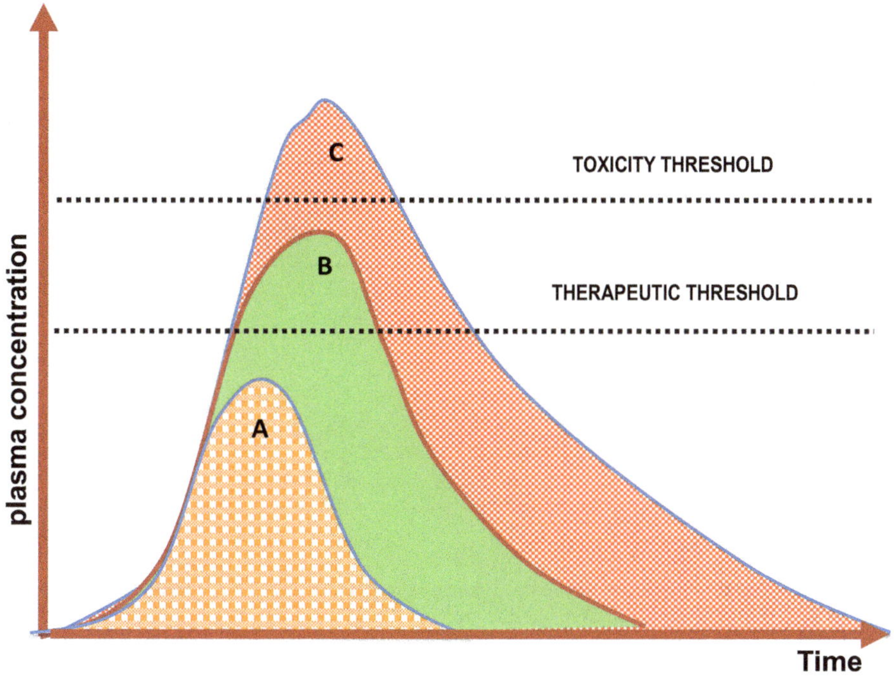

Fig. 8.7 Scheme of plasma concentration *versus* time curve in patients with different genotype of metabolic enzymes

suspended. In the case of voriconazole, one of the most serious adverse reactions is liver damage, leading to hepatic failure. This could be avoided with correct dosage and monitoring. Differences in the final phase of the curve can also be observed. The slope of the log transformation of the plasma drug concentration-time profile represents the elimination rate; from it, the elimination half-life could be calculated. Patient C eliminates slower than patients A and B, driving to a longer half-life.

Other relevant situation to be analyzed is when the administered drug is not active but its metabolite is (as in the case of prodrugs). In that scenario, a poor metabolism is associated with lower bio-activation and, as a consequence, to a lower or no therapeutic effect. In this respect, Fig. 8.8 shows plasmatic levels for patients who received an oral dose of a prodrug, for example, codeine, which is biotransformed by CYP2D6 to morphine in the liver. Patient B carries the wild-type allele for CYP2D6 enzyme in homozygous way, so it is an extensive metabolizer of codeine. Patient A carries an extra copy of the gene CYP2D6 in one allele, so it is an ultrarapid metabolizer of codeine. As can be appreciated through the low plasmatic level reached, this patient reached a lower peak than the other two because when codeine was absorbed, it was rapidly metabolized and the concentration-time curve fell faster. Patient C carries the genotype CYP2D6*5/*5; therefore, it is a poor metabolizer of codeine, so the peak of codeine should be higher and the elimination

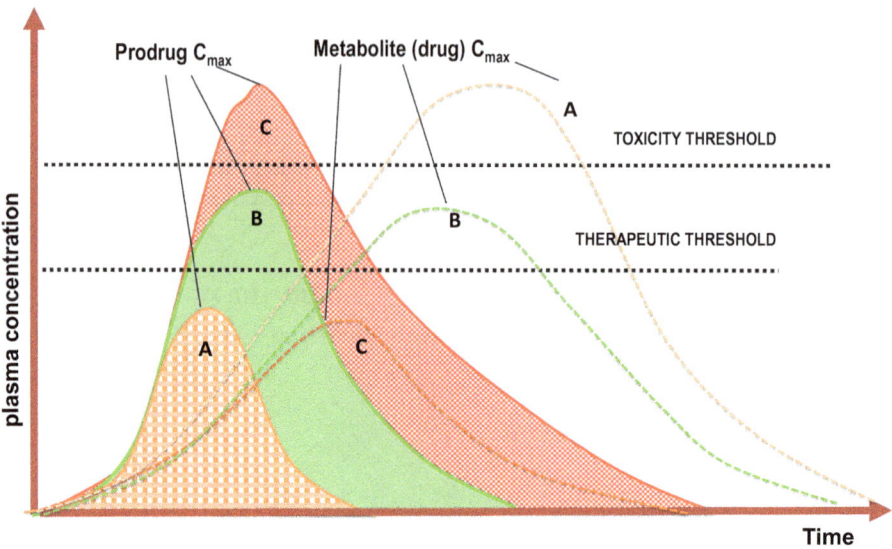

Fig. 8.8 Scheme of plasma concentration *versus* time curve for an administration of a prodrug, in patients with different genotype of metabolic enzymes

slower than for subjects A and B. The metabolite (e.g., morphine) will appear in plasma after biotransformation of codeine.

Note that the absorption phase is similar in all patients, but the elimination phase and peak (Cmax) are different for the prodrug and the metabolite. However, this is not always observed. The metabolite appearance in plasma could be different and the elimination rate as well. In this case the therapeutic threshold is associated only with the metabolite and the toxic threshold could be related to both compounds. In the case of codeine, the therapeutic activity is not related to the prodrug (codeine) but to the "activated" drug, which is the metabolite (morphine). In this sense, using our example, we can see that patient B achieves the therapeutic threshold, but does not surpass the toxicity level because habitual dosage is useful and secure in wild-type CYP2D6 carriers. However, in patient A, the level of morphine is higher due to a faster biotransformation reaching the toxicity threshold (possible adverse reactions). On the other hand, patient C displays the highest level of codeine and the lowest level of morphine because his poor metabolism fails to transform codeine to morphine; thus, this patient will not have a therapeutic success. Note that the ascendant curve for metabolites in Fig. 8.8 is not due to absorption, but to biotransformation from codeine, thus the start point is not zero time. For patient A, the bio-activation is faster than for the remaining two (as seen in the higher slope) due to an extra copy of the gene that encodes the biotransformation enzyme.

In other cases, as, for example, tamoxifen, both the drug (tamoxifen) and its metabolites (4-hydroxytamoxifen, N-demethyltamoxifen, and endoxifen) have

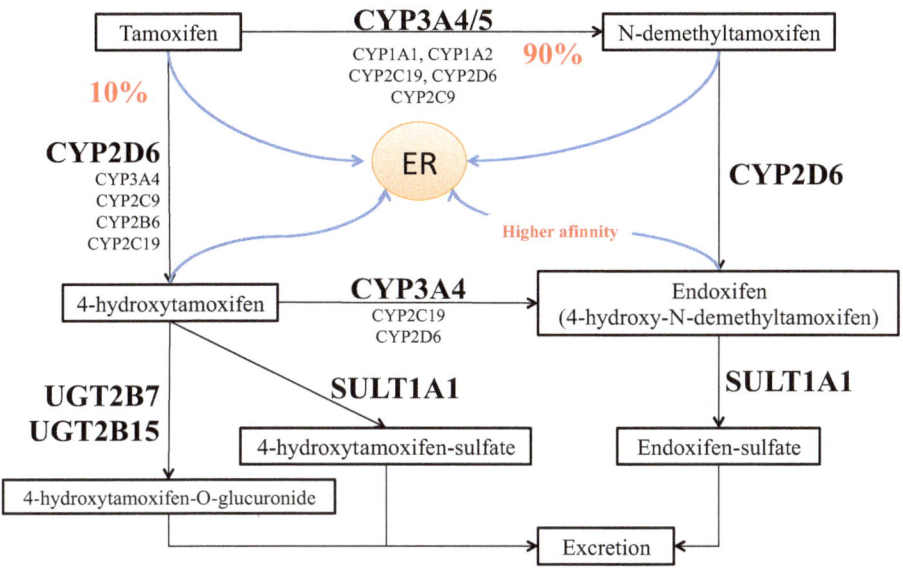

Fig. 8.9 Metabolic pathways of tamoxifen. (Adapted from Miranda 2016). *ER, estrogen receptor; CYP, cytochrome P450; UGT, UDP-glucuronosyltransferase; SULT, sulfotransferase*

therapeutic activity, even though it is suggested that endoxifen is more active (Fig. 8.9) (Skaar and Desta 2018; Neven et al. 2018).

8.3.4 Renal Excretion

Drug excretion may be a pharmacogenomically relevant process whenever the drug is not widely metabolized, and the elimination not only involves glomerular filtration but active secretion in the renal tubule as well. This is the case of β-lactam antibiotics, as penicillin and cephalosporins. These drugs are poorly metabolized, so their excretion is principally performed by excretion of the unaltered drug, to which transporters do contribute.

In renal proximal tubule, there are three transporters implied in the active secretion of β-lactam drugs, belonging to the family of organic anion transporter (OAT): OAT1, encoded by *SLC22A6* and located in the basolateral membrane together with OAT3, encoded by *SLC22A8*, and in the apical membrane OAT4, encoded by *SLC22A11* gene (Fig. 8.10) (Lee et al. 2006; Roth et al. 2012).

It has been described that all these genes have SNPs affecting the normal activity of transporters. The most studied SNPs are rs11568626 (149C>T) in OAT1, rs11568482 in OAT3 (62763264T>A), and rs11231809 (64302950T>A) in OAT4. In patients carrying some of these polymorphic alleles, the elimination process will be slower favoring the incidence of adverse effects. On the other

Fig. 8.10 Location of SLC family transporters in kidney proximal tubule. *OAT, organic anion transporter*

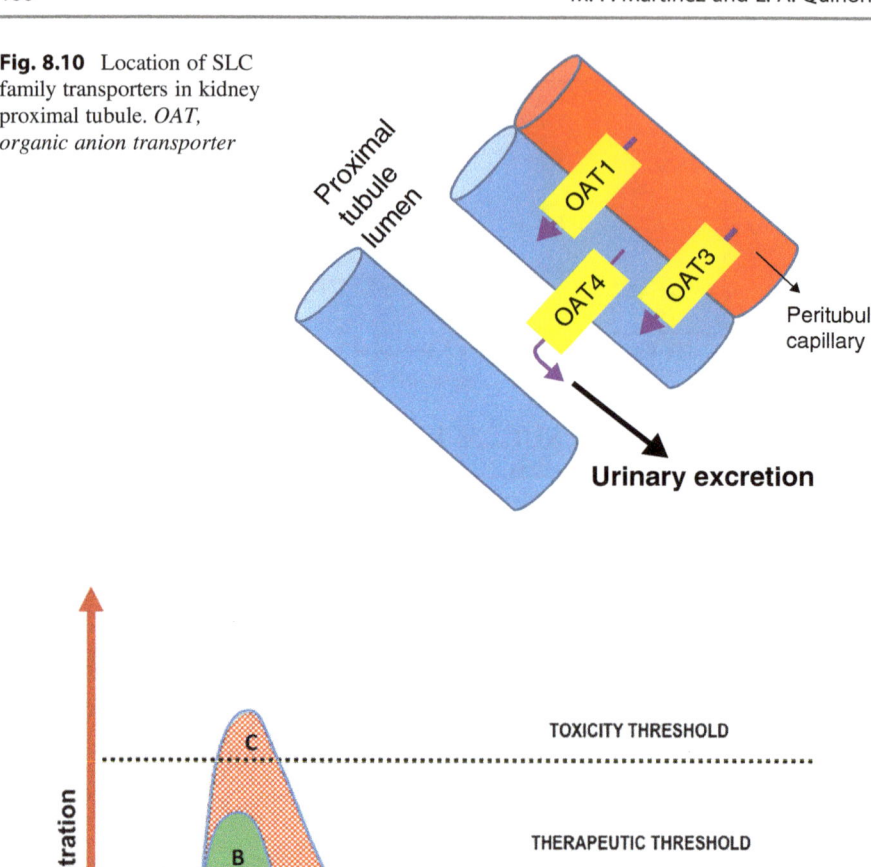

Fig. 8.11 Scheme of plasma concentration *versus* time curve of a drug (e.g., cefotaxime) in patients with different genotypes in OAT3 gene. *In the case of cefotaxime, therapeutic threshold corresponds to the minimum inhibitory concentration*

hand, an overexpression of genes encoding these excretion transporters will result in lower drug plasma levels, and the therapeutic effect would be reduced.

As an example, cefotaxime is a third-generation cephalosporin antibiotic, and its elimination from the body is mediated by OAT3. Figure 8.11 shows plasmatic levels after an intravenous dose of cefotaxime in patients with different genotypes for OAT3 (rs11568482). Patient C would be homozygous for rs11568482 leading to a reduced clearance (Yee et al. 2013), patient B is heterozygous having a reduced clearance too, and patient A has a wild-type genotype.

Note that the ascendant phase is quite fast, because the example does not involve intestinal absorption, but an intravenous administration, which is similar for three patients, the Tmax being also similar. For cefotaxime, the therapeutic threshold is now called minimum inhibitory concentration (MIC), that is, the minimum plasma level of a drug required to inhibit bacterial grow. In other antibiotics, as, for example, cephalosporine, it is important that the time concentration remains over the MIC and not only to reach the minimum level.

Patient A has the fastest elimination process and a lower Cmax. This patient has a wild-type genotype for the variant rs11568482 in *OAT3* gene, and probably the infection will not be solved with this dose of cefotaxime. Patient B is heterozygote for this variant and has a better pharmacokinetic profile, i.e., reach the MIC during enough time over MIC, without surpassing the toxic threshold. Patient C, homozygote for the variation, will have a very good activity against bacteria, but will probably develop adverse effects, in the case of cefotaxime, nausea, diarrhea, or kidney deterioration due to a very slow elimination of the drug.

8.3.5 Drug Toxicity (Safety)

The first historical report of pharmacogenetics was in 510 B.C. when Pythagoras observed that some people suffered a potentially fatal reaction after ingestion of fava beans (Pirmohamed 2001). Later, this reaction was related to a deficiency of the enzyme glucose-6-phosphate dehydrogenase (G6PD). Now, this deficiency is related to several adverse effects of some drugs, for example, rasburicase (a drug used for hyperuricemia). Patients with a G6PD deficiency could experience potentially fatal hemolytic anemia (Nguyen and Ness 2014). Another example of this adverse reaction is the use of fluoroquinolone antibiotics which should be avoided in those patient with G6PD deficiency (Bensouda et al. 2002; Sansone et al. 2010).

Another good example of the use of a polymorphism as toxicity marker is the HLA-B*5701 genotype. Patients using abacavir, an HIV reverse transcriptase inhibitor, need to be studied for HLA-B*5701 genotypes, because the patients who carry this variant have a high risk of hypersensitivity reactions, with cutaneous expression and life risk (Mallal et al. 2008).

8.4 Dosing Modifications and Drug Selection

The Clinical Pharmacogenetics Implementation Consortium (CPIC) is an international organization interested in facilitating the use of pharmacogenetics in the clinical practice. CPIC have created some clinical guidelines for different gene/drug pairs. These guidelines have indications of drug dosage or drug selection, to improve the response of patients.

Consensual CPIC recommendations are shown in Table 8.1. Recommendations are related to dose adjustment and drug selection or alternative use of other drugs.

Table 8.1 CPIC recommendation according to drug and polymorphism[a]

Drug	Gen	Polymorphism	CPIC recommendation
Abacavir (Martin et al. 2012, 2014)	HLA-B	Presence of at least one *57:01 allele	Abacavir is not recommended. High risk of hypersensitivity (~6% of patients)
Allopurinol (Hershfield et al. 2013; Saito et al. 2016)	HLA-B	Carrier of HLA-B*5801	Significantly increased risk of allopurinol-induced severe cutaneous adverse reaction
Amitriptyline Clomipramine Doxepin Imipramine Trimipramine (Hicks et al. 2017)	CYP2D6	*1/*1 × N, *1/*2 × N, *2/*2 × Nc (ultrarapid metabolizer)	Avoid tricyclic use due to potential lack of efficacy. Consider alternative drug not metabolized by CYP2D6
		*4/*10, *4/*41, *5/*9 (intermediate metabolizer)	Consider a 25% reduction of recommended starting dose. Utilize therapeutic drug monitoring to guide dose adjustments
		*3/*4, *4/*4, *5/*5, *5/*6 (poor metabolizer)	Avoid tricyclic use due to potential for side effects. Consider alternative drug not metabolized by CYP2D6. If a TCA is warranted, consider a 50% reduction of recommended starting dose
	CYP2C19	*1/*17, *17/*17 (rapid and ultrarapid metabolizer)	Avoid tertiary amine use due to potential for sub-optimal response. Consider alternative drug not considerably metabolized by CYP2C19 (e.g., nortriptyline and desipramine). If a tertiary amine is warranted, utilize therapeutic drug monitoring to guide dose adjustments
		*2/*2, *2/*3, *3/*3 (poor metabolizer)	Avoid tertiary amine use due to potential for sub-optimal response. Consider alternative drug not metabolized by CYP2C19. TCAs without major CYP2C19 metabolism include the secondary amines nortriptyline and desipramine. For tertiary amines, consider a 50% reduction of the recommended starting dose

(continued)

Table 8.1 (continued)

Drug	Gen	Polymorphism	CPIC recommendation
Atazanavir Gammal et al. 2016)	UGT1A1	*28/*28, *28/*37, *37/*37, rs887829T/T(*80/*80), *6/*6 (poor metabolizer)	High likelihood of bilirubin-related discontinuation of the drug. Consider an alternative agent particularly where jaundice would be of concern to the patient
Capecitabine Fluorouracil Tegafur (Amstutz et al. 2018)	DPYD	c.[1905 + 1G>A]/wt, c.[1679T>G]/wt,c.[2846A>T]/wt; c.[1129–5923C>G]/wt; c.[1129–5923C>G]/[1129–5923C>G], c.[2846A>T]/[2846A>T] (intermediate metabolizer)	Reduce starting dose based on activity score followed by titration of dose based on toxicity
		c.[1905 + 1G>A]/[1905 + 1G>A], c.[1679T>G]/[1679T>G], c.[1905 + 1G>A]/[2846A>T] c.[1905 + 1G>A]/[1129–5923C>G] (poor metabolizer)	Avoid use of 5-fluorouracil or 5-fluorouracil prodrug-based regimens
Carbamazepine (Leckband et al. 2013)	HLA-B	Carrier of HLA-B*15:02	If patient is carbamazepine-naive, do not use carbamazepine. Increased risk of carbamazepine induced Stevens-Johnson syndrome/toxic epidermal necrolysis. If patient has previously used carbamazepine for longer than 3 months without incidence of cutaneous adverse reactions, cautiously consider use of carbamazepine
citalopram escitalopram Paroxetine (Hicks et al. 2015)	CYP2C19	*17/*17, *1/*17 (rapid and ultrarapid metabolizer)	Select alternative drug not predominantly metabolized by CYP2D6
		*2/*2, *2/*3, *3/*3 (poor metabolizer)	Select alternative drug not predominantly metabolized by CYP2D6b or if drug use warranted, consider a 50% reduction of recommended starting dose and titrate to response
Clopidogrel (Scott et al. 2013)	CYP2C19	*1/*2, *1/*3, *2/*17 (intermediate metabolizer)	Alternative antiplatelet therapy (if no contraindication), e.g., prasugrel, ticagrelor
		*2/*2, *2/*3, *3/*3 (Poor metabolizer)	Alternative antiplatelet therapy (if no contraindication), e.g., prasugrel, ticagrelor

(continued)

Table 8.1 (continued)

Drug	Gen	Polymorphism	CPIC recommendation
Codeine (Crews et al. 2014)	CYP2D6	*1/*1 × N, *1/*2 × N (ultrarapid)	Avoid codeine use due to potential for toxicity
		*4/*4, *4/*5, *5/*5, *4/*6 (poor metabolizer)	Avoid codeine use due to lack of efficacy
Desipramine Nortriptyline ((Hicks et al. 2017)	CYP2D6	(*1/*1) × N,(*1/*2) × N, (*2/*2) × N (ultrarapid metabolizer)	Avoid tricyclic use due to potential lack of efficacy. Consider alternative drug not metabolized by CYP2D6
		*4/*41, *5/*9, *4/*10 (intermediate metabolizer)	Consider a 25% reduction of recommended starting dose. Utilize therapeutic drug monitoring to guide dose adjustments
		*4/*4, (*4/*4) × N,*3/*4,*5/*5,*5/*6 (poor metabolizer)	Avoid tricyclic use due to potential for side effects. Consider alternative drug not metabolized by CYP2D6
Fluvoxamine (Hicks et al. 2015)	CYP2D6	*3/*4, *4/*4, *5/*5, *5/*6 (poor metabolizer)	Consider a 25–50% reduction of recommended starting dose and titrate to response or use an alternative drug not metabolized by CYP2D6
Ivacaftor (Clancy et al. 2014)	CFTR	Homozygous or heterozygous G551D-CFTR – e.g., G551D/ F508del, G551D/G551D, rs75527207 genotype AA or AG	Use ivacaftor according to the product label (e.g., 150 mg every 12 h for patients aged 6 years and older without other diseases; modify dose in patients with hepatic impairment)
		Noncarrier of G551D-CFTR – e.g., F508del/R553X, rs75527207 genotype GG	Ivacaftor is not recommended
		Homozygous for F508del-CFTR (F508del/F508del), rs113993960, or rs199826652 genotype del/de	Ivacaftor is not recommended
Mercaptopurine Tioguanina Azathioprine (Relling et al. 2011)	TPMT	*1/*1 (extensive metabolizer)	Start with normal starting dose (e.g., 75 mg/m2/day or 1.5 mg/kg/day) and adjust doses of MP (and of any other myelosuppressive therapy) without any special emphasis on MP compared to other agents. Allow 2 weeks to reach steady state after each dose adjustment
			Start with reduced doses (start at 30–70% of full dose: e.g.,

(continued)

Table 8.1 (continued)

Drug	Gen	Polymorphism	CPIC recommendation
		*1/*2, *1/*3A, *1/*3B, *1/*3C, *1/*4 (Intermediate metabolizer)	at 50 mg/m2/day or 0.75 mg/kg/day) and adjust doses of MP based on degree of myelosuppression and disease-specific guidelines. Allow 2–4 weeks to reach steady state after each dose adjustment. In those who require a dosage reduction based on myelosuppression, the median dose may be ~40% lower (44 mg/m^2) than that tolerated in wild-type patients (75 mg/m^2). In setting of myelosuppression, and depending on other therapy, emphasis should be on reducing MP over other agents
		*3A/*3A, *2/*3A, *3C/*3A, *3C/*4, *3C/*2, *3A/*4 (poor metabolizer)	Start with drastically reduced doses (reduce daily dose by tenfold and reduce frequency to thrice weekly instead of daily, e.g., 10 mg/m^2/day given just 3 days/week) and adjust doses of MP based on degree of myelosuppression and disease-specific guidelines. Allow 4–6 weeks to reach steady state after each dose adjustment. In setting of myelosuppression, emphasis should be on reducing MP over other agents. For nonmalignant conditions, consider alternative nonthiopurine immunosuppressant therapy
Ondansetron Tropisetron (Bell et al. 2017)	CYP2D6	*1/*1 × N, *1/*2 × N, *2/*2 × N (ultrarapid metabolizer)	Select alternative drug not predominantly metabolized by CYP2D6 (i.e., granisetron)
		*3/*4, *4/*4, *5/*5, *5/*6 (poor metabolizer)	Insufficient evidence demonstrating clinical impact based on CYP2D6 genotype. Initiate therapy with no recommendation recommended starting dose

(continued)

Table 8.1 (continued)

Drug	Gen	Polymorphism	CPIC recommendation
Peginterferon alfa-2a Peginterferon alfa-2b Ribavirin (Muir et al. 2014)	IFNL3	rs12979860 CC (favorable responsive genotype)	Approximately 90% chance for sustained virologic response (defined by undetectable serum viral RNA 12–24 weeks after the end of treatment) after 24–48 weeks of treatment. Approximately 80–90% of patients are eligible for shortened therapy (24–28 weeks vs. 48 weeks). Weighs in favor of using PEG-IFN-α- and ribavirin-containing regimens
		rs12979860 CT or TT (unfavorable responsive genotype)	Approximately 60% chance of sustained virologic response after 24–48 weeks of treatment. Approximately 50% of patients are eligible for shortened therapy regimens (24–28 weeks). Consider implications before initiating PEG-IFN-α- and ribavirin-containing regimens
Phenytoin (Caudle et al. 2014)	CYP2C9	*1/*3, *1/*2 (Intermediate metabolizer)	Consider 25% reduction of recommended starting maintenance dose. Subsequent maintenance doses should be adjusted according to therapeutic drug monitoring and response
		*2/*2, *3/*3, *2/*3 (poor metabolizer)	Consider 50% reduction of recommended starting maintenance dose
	HLA-B	HLA-B*15:02 noncarrier. No *15:02 alleles reported	Initiate therapy following CYP2C9 genotype
		HLA-B*15:02 carrier. One or two *15:02 alleles	If patient is phenytoin naive, do not use phenytoin/ fosphenytoin
Rasburicase (Relling et al. 2014)	G6PD	Deficient (<10–60% of normal enzyme activity)	Rasburicase is contraindicated; alternatives include allopurinol
		Deficient with chronic nonspherocytic hemolytic anemia (Severe enzyme deficiency (<10% activity)	Rasburicase is contraindicated; alternatives include allopurinol
		Variable (normal or deficient enzyme activity) (due to X-linked mosaicism, females	To ascertain that G6PD status is normal, enzyme activity must be measured;

(continued)

Table 8.1 (continued)

Drug	Gen	Polymorphism	CPIC recommendation
		heterozygous may display a normal or a deficient phenotype. It is therefore difficult to predict the phenotype of these individuals)	alternatives include allopurinol
Sertraline (Hicks et al. 2015)	CYP2C19	*17/*17, *1/*17 (rapid and ultrarapid metabolizer)	Initiate therapy with recommended starting dose. If patient does not respond to recommended maintenance dosing, consider alternative drug not predominantly metabolized by CYP2C19
		*2/*2, *2/*3, *3/*3 (poor metabolizer)	Consider a 50% reduction of recommended starting dose and titrate to response or select alternative drug not predominantly metabolized by CYP2C19
Simvastatina (Ramsey et al. 2014)	SLCO1B1	rs4149056 CT (Intermediate function)	Prescribe a lower dose or consider an alternative statin (e.g., pravastatin or rosuvastatin); consider routine CK surveillance
		rs4149056 CC (low function)	Prescribe a lower dose or consider an alternative statin (e.g., pravastatin or rosuvastatin); consider routine CK surveillance
Tacrolimus (Birdwell et al. 2015)	CYP3A5	*1/*1 (extensive metabolizer)	Increase starting dose 1.5–2 times recommended starting dose. Total starting dose should not exceed 0.3 mg/kg/day
		*1/*3, *1/*6, *1/*7 (intermediate metabolizer)	Increase starting dose 1.5–2 times recommended starting dose. Total starting dose should not exceed 0.3 mg/kg/day
Voriconazole (Moriyama et al. 2017)	CYP2C19	*17/*17, *1/*17 (rapid and ultrarapid metabolizer)	Choose an alternative agent that is not dependent on CYP2C19 metabolism as primary therapy in lieu of voriconazole. Such agents include isavuconazole, liposomal amphotericin B, and posaconazole

(continued)

Table 8.1 (continued)

Drug	Gen	Polymorphism	CPIC recommendation
		(*2/*2, *2/*3, *3/*3 (poor metabolizer)	Choose an alternative agent that is not dependent on CYP2C19 metabolism as primary therapy in lieu of voriconazole. Such agents include isavuconazole, liposomal amphotericin B, and posaconazole
Warfarin (Johnson et al. 2017)	CYP2C9 and VKORC1	VKORC1-1639 G>A and CYP2C9*2 or *3 available	Calculate dose based on validated published pharmacogenetic algorithms
		VKORC1-1639 G>A and CYP2C9*2 or *3 Not available	Dose clinically
	CYP4F2	Carriers of rs2108622 T allele	Increase dose by 5–10%
	CYP2C9	Carriers of *5, *6, *8, *11 variant alleles	Decrease calculated dose by 15–30%
	CYP2C cluster	rs12777823 A carriers	Decrease dose by 10–25%

[a]Some genotypes have not been included, because there is no recommendation of modification based in pharmacogenetics (principally extensive metabolizer genotypes)

Main phase I enzymes involved in the guidelines are CYP2C9, CYP2C19, CYP2D6, CYP3A5, and CYP4F2. Main phase II enzymes are UGT1A1 and TPMT. Other polymorphic enzymes/proteins involved are DPYD, VKORC1, G6PD, HLA-B, IFNL3, and CFTR; the only transporter with a clinical recommendation is so far SLCO1B1.

Besides CPIC, there are other institutions working on clinical implementation of pharmacogenetics around the world. The *Dutch Pharmacogenetics Working Group (DPWG)* and the *Canadian Pharmacogenomics Network for Drug Safety (CPNDS)* are good examples. DPWG have implemented guidelines for aripiprazole, atomoxetine, clozapine, duloxetine, flecainide, flupenthixol, haloperidol, metoprolol, mirtazapine, olanzapine, oxycodone, propafenone, risperidone, tamoxifen, tolbutamide, tramadol, venlafaxine, and zuclopenthixol. Every one of them is in relation with CYP2D6 genotype. Esomeprazole and omeprazole are related to CYP2C19, glibenclamide and tolbutamide to CYP2C9, irinotecan to UGT1A1. CPNDS has guidelines for daunorubicin and doxorubicin with RARG, SLC8A3, and UGT1A6 polymorphism and cisplatin with polymorphism in TPMT, among others.

In Latin America, the pharmacogenetics and pharmacogenomics areas are recently emerging fields, and the main focus of the research is to evaluate ethnic differences to apply adapted guidelines to manage personalized pharmacotherapy. Large differences are presumed among countries, as well as potential benefits from the use of Pharmacogenomics testing, which are not extensively implemented to date. The RIBEF (*Red Iberoamericana de Farmacogenética y Farmacogenómica*) and SOLFAGEM (*Sociedad Latinoamericana de Farmacogenómica y Medicina Personalizada*) networks are focused on studying and adapting pharmacogenomic guidelines to the region. A recent book collected the present efforts to investigate variability in drug response using molecular approaches and considering the limitations to apply pharmacogenomics tests in clinical centers and hospitals of Latin America (Quiñones 2017).

8.5 Integrative Pharmacogenomic Clinical Cases

Clinical Case 1

A 55-year-old male patient is diagnosed with depression, with no other clinically relevant antecedent. The psychiatrist has prescribed amitriptyline in the recommended starting dose and a control 2 months later. The patient reported that he has not felt better and the physician suspected lack of efficacy; thus he solicited a genetic test for CYP2D6 and CYP2C19. The result of the genotypic analysis was CYP2D6 *1/*1 and CYP2C19 *17/*17.

As the patient is an extensive metabolizer for CYP2D6, the conclusion is that the therapeutic failure is not due to this enzyme; therefore, the new chosen treatment could be a drug metabolized by CYP2D6. On the other hand, CYP2C19 genotyping showed an ultrarapid metabolizer phenotype; therefore, the therapeutic failure probably is due this accelerated metabolism; thus, the chosen medication should not be metabolized by CYP2C19 to avoid failure due to a high elimination.

In conclusion, if the psychiatrist wanted to keep tricyclic antidepressant (TCA) treatment, he had two good options, desipramine and nortriptyline, which are metabolized by CYP2D6 and have no mayor metabolism by CYP2C19. However, therapeutic drug monitoring must be carried out to ensure the therapeutic success.

Clinical Case 2

A 68-year-old female patient with antecedents of hypercholesterolemia, acute myocardial infarction, depression, and epilepsy is using simvastatin, aspirin, warfarin, clopidogrel, sertraline, and phenytoin, every one of them at the usual doses, with unsatisfactory results. To improve the pharmacological therapy, genotyping is required. It was proposed to genotype SLCO1B1 for simvastatin, CYP2C19 for clopidogrel and sertraline, CYP2C9 for phenytoin and warfarin, HLA-B for phenytoin, and VKORC1 for warfarin.

(continued)

The result of genotyping was:

- SLCO1B1 rs4149056: CT
- CYP2C19: *2/*2
- CYP2C9: *1/*3
- HLA-B 15:02: noncarrier
- VKORC1: -1639 GG

SLCO1B1 rs4149056 CT genotype means that the transporter has an intermediate function for simvastatin transport, so the recommendation should reduce the dose or change the medication for hypercholesterolemia. Pravastatin could be a good option in this case. CYP2C19*2/*2 genotype corresponds to a poor metabolizer; therefore, the dose of sertraline must be decreased in at least a 50% and clopidogrel will not be effective, because it is a prodrug bioactivated by CYP2C19. The recommendation should be a change of drug. Prasugrel is a good option in this case. In both cases monitoring is necessary to ensure the therapeutic result.

CYP2C9 *1/*3 genotype speaks of an intermediate metabolizer phenotype and together with VKORC1 -1639 GG (wild-type genotype) indicated that probably this patient needs a lower dose than the usual. Warfarin dosage must be based on validated published Pharmacogenetics algorithms considering other nongenetic factors.

Similarly, considering the intermediate metabolizer phenotype of CYP2C9 and the noncarrier HLA-B15:01 genotype, the use of phenytoin does not drive to the risk for hypersensitivity reaction; however, the dose should be reduced, starting with a dose 25% lower than the usual and adjusting according to response, to improve the therapeutic result.

Clinical Case 3

A Chilean young man (24 years old), 1.60 m height and 54 kg weight, suffered the torsion of his right ankle. He went to the traumatologist who diagnosed sprain and indicates immobilization and NSAIDs. Several days after that visit, he started with pain in the popliteal zone, nearby the right knee; thus, he consulted another traumatologist, who suspects deep venous thrombosis (DVT) and requests Eco Doppler analysis confirming DVT. Patient is referred to internist, who indicates warfarin 5 mg/day and control in 15 days. It also indicates not to consume green vegetables. To control his treatment, he decided to evaluate his coagulation at the fourth day, advised by a physician. The INR value was 5.6; therefore the doctor decided to decrease the weekly

(continued)

dose to a half tablet from Monday to Friday and one tablet on Saturday and Sunday. After 1 week the INR value was 3.9; therefore, the new dose adjustment was half tablet from Monday to Saturday and one tablet on Sunday; after this adjustment, INR was between 2.0 and 3.0, as is recommended.

Genotypic analysis after his therapeutic start was carried out. The result was:

- VKORC1-1639: G/A (heterozygous)
- CYP2C9*2: C/C (homozygous wild type)
- CYP2C9*3: C/C (homozygous variant)

Considering the previous antecedents, if warfarin dosage for patient's genetics should be different, there are two tools to choose the correct warfarin dose: the table in the warfarin product insert approved by the US Food and Drug Administration (FDA) (Table 8.2) and online warfarin dosage algorithm (www.wafarindosing.org).

According to the Table 8.2, the patient has a heterozygous genotype for VKORC1-1639 polymorphism, thus we must follow the second row until the CYP2C9 genotype (CYP2C9*1/*3), so the recommended daily dose for this patient is between 3 and 4 mg of warfarin. This method does not consider other factors as the age, sex, height, or weight.

The online algorithm considers genetic and not genetic factors (as demographic, pharmacologic, and smoking) to calculate the dose. If we include the factors in the field, the daily recommended dose is 2.3 mg of warfarin to achieve an INR of 2.0. This dose is very close to the real stable dose. If the algorithm had been used from the beginning, the patient would not have been at risk of hemorrhage with INR values over 3.0.

Table 8.2 Recommended daily warfarin doses (mg/day) to achieve a therapeutic INR based on *CYP 2C9* and *VKOR C1* genotype using the warfarin product insert approved by the US Food and Drug Administration

VKORC1: −1639G>A	CYP2C9 *1/*1	CYP2C9 *1/*2	CYP2C9 *1/*3	CYP2C9 *2/*2	CYP2C9 *2/*3	CYP2C9 *3/*3
GG	5–7	5–7	3–4	3–4	3–4	0.5–2
GA	5–7	3–4	3–4	3–4	0.5–2	0.5–2
AA	3–4	3–4	0.5–2	0.5–2	0.5–2	0.5–2

Reproduced from updated warfarin product label

8.6   Conclusions

- Pharmacokinetics in patients varies due to genetic factors, and every stage in ADME process is potentially susceptible to be modified by them, giving rise to variable therapeutic response.
- Pharmacogenetics and pharmacogenomic tools can help improving the clinical outcome of drug therapy, considering both drug efficacy and safety, and should be considered at the beginning of patient treatment.
- Not only ADME processes-related polymorphisms should be associated with drug selection. Factors as HLA-B, related to immune response, and others should be considered to improve efficacy and prevent adverse reactions for some drugs.
- Genetic factors must be considered together with clinical and demographic factors to choose the correct medication and dosage.
- As pharmacogenetics is a discipline in present development, new studies must be carried out to increase the knowledge on relationships between genetic variants and pharmacological response, particularly in mixed populations. Clinical trials considering genotypes will be useful to favor the application in daily clinical practice in health centers.

References

Amstutz U, Henricks LM, Offer SM et al (2018) Clinical Pharmacogenetics Implementation Consortium (CPIC) guideline for dihydropyrimidine dehydrogenase genotype and fluoropyrimidine dosing: 2017 update. Clin Pharmacol Ther 103:210–216

Annalora AJ, Marcus CB, Iversen PL (2017) Alternative splicing in the cytochrome P450 super-family expands protein diversity to augment gene function and redirect human drug metabolism. Drug Metab Dispos 45:375–389

Bell GC, Caudle KE, Whirl-Carrillo M et al (2017) Clinical Pharmacogenetics Implementation Consortium (CPIC) guideline for CYP2D6 genotype and use of ondansetron and tropisetron. Clin Pharmacol Ther 102:213–218

Bensouda L, Jarry C, Jonville-Béra A et al (2002) Risk medications in case of glucose-6-phosphate dehydrogenase deficiency. Arch Pediat 9:316–319

Birdwell KA, Decker B, Barbarino JM et al (2015) Clinical Pharmacogenetics Implementation Consortium (CPIC) guidelines for CYP3A5 genotype and tacrolimus dosing. Clin Pharmacol Ther 98:19–24

Bolleddula J, DeMent K, Driscoll JP et al (2014) Biotransformation and bioactivation reactions of alicyclic amines in drug molecules. Drug Metab Rev 46:379–419

Brookes AJ (1999) The essence of SNPs. Gene 234:177–186

Brunton L, Knollman B, Hilal-Dandan R (2017) Goodman and Gilman's the pharmacological basis of therapeutics, 13th edn. McGraw Hill Professional, New York

Cappellini MD, Fiorelli G (2008) Glucose-6-phosphate dehydrogenase deficiency. Lancet 371:64–74

Caudle KE, Rettie AE, Whirl-Carrillo M et al (2014) Clinical Pharmacogenetics Implementation Consortium guidelines for CYP2C9 and HLA-B genotypes and phenytoin dosing. Clin Pharmacol Ther 96:542–548

Clancy J, Johnson S, Yee S et al (2014) Clinical Pharmacogenetics Implementation Consortium (CPIC) guidelines for ivacaftor therapy in the context of CFTR genotype. Clin Pharmacol Ther 95:592–597

Colas C, Ung PM-U, Schlessinger A (2016) SLC transporters: structure, function, and drug discovery. Med Chem Comm 7:1069–1081

Crews KR, Gaedigk A, Dunnenberger HM et al (2014) Clinical Pharmacogenetics Implementation Consortium guidelines for cytochrome P450 2D6 genotype and codeine therapy: 2014 update. Clin Pharmacol Ther 95:376–382

Edwards IR, Aronson JK (2000) Adverse drug reactions: definitions, diagnosis, and management. Lancet 356:1255–1259

Evans WE, McLeod HL (2003) Pharmacogenomics—drug disposition, drug targets, and side effects. N Engl J Med 348:538–549

Ferrell PB, McLeod HL (2008) Carbamazepine, HLA-B* 1502 and risk of Stevens–Johnson syndrome and toxic epidermal necrolysis: US FDA recommendations. Pharmacogenomics 9 (10):1543–1546

Froehlich TK, Amstutz U, Aebi S et al (2015) Clinical importance of risk variants in the dihydropyrimidine dehydrogenase gene for the prediction of early-onset fluoropyrimidine toxicity. Int J Cancer 136:730–739

Fromm MF (2002) The influence of MDR1 polymorphisms on P-glycoprotein expression and function in humans. Adv Drug Deliv Rev 54:1295–1310

Fujikura K, Ingelman-Sundberg M, Lauschke VM (2015) Genetic variation in the human cytochrome P450 supergene family. Pharmacogenet Genomics 25:584–594

Gallagher EP, Gardner JL, Barber DS (2006) Several glutathione S-transferase isozymes that protect against oxidative injury are expressed in human liver mitochondria. Biochem Pharmacol 71:1619–1628

Gammal RS, Court MH, Haidar CE et al (2016) Clinical Pharmacogenetics Implementation Consortium (CPIC) guideline for UGT1A1 and atazanavir prescribing. Clin Pharmacol Ther 99:363–369

Group SC (2008) SLCO1B1 variants and statin-induced myopathy—a genomewide study. N Engl J Med 2008:789–799

Hayes JD, Strange RC (1995) Invited commentary potential contribution of the glutathione S-transferase supergene family to resistance to oxidative stress. Free Radic Res 22:193–207

Hershfield M, Callaghan J, Tassaneeyakul W et al (2013) Clinical Pharmacogenetics Implementation Consortium guidelines for human leukocyte antigen-B genotype and allopurinol dosing. Clin Pharmacol Ther 93:153–158

Hicks JK, Bishop JR, Sangkuhl K et al (2015) Clinical Pharmacogenetics Implementation Consortium (CPIC) guideline for CYP2D6 and CYP2C19 genotypes and dosing of selective serotonin reuptake inhibitors. Clin Pharmacol Ther 98:127–134

Hicks JK, Sangkuhl K, Swen JJ et al (2017) Clinical Pharmacogenetics Implementation Consortium guideline (CPIC) for CYP2D6 and CYP2C19 genotypes and dosing of tricyclic antidepressants: 2016 update. Clin Pharmacol Ther 102:37. https://doi.org/10.1002/cpt.597

Hoffmeyer S, Burk O, Von Richter O et al (2000) Functional polymorphisms of the human multidrug-resistance gene: multiple sequence variations and correlation of one allele with P-glycoprotein expression and activity in vivo. Proc Natl Acad Sci U S A 97:3473–3478

Hung SI, Chung WH, Liou LB et al (2005) HLA-B* 5801 allele as a genetic marker for severe cutaneous adverse reactions caused by allopurinol. Proc Natl Acad Sci U S A 102:4134–4139

Ikeda Y, Umemura K, Kondo K et al (2004) Pharmacokinetics of voriconazole and cytochrome P450 2C19 genetic status. Clin Pharmacol Ther 75:587–588

Ingelman-Sundberg M, Gomez A (2010) The past, present and future of pharmacoepigenomics. Pharmacogenomics 11:625–627

Jancova P, Anzenbacher P, Anzenbacherova E (2010) Phase II drug metabolizing enzymes. Biomed Pap 154:103–116

Johnson JA, Caudle KE, Gong L et al (2017) Clinical Pharmacogenetics Implementation Consortium (CPIC) guideline for pharmacogenetics-guided warfarin dosing: 2017 update. Clin Pharmacol Ther 102:397–404

Kameyama Y, Yamashita K, Kobayashi K et al (2005) Functional characterization of SLCO1B1 (OATP-C) variants, SLCO1B1* 5, SLCO1B1* 15 and SLCO1B1* 15+ C1007G, by using transient expression systems of HeLa and HEK293 cells. Pharmacogenet Genomics 15:513–522

Kalow W (1962) Pharmacogenetics, heredity and the response to drugs, 1st edn. W. B. Saunders Co, Philadelphia

Klaassen CD, Watkins J (2013) Casarett and Doull's toxicology: the basic science of poisons, 8th edn. McGraw-Hill Education, New York

Leckband S, Kelsoe J, Dunnenberger H et al (2013) Clinical Pharmacogenetics Implementation Consortium guidelines for HLA-B genotype and carbamazepine dosing. Clin Pharmacol Ther 94:324–328

Lee AM, Shi Q, Pavey E et al (2014) DPYD variants as predictors of 5-fluorouracil toxicity in adjuvant colon cancer treatment (NCCTG N0147). J Natl Cancer Inst 106(12):1–2

Lee WK, Jung S-M, Kwak J-O et al (2006) Introduction of organic anion transporters (SLC22A) and a regulatory mechanism by caveolins. Electrolyte Blood Press 4:8–17

Li X, Yu C, Wang T et al (2016) Effect of cytochrome P450 2C19 polymorphisms on the clinical outcomes of voriconazole: a systematic review and meta-analysis. Eur J Clin Pharmacol 72:1185–1193

Lin JH, Lu AY (2001) Interindividual variability in inhibition and induction of cytochrome P450 enzymes. Annu Rev Pharmacol Toxicol 41:535–567

Lin L, Yee SW, Kim RB et al (2015) SLC transporters as therapeutic targets: emerging opportunities. Nat Rev Drug Discov 14:543–560

Mallal S, Phillips E, Carosi G et al (2008) HLA-B* 5701 screening for hypersensitivity to abacavir. N Engl J Med 358:568–579

Martin M, Klein T, Dong B et al (2012) Clinical Pharmacogenetics Implementation Consortium guidelines for HLA-B genotype and abacavir dosing. Clin Pharmacol Ther 91:734–738

Martin M, Hoffman J, Freimuth R et al (2014) Clinical Pharmacogenetics Implementation Consortium guidelines for HLA-B genotype and abacavir dosing: 2014 update. Clin Pharmacol Ther 95:499–500

Mason PJ, Bautista JM, Gilsanz F (2007) G6PD deficiency: the genotype-phenotype association. Blood Rev 21:267–283

Meulendijks D, Henricks LM, Sonke GS et al (2015) Clinical relevance of DPYD variants c. 1679T> G, c. 1236G> A/HapB3, and c. 1601G> A as predictors of severe fluoropyrimidine-associated toxicity: a systematic review and meta-analysis of individual patient data. Lancet Oncol 16:1639–1650

Meyer UA (2004) Pharmacogenetics–five decades of therapeutic lessons from genetic diversity. Nat Rev Genet 5:669–676

Meyer UA (2000) Pharmacogenetics and adverse drug reactions. Lancet 356:1667–1671

Miranda C (2016) Estudio de la asociación entre polimorfismos genéticos y la respuesta clínica a Tamoxifeno en pacientes con Cáncer de Mama. Universidad de Chile, Santiago/Chile

Moriyama B, Obeng AO, Barbarino J et al (2017) Clinical Pharmacogenetics Implementation Consortium (CPIC) guidelines for CYP2C19 and voriconazole therapy. Clin Pharmacol Ther 102:45. https://doi.org/10.1002/cpt.583

Muir A, Gong L, Johnson S et al (2014) Clinical Pharmacogenetics Implementation Consortium (CPIC) guidelines for IFNL3 (IL28B) genotype and PEG interferon-α–based regimens. Clin Pharmacol Ther 95:141–146

Munro AW, McLean KJ, Grant JL, Makris TM (2018) Structure and function of the cytochrome P450 peroxygenase enzymes. Biochem Soc Trans 46:183–196

Nelson DR (2009) The cytochrome P450 homepage. Hum Genomics 4:59–65

Neven P, Jongen L, Lintermans A et al (2018) Tamoxifen metabolism and efficacy in breast cancer: a prospective multicenter trial. Clin Cancer Res 24:2312–2318

Nguyen AP, Ness GL (2014) Hemolytic anemia following rasburicase administration: a review of published reports. J Pediatr Pharmacol Ther 19:310–316

Niemi M, Schaeffeler E, Lang T et al (2004) High plasma pravastatin concentrations are associated with single nucleotide polymorphisms and haplotypes of organic anion transporting polypeptide-C (OATP-C, SLCO1B1). Pharmacogenet Genomics 14:429–440

Nkhoma ET, Poole C, Vannappagari V et al (2009) The global prevalence of glucose-6-phosphate dehydrogenase deficiency: a systematic review and meta-analysis. Blood Cells Mol Dis 42:267–278

Orellana M, Guajardo V (2004) Cytochrome P450 activity and its alteration in different diseases. Rev Med Chil 132:85–94

Pasanen M, Fredrikson H, Neuvonen P et al (2007) Different effects of SLCO1B1 polymorphism on the pharmacokinetics of atorvastatin and rosuvastatin. Clin Pharmacol Ther 82:726–733

Peedicayil J (2008) Pharmacoepigenetics and pharmacoepigenomics. Pharmacogenomics 9:1785–1786

Pirmohamed M (2001) Pharmacogenetics and pharmacogenomics. Br J Clin Pharmacol 52:345–347

Preissner SC, Hoffmann MF, Preissner R et al (2013) Polymorphic cytochrome P450 enzymes (CYPs) and their role in personalized therapy. PLoS One 8:e82562

Quiñones L (2017) Pharmacogenomics in Latin America: challenges and opportunities, 1st edn. Nova Science, New York

Quiñones L, Rosero M, Roco Á et al (2008) Papel de las enzimas citocromo p450 en el metabolismo de fármacos antineoplásicos: Situación actual y perspectivas terapéuticas. Rev Med Chil 136:1327–1335

Ramsey LB, Johnson SG, Caudle KE et al (2014) The Clinical Pharmacogenetics Implementation Consortium guideline for SLCO1B1 and simvastatin-induced myopathy: 2014 update. Clin Pharmacol Ther 96:423–428

Relling MV, McDonagh EM, Chang T et al (2014) Clinical Pharmacogenetics Implementation Consortium (CPIC) guidelines for rasburicase therapy in the context of G6PD deficiency genotype. Clin Pharmacol Ther 96:169–174

Relling M, Gardner E, Sandborn W et al (2011) Clinical Pharmacogenetics Implementation Consortium guidelines for thiopurine methyltransferase genotype and thiopurine dosing. Clin Pharmacol Ther 89:387–391

Roses AD (2000) Pharmacogenetics and the practice of medicine. Nature 405:857–865

Roth M, Obaidat A, Hagenbuch B (2012) OATPs, OATs and OCTs: the organic anion and cation transporters of the SLCO and SLC22A gene superfamilies. Br J Pharmacol 165:1260–1287

Saito Y, Stamp LK, Caudle KE et al (2016) Clinical Pharmacogenetics Implementation Consortium (CPIC) guidelines for human leukocyte antigen B (HLA-B) genotype and allopurinol dosing: 2015 update. Clin Pharmacol Ther 99:36–37

Sansone S, Rottensteiner J, Stocker J et al (2010) Ciprofloxacin-induced acute haemolytic anaemia in a patient with glucose-6-phosphate dehydrogenase Mediterranean deficiency: a case report. Ann Hematol 89:935–937

Sear J (2004) In: Evers AS, Maze M (eds) Anesthetic pharmacology: physiologic principles and clinical practice. Oxford University Press, Churchill Livingstone, Philadelphia

Scott SA, Sangkuhl K, Stein C et al (2013) Clinical Pharmacogenetics Implementation Consortium guidelines for CYP2C19 genotype and clopidogrel therapy: 2013 update. Clin Pharmacol Ther 94:317–323

Shalan H, Kato M, Cheruzel L (2019) Keeping the spotlight on cytochrome P450. Biochim Biophys Acta 1866:80–87

Sim SC, Risinger C, Dahl ML et al (2006) A common novel CYP2C19 gene variant causes ultrarapid drug metabolism relevant for the drug response to proton pump inhibitors and antidepressants. Clin Pharmacol The 79:103–113

Skaar TC, Desta Z (2018) CYP2D6 and endoxifen in tamoxifen therapy: a tribute to David A. Flockhart. Clin Pharmacol Ther 103:755–757

Strange RC, Spiteri MA, Ramachandran S et al (2001) Glutathione-S-transferase family of enzymes. Mutation Res 482:21–26

Tirona RG, Leake BF, Merino G et al (2001) Polymorphisms in OATP-C identification of multiple allelic variants associated with altered transport activity among European-and African-Americans. J Biol Chem 276:35669–35675

Vasiliou V, Vasiliou K, Nebert DW (2009) Human ATP-binding cassette (ABC) transporter family. Hum Genomics 3:281–290

Vogel F (1959) Moderne probleme der humangenetik. In: Ergebnisse der inneren medizin und kinderheilkunde. Springer, Berlin, pp 52–125

Walsh JS, Miwa GT (2011) Bioactivation of drugs: risk and drug design. Annu Rev Pharmacol Toxicol 51:145–167

Wijnen P, Op Den Buijsch R, Drent M et al (2007) The prevalence and clinical relevance of cytochrome P450 polymorphisms. Aliment Pharmacol Ther 26:211–219

Wilkinson GR (2005) Drug metabolism and variability among patients in drug response. N Engl J Med 352:2211–2221

World Health Organization (2002) The world health report 2002: reducing risks, promoting healthy life. World Health Organization, Geneva

Xie H-G, Frueh FW (2005) Pharmacogenomics steps toward personalized medicine. Pers Med 2:325–337

Yee SW, Nguyen AN, Brown C et al (2013) Reduced renal clearance of cefotaxime in asians with a low-frequency polymorphism of OAT3 (SLC22A8). J Pharm Sci 102:3451

Zhou S-F, Ming Di Y, Chan E et al (2008) Clinical pharmacogenetics and potential application in personalized medicine. Curr Drug Metab 9:738–784

Zhou S-F, Liu J-P, Chowbay B (2009) Polymorphism of human cytochrome P450 enzymes and its clinical impact. Drug Metab Rev 41:89–295

Zhu Y, Shennan M, Reynolds KK et al (2007) Estimation of warfarin maintenance dose based on VKORC1 ($-$ 1639 G$>$ A) and CYP2C9 genotypes. Clin Chem 53:1199–1205

The Relationship Between Pharmacogenomics and Pharmacokinetics and Its Impact on Drug Choice and Dosing Regimens in Pediatrics

9

Venkata K. Yellepeddi, Jessica K. Roberts, Leslie Escobar, Casey Sayre, and Catherine M. Sherwin

9.1 Introduction

▶ **Important** Humans are a very heterogeneous population. As such, it is nearly impossible to include every version of a person in a pharmacokinetic clinical trial. Clinical trials do their best to study as many different people as possible, but there are always those within the population that

V. K. Yellepeddi
Division of Clinical Pharmacology, Department of Pediatrics, University of Utah, Salt Lake City, UT, USA

Department of Pharmaceutics & Pharmaceutical Chemistry, College of Pharmacy, University of Utah, Salt Lake City, UT, USA

J. K. Roberts
Division of Clinical Pharmacology, Department of Pediatrics, University of Utah, Salt Lake City, UT, USA

L. Escobar
Department of Pediatrics and Infant Surgery, Faculty of Medicine, University of Chile, Santiago, Chile

C. Sayre
College of Pharmacy, Roseman University of Health Sciences, South Jordan, UT, USA

C. M. Sherwin (✉)
Division of Clinical Pharmacology, Department of Pediatrics, University of Utah, Salt Lake City, UT, USA

Department of Pharmaceutics & Pharmaceutical Chemistry, College of Pharmacy, University of Utah, Salt Lake City, UT, USA

Department of Pharmacotherapy, College of Pharmacy, University of Utah, Salt Lake City, UT, USA
e-mail: catherine.sherwin@hsc.utah.edu

© Springer Nature Switzerland AG 2018
A. Talevi, P. A. M. Quiroga (eds.), *ADME Processes in Pharmaceutical Sciences*,
https://doi.org/10.1007/978-3-319-99593-9_9

are outliers. Many factors contribute to the pharmacokinetic hetero-geneity, including the genetic makeup of each individual.

Pharmacogenomics is a relatively new field of pharmaceutical research that covers both somatic and germ line mutations.

▶ **Definition** Pharmacogenomics and pharmacogenetics are often used inter-changeably; however, this is not always appropriate. According to PharmGKB, pharmacogenetics generally refers to the impact of variation in a single gene has on the response to a drug. Pharmacogenomics, on the other hand, includes how the whole genome affects the reaction (PharmGKB 2018). Generally, when discussed regarding a single gene, pharmacogenetics is the most appropriate term to be used and will be used throughout this chapter accordingly.

Pharmacogenetic studies are often conducted to avoid toxicity in specific patient populations. For example, a study completed by Xu et al. found that the acylphosphatase-2 (ACYP2) risk variant predisposed children with brain tumors treated with cisplatin to ototoxicity (Xu et al. 2015). However, studies can also be completed to demonstrate specific populations have pharmacological benefit due to the expression of variants. A study completed by Stockmann et al. (2013) evaluated fluticasone propionate for the treatment of asthma in children and found that patients with the CYP3A4*22 variant had increased asthma control as compared to those with other variants (Brown et al. 2015).

There are a couple of ways to begin considering a pharmacogenetic study. Pharmacogenetic studies are often conducted with the pharmacology of the drug in mind. The metabolizing enzyme, target enzyme, or other drug responses are used as a guide to decide which genes to consider when studies are undertaken. This was true in the review by Stockmann et al. (2013). Additionally, more complex analyses, such as genome-wide association studies (GWAS), will use previously identified genetic polymorphisms to probe for varying responses to drugs in particular populations, as completed by Xu et al. (2015). While there are examples of pediatric-specific pharmacogenetic studies, the majority of pharmacogenetic studies are conducted in adults.

▶   **Important** It is important to note that adult pharmacogenetic studies do not translate to pediatrics (Kearns et al. 2003), resulting in the majority of pediatric pharmacotherapy being off-label (Blumer 1999). In pediatrics due to maturational changes in the expression of drug-metabolizing enzymes (DMEs) (CYP, UGT, NAT2, *TPMT*, etc.), transporters (P-gp, OATP, etc.), and target proteins, pharmacogenetics of drugs is much complex. These types of changes and their interplay with pharmacogenetics are discussed in detail in Sect. 9.3.

Personalized medicine, or precision medicine, in pediatric patients is still in its infancy. Almost every pediatric-specific disease is considered an orphan disease because the population sizes of each are so small. This makes it very difficult to study a drug effect in a pediatric population, including the pharmacogenetics. There have also been ethical considerations that have had to be overcome. Historically, pediatric-specific pharmacokinetic trials, including pharmacogenetics, were not completed because it was not considered ethical to conduct research on such a vulnerable population. However, clinical research demonstrating pediatric patients are not small adults has highlighted the need for pediatric-specific studies (Kearns et al. 2003). The changing of the research paradigm was begun by the first piece of legislation passed in 1994: the Pediatric Labeling Rule (Center for Drug Evaluation and Research 1994). This allowed companies to extrapolate adult pharmacokinetic and pharmacodynamic data to pediatric patients for labeling purposes if the disease state was similar enough. This act was soon followed by the Pediatric Rule, which began in 1997 and was finalized in 1998. Because the legislation passed in 1994 was voluntary, a more concerted effort was made to require companies to comply. This required that labeling information that was relevant to pediatric patients be included for new pharmaceuticals unless a waiver could be obtained demonstrating the use in this specific population would be unlikely or unsafe. The rule required companies to submit safety and efficacy studies before or soon after approval for pediatric patients. Furthermore, in 1997 the Food and Drug Administration Modernization Act (FDAMA 1997) aimed to provide a process to make a list of drugs where additional pediatric information would be beneficial. These would be discussed, and a written request would be made to the sponsor to complete pediatric studies that were deemed necessary. As an incentive for completing the studies, sponsors would be granted an additional 6 months of marketing exclusivity (Food and Drug Administration Modernization Act of 1997 1997). The Best Pharmaceuticals for Children Act (BPCA) was introduced in 2002, renewed the exclusivity of FDAMA, provided an avenue for on- and off-patent agents, and required disclosure of the studies that were being completed in pediatric patients. In 2003 the Pediatric Research Equity Act (PREA) made the Pediatric Rule legislation, now requiring pediatric assessment unless waivers were obtained. Continuing to advance pediatric studies, the Food and Drug Administration Amendments Act of 2007 renewed PREA and BPCA and introduced the first Pediatric Review Committee to oversee the review of pediatric-specific studies. The FDA Safety and Innovation Act in 2012 renewed BPCA yet again. The most recent FDA Reauthorization Act (FDARA) of 2017 was officially signed into law and extends BPCA programs for testing off-patent drugs for pediatric indications through 2022. A timeline of the legislature specific to pediatric patients can be found in Fig. 9.1. Also, the National Pediatric Research Network Act will continue to implement pediatric-specific studies. As a result, many more pediatric-specific studies will be conducted, and some of these will more than likely contain pharmacogenetic considerations. Also, as the tools required for pharmacogenetic testing become more cost-effective (genetic sequencing, etc.), these studies will become more feasible, and precision medicine for a pediatric patient will become more mainstream.

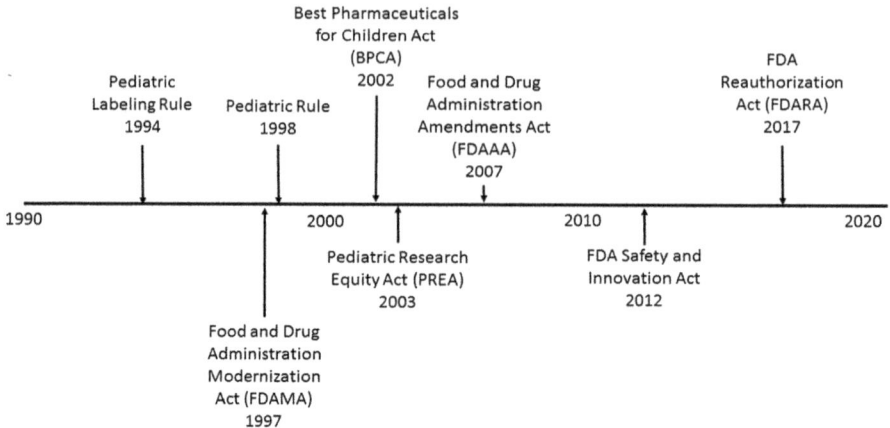

Fig. 9.1 A timeline of the legislature specific to pediatric patients

Despite the current limitations of pediatric pharmacogenetic testing, there have been many pediatric-specific pharmacogenetic studies completed in diseases such as cancer (Ross et al. 2009; Pussegoda et al. 2013; Brown et al. 2015; Lanvers-Kaminsky et al. 2015; Moriyama et al. 2017a, b), asthma (Stockmann et al. 2015, 2013; Zuurhout et al. 2013; Palmer et al. 2006; Giubergia et al. 2013; Turner et al. 2016; Lipworth et al. 2013), and thrombosis (Nowak-Gottl et al. 2010; Biss et al. 2012; Kato et al. 2011; Shaw et al. 2014; Moreau et al. 2012; Vear et al. 2014), to name a few. While pediatric-specific pharmacogenetic studies are on the rise, the clinical implications for many of the findings have not been elucidated. For example, whole genome sequencing of patients has resulted in large data sets with information for genes that have no current meaning. Additional scientific research needs to be performed to elucidate the essence of all the data collected for implementation in the clinic. Additionally, though there appear to be studies that implicate specific genes, some studies have found that children who express variants do not always require action in the changing of the treatment with particular agents. Therefore, it is essential to understand the dynamics between the cost of genetic testing and the implications of these studies when treating pediatric patients. Specific examples, including *TPMT, CYP3A, FCER2, CYP2D6, CYP2C19*, etc., will be discussed in Sect. 9.3.

▶ **Important** Pediatric pharmacogenetic considerations are continuing to come to the forefront of clinical treatment as precision medicine continues to gain importance in a clinical setting. Therefore, this chapter aims to discuss the challenges that are associated with pediatric pharmacotherapy, the role of pharmacogenetics in pediatric dosing, and the impact of specific genetic polymorphisms on the pharmacokinetics of drugs in pediatric patients.

9.2 Challenges in Pediatric Pharmacotherapy

9.2.1 Lack of Evidence in Pediatric Pharmacotherapy

There is a significant and generalized lack of information and evidence in pediatric pharmacotherapy (Sachs et al. 2012). This has hindered optimal provision of care for pediatric patients, as pharmacotherapeutic decision-making for prescribing, dosing, efficacy, and current and high-quality scientific evidence typically backs toxicity in adults. Consequently, pediatric information in drug monographs is also frequently absent in a multitude of commercially available pharmaceuticals. In pediatric pharmacotherapy, due to a lack of knowledge in drug monographs, there is an increase in the frequency of medication errors (Kalra and Goindi 2014; Frattarelli et al. 2014; Preventing errors relating to commonly used anticoagulants 2008). Additionally, these limitations in pharmacotherapy practice have been the cause of poor health outcomes and tragedy. Benzyl alcohol poisoning in infants occurred in the 1980s due to the lack of knowledge on the effects of this preservative in pediatric patients (Gershanik et al. 1982).

Current pediatric pharmacotherapy practice, however, now benefits from a substantial increase in pediatric drug information and evidence. Several legislative changes resulted in the Pediatric Exclusivity Program, the 1998 Pediatric Rule, the 2003 Research Equity Act, and the Best Pharmaceuticals for Children Act (Kearns 2015; Brown et al. 2015). These new regulations allowed the FDA to require that drug manufacturers conduct clinical trials that included a significant number of patients. They also provided some incentives for companies to pursue development of pediatric formulations of both new and older off-patent drugs (Korbel et al. 2014).

Since the implementation of these legislative changes, the number of pediatric clinical trials has substantially increased as well as the number of patent extension applications. As of March 31, 2018, there have been 731 pediatric labeling changes with 664 that incorporated new pediatric studies (Food and Drug Administration 2018, March 31). Additionally, dosing and toxicity information has been added to the prescribing information of many drugs. Access to this information has also increased (Mukattash et al. 2014). However, it will take many more years for the evidence and information for pediatric pharmacotherapy to reach the standards relative to that of adult patients.

9.2.2 Off-Label Use of Medications and Potential Risks Associated with it

▶ **Important** Even with the significant increase in pediatric pharmacotherapy information, an unavoidable consequence of the persisting lack of pediatric drug information and evidence is widespread off-label use of medication. The practice of off-label use is both prevalent and necessary in all parts of the world (Palmaro et al. 2015; Frattarelli et al. 2014;

Gore et al. 2017). Off-label prescribing, though legal and often necessary, can have significant risks. It is often the case that when a drug is used for a non-FDA-approved indication, the risk for adverse drug events (ADRs) increases (Cuzzolin et al. 2006).

There are physiological differences between pediatric and adult patients that can complicate the therapeutic assumptions made by clinicians when prescribing drugs off-label. This can make things like finding the correct dose difficult. For example, the maintenance dose of digoxin is much higher in infants than in adults. This is due in part to a relatively decreased binding affinity at receptors in the heart in infants (Kearin et al. 1980). Rate and extent of drug absorption are also widely variable between pediatric patients and adults (Kearns et al. 2003; Debotton and Dahan 2014). Gastric emptying time and pH are significant contributors to oral drug absorption and are different in pediatric patients (Harries and Fraser 1968; Anderson et al. 2002). In neonates, gastric pH can rise above 4 resulting in a more significant rate and extent of absorption of acid-labile drugs like penicillin G, ampicillin, and erythromycin (Agunod et al. 1969; Brown and Campoli-Richards 1989). Conversely, weakly acid drugs like phenytoin, phenobarbital, and ganciclovir can be absorbed to a lesser extent due to the ionization state induced by the elevated pH (Anderson and Lynn 2009).

9.2.3 Lack of Appropriate Formulations

Similar to the general lack of pediatric drug information or evidence, there is also a lack of commercially available dosage forms for pediatric administration. This leads to several complications in pediatric pharmacotherapy. The first is the need to make adjustments to existing dosage forms for pediatric administration. One modification is frequently needed to create a new concentration of a drug from a commercially available one, particularly the dilution of concentrated dosage forms. The range of volume measurements required to make a correct concentration for administration can be broad, increasing the potential for errors. Digoxin and morphine toxicity has occurred in infants following some well-known examples of pediatric dilution errors (Berman et al. 1978). Some organizations are addressing this by issuing standardized concentrations for compounded liquid medications that are disseminated and adopted into pharmacy practice (Engels et al. 2016).

The second is the conversion of a commercially available dosage designed for adults into one that can be administered to a pediatric patient. Many drugs in pediatric pharmacotherapy require oral administration. Since pediatric patients can have difficulty swallowing and may refuse medications due to poor taste, texture, or smell, it is common to convert a solid oral dosage form (tablet or capsule) into a liquid oral dosage form. This manipulation can occur in the home or the hospital via a caregiver crushing a tablet or opening a capsule and mixing the contents into applesauce. In some instances, it can happen in a compounding pharmacy utilizing raw ingredients and a validated extemporaneous formulation. In either case, the manipulation is required to administer the drug safely and effectively. Whether the alteration is for an orally or parenterally administered dosage form, there are multiple complications and potential pitfalls. Among these are lack of stability and compatibility studies and, as previously mentioned, the propensity for error in measurement and other techniques in the compounding process (Rood et al. 2014; Lam 2011). As drug manufacturers continue to increase their development of pediatric dosage forms, there is a need for an increase in research regarding the stability, compatibility, and quality of compounded pediatric dosage forms.

9.3 Role of Pharmacogenomics and Pharmacogenetics in Pediatric Dosing

9.3.1 Role of Ontogeny in Pediatric Pharmacogenomics

The ontogeny of DMEs (phases 1 and 2), transporters, and target proteins is crucial to understand the role of pharmacogenomics in pediatric dosing. For example, the major DMEs such as cytochrome P450 (CYP) have shown to significantly change their activity based on the developmental pattern (Mlakar et al. 2016). CYP3A4 is the most abundantly expressed CYP in the liver and small intestine and was shown to have extremely low activity at birth, reaching 30–40% of adult activity by the first month and complete adult activity by the sixth month followed by exceeding adult activity (~120%) between 1 and 4 years of age and decreasing to adult levels after puberty (Leeder and Kearns 1997). CYP2D6 ontogeny was reported by in vitro analysis, and the results have shown that the CYP2D6 protein activity remains relatively constant after 1 week of age up to 18 years (Stevens et al. 2008).

The glucuronosyltransferases (UGTs) are enzymes predominantly involved in phase 2 biotransformation of drugs used in pediatrics. UGTs catalyze the conjugation of glucuronic acid to drugs including morphine, acetaminophen, nonsteroidal anti-inflammatory drugs (NSAIDs), and benzodiazepines. By in vitro tests, it was shown that UGT activity toward bilirubin (a probe substrate for UGT) was nearly undetectable in fetal liver, and the activity increases immediately after birth, reaching adult levels around 3–6 months of age (Hume et al. 1995, 1996). Similarly, UGT2B7, an essential isoform of UGT enzyme family, was shown to have 20% of levels of older children (1 to 16 years) during the first 2 weeks of life (de Wildt et al. 1999). UGT1A6 activity was reported by in vitro measurements and was shown to have 1–10% of adult levels during fetal life, followed by slowly increasing to 50% of adult activity after birth (de Wildt et al. 1999).

Arylamine N-acetyltransferase (NAT2) is involved in acetylation of many drugs used in pediatrics. By utilizing caffeine as an in vivo phenotyping probe, its ontogeny was reported in pediatrics. All infants 0–55 days of age appear to be phenotypically slow acetylators, whereas 50% and 62% of infants 122–224 and 225–342 days of age were characterized as fast acetylators (Evans et al. 1989). Thiopurine S-methyltransferase (*TPMT*) catalyzes the *S*-methylation of aromatic and heterocyclic drugs used in pediatrics. Examples of drugs that are metabolized by

TPMT that are used in pediatrics include 6-mercaptopurine (6-MP), azathioprine, and 6-thioguanine. The ontogeny data of *TPMT* currently available in the reported literature is conflicting. In one study involving newborn infants, peripheral blood *TPMT* activity was reported to be 50% greater than in adults (McLeod et al. 1995). However, another study reported no significant difference in *TPMT* activity between cord blood and only a slightly lower *TPMT* activity in children than when compared with adults (Ganiere-Monteil et al. 2004).

In addition to DMEs, drug transporters that are expressed at numerous epithelial barriers influence drug disposition in pediatrics. The ontogeny of ATP-binding cassette (ABC) protein transporter, P-glycoprotein (P-gp), was characterized in human enterocytes. The P-gp activity increased rapidly during the first 3–6 months of life, reaching adult levels by approximately 2 years of life (Johnson and Thomson 2008). The ontogeny of organic anion-transporting polypeptides (OATPs) showed that the mRNA for the OATP isoforms OATP1B1, OATP1B3, and OATP2B1 was detectable in fetal hepatocytes by gestational weeks 18–23 and was significantly higher in adults compared to fetal livers (Sharma et al. 2013).

▶ **Important** There is a multitude of reports highlighting the importance of developmental differences in pharmacologic targets for drug action. Similar to the ontogeny of DMEs and transporters, ontogeny of drug receptors can result in age-dependent differences in the pharmacokinetics of drugs leading to changes in dose exposure-response relationship. For example, in an animal model, it was demonstrated that mu and kappa receptors were predominantly present postnatally, whereas delta receptors are expressed in later stages of development (Neville et al. 2011).

9.3.2 Pharmacogenomic Testing in Pediatric Practice

▶ **Important** As discussed in the previous section, the developmental changes in the expression of various enzymes, transporters, and receptors influence the pharmacokinetics of drugs in pediatrics. Therefore, it is critical to account for changes in gene expression during development to understand the genotype-phenotype relationships in pediatric populations. Despite the paucity of data of ontogeny in the pediatric population, the scope of pharmacogenomic testing in the pediatric clinical setting is expanding. Pharmacogenetic or pharmacogenomic testing leverages variations in individual genes relevant to medication metabolism or targets to predict treatment response and may guide treatment selection (Wehry et al. 2018). Furthermore, knowledge of pharmacogenomic variants that impact risk of adverse drug reactions or pharmacologic activity can significantly improve the health outcomes in pediatrics.

The Food and Drug Administration (FDA) recommends genetic testing for the enzyme *TPMT* before initiating mercaptopurine (6-MP) therapy. Clinicians should consider lower doses of 6-MP in their patients who are carriers of a variant *TPMT* allele and should consider treatments other than 6-MP for ALL patients that are homozygous for the *TPMT* variant allele (Korbel et al. 2014). As 6-MP is used in pediatrics for the treatment of acute lymphoblastic leukemia (ALL) and inflammatory bowel disease, it is vital to test *TPMT* levels to ensure optimal dosing of 6-MP (Kirschner 1998; Schmiegelow et al. 2014). Tacrolimus, a calcineurin inhibitor, is used to prevent allograft rejection and is commonly used in pediatric heart and liver transplant patients. However, the FDA black box warning recommends dose adjustments when tacrolimus is taken concurrently with other medications inhibiting or including CYP3A metabolism (Korbel et al. 2014). CYP3A5*3, the most common nonfunctional variant allele of CYP3A5, and CYP3A4*22, a common reduced-function CYP3A4 allele, are associated with increased concentrations of tacrolimus resulting in toxicity at standard doses (Elens et al. 2013). Therefore, there is a need to incorporate CYP3A genetic testing information for tacrolimus dosing recommendations.

Corticosteroids are commonly used in pediatrics to reduce airway inflammation, improve lung function, improve asthma symptoms, and reduce asthma exacerbations. However, a meta-analysis study showed that in children carrying a single nucleotide polymorphism (SNP) in the Fc fragment of IgE receptor II (FCER2) receptor gene resulted in lower expression and increased hospital visits for pediatric asthma patients receiving inhaled corticosteroids (Maitland-van der Zee and Raaijmakers 2012). The Clinical Pharmacogenetics Implementation Consortium (CPIC) recently published guidelines regarding dosing recommendations for five commonly prescribed selective serotonin reuptake inhibitors (SSRIs): fluvoxamine, paroxetine, citalopram, escitalopram, and sertraline based on CYP2D6 and CYP2C19 polymorphisms (Wehry et al. 2018). As psychiatric illnesses are common in pediatrics, and many of the drugs mentioned above are prescribed in pediatrics, incorporating genetic testing information for CYP2D6 and CYP2C19 in pediatric dosing recommendations will be valuable in avoiding severe toxicity. A complete discussion of drugs that have pharmacogenetic testing information is out of the scope of this book chapter. However, Table 9.1 provides a list of drugs that are used in pediatrics with pharmacogenomics and pharmacogenetic testing information.

9.4 Impact of Genetic Polymorphisms on the Pharmacokinetics of Drugs in Pediatrics

▶ **Definition** The term polymorphism derives from "polymorph," a word built with the Greek root "poly" (many) and "morphe" (form/shape). Typically, a polymorph is more related to a different crystalline form of a substance, which provides different physicochemical properties. In pharmacogenomics, a single nucleotide polymorphism (SNP) throughout the human genome contains a different genetic response that can be applied to understand the differences in drug response.

Table 9.1 Table of drugs with pharmacogenomic testing information included in FDA labeling. The table was prepared from the reference (Food and Drug Administration 2018)

S. No	Drug	Therapeutic category	Pharmacogenomic biomarker
1	Arformoterol	Pediatric asthma	UGT1A1, CYP2D6
2	Atomoxetine	Pediatric ADHD	CYP2D6
3	Celecoxib	Juvenile rheumatoid arthritis	CYP2C9
4	Chloroquine	Malaria	G6PD
5	Codeine	Cough	CYP2D6
6	Daclatasvir	HCV/HIV	IFNL3
7	Efavirenz	HIV	CYP2B6
8	Esomeprazole	Gastric reflux and ulcer	CYP2C19
9	Everolimus	Oncology	ERBB2
10	Irinotecan	Oncology	UGT1A1
11	Ivacaftor and lumacaftor	Cystic fibrosis	CFTR
12	Lansoprazole	Gastric reflux and ulcer	CYP2C19
13	Mycophenolic acid	Transplantation	HPRT1
14	Obinutuzumab	Oncology	MS4A1
15	Ondansetron	Anti-nausea	CYP2D6

It is known that most medicinal drugs do not work effectively in all patients at similar magnitude, and also heterogeneity of the drug effects is quite common. Some of them have an inappropriate response, including toxicity or even lack of effect. Nevertheless, pharmacogenetics, which can be described as the study of how genes and genetics influence the action of a drug or how patients respond to therapy, is often forgotten to explain these variabilities (Drew 2016). Some genetic polymorphisms can affect the efficacy of drugs or even contribute to adverse reactions. In adults information on genetic polymorphism and their relationship to drug pharmacokinetics can be obtained from clinical trials. However, in pediatric patients, these conditions are less described. When the use of drugs in the pediatric population is challenging, pharmacogenomic issues must be taken into account in the highly variable scenario of age-related changes in pharmacokinetics.

Various organizations are working to construct guidelines for medication use considering genetics in drug therapy. Clinical Pharmacogenetics Implementation Consortium (CPIC), Pharmacogenomic Research Network (PGRN), PGEN4Kids, International HapMap Project, 1000 Genomes Project, Pharmacogenomics Knowledgebase (pharmGKB), genome-wide association studies (GWAS), and Pharmacogenomic Resource for Enhanced Decisions in Care and Treatment (PREDICT), among others, have contributed in our understanding of human genetic variability with the optimist idea of personalized medicine (Quiñones et al. 2017; Lavertu et al. 2018).

The pharmacokinetic and pharmacodynamic studies in the pediatric population are increasing, but they remain insufficient to describe drug behavior in specific conditions. In pediatric patients, it is well-known that high inter- and intraindividual variation on pharmacokinetics is because of age-dependent changes in body composition and organ function. Besides, pharmacogenetic determinants of drug actions contribute to the age-dependent differences in pharmacological treatment of a child (Kearns et al. 2003; Roberts et al. 2014).

Pharmacogenomic studies have been conducted to find or explain biomarkers related to diagnosis, prognosis, susceptibility, and treatment outcomes. Clinical impact of them is more in the identification of polymorphisms of a protein involved in one or more steps of ADME process (drug absorption, distribution, metabolism, and/or excretion).

▶   **Important** Drug metabolism is the most common pharmacokinetic step studied with genetic variation in enzymes. The primary goal of metabolism or biotransformation is to transform a drug into an easily clearable substance. This new substance created could be pharmacologically active or inactive metabolite. Frequently, drug metabolism occurs in two reaction processes: phase I (oxidation, reductions, and hydrolysis) involving cytochrome P450 system (CYP450) and phase II (conjugation reactions: glucuronidation, acetylation, sulfation, and methylation).

   Depending on the capability to metabolize substance and the enzymes involved, it is possible to obtain a reduction, increase, or absence of enzyme function. Patients can be classified into four categories according to their genotype metabolizer activity to perform the drug biotransformation: as poor/slow, intermediate, extensive, or ultrarapid metabolizer. This means that drug plasma concentration in each patient will be different at the same dose; hence the difference in pharmacological response is observed.

According to the Food and Drug Administration, pharmacogenomics can play an essential role in identifying responders and nonresponders to medications, avoiding adverse events, and optimizing drug dose. Currently, over 160 drugs require pharmacogenomic testing and information in their labels to show the variability in clinical response, adverse events, and dosing according to genotype (Constance et al. 2017). Drug labeling includes specific actions to be taken based on the biomarker information: describing drug exposure and clinical response variability, the risk for adverse events, genotype-specific dosing, mechanisms of drug action, polymorphic drug target and disposition genes, and trial design features.

We are looking forward to seeing pharmacogenetics become a standard in clinical practice to adjust medication doses in pediatric patients. In the literature, there are investigations about the impact of pharmacogenomics on the pharmacokinetics of some examples of drugs investigated in pediatrics.

9.4.1 Codeine

Codeine is included in the list of medications with the pharmacogenomic informa-tion requested in its label. In normal metabolizers, around 10% of codeine is metabolized by CYP2D6 isoenzymes to morphine, its active metabolite (Crews et al. 2014) (Fig. 9.2). This biotransformation is responsible for the analgesic potency of codeine, conditioning the effective morphine plasma concentration according to the metabolizer capability of the patient or by drug-drug interaction

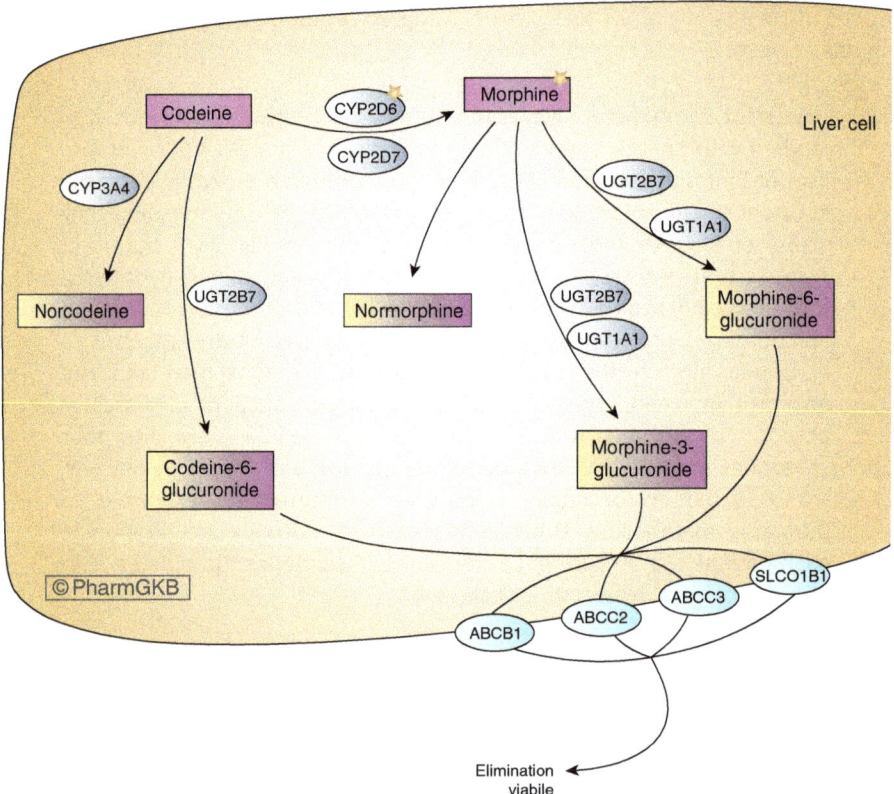

Fig. 9.2 Representation of the candidate genes involved in the metabolism of codeine and morphine. The principal pathways for metabolism of codeine occur in the liver, although some metabolism occurs in the intestine and brain. The most commonly studied gene with respect to this pathway is CYP2D6. The CYP2D6 gene is highly polymorphic with many essential SNPs, haplotypes, and copy number variants (see CYP2D6 VIP annotation for further details). Simplisti-cally, CYP2D6 variants can be categorized into poor metabolizers (low-activity variants), extensive metabolizers (high- or normal-activity variants), and ultrarapid metabolizers (multiple gene copy variants). Poor metabolizers are unable to convert codeine to morphine efficiently and as a consequence may not experience pain relief. Ultrarapid metabolizers may metabolize codeine too efficiently leading to morphine intoxication. (The figure was reprinted with permission from PharmGKB and Stanford University)

with the presence of another drug that is metabolized by CYP2D6 (Crews et al. 2014). In patients with ultrarapid CYP2D6 genotype or poor metabolizers of CYP2D6, it is likely to have high variability of morphine plasma concentration. As a result, in pediatric patients, the pain relief response with codeine is unpredictable, with an increased risk of rare but life-threatening adverse event or death due to morphine toxicity. Codeine use in children must be avoided (Constance et al. 2017; Crews et al. 2014).

9.4.2 Celecoxib

Celecoxib is an inhibitor of cyclooxygenase 2 (COX-2) and is metabolized by CYP2C9. In children, the pharmacokinetics of celecoxib is different compared to adults, with high clearance and short elimination half-life which can be related to the short duration of effects in this population. People with CYP2C9*3 have a reduction in the metabolism of celecoxib (Kirchheiner et al. 2003). This characteristic tends to be greater in *3 homozygotes than *3 heterozygotes (Murto et al. 2015). The impact of this genetically polymorphic CYP enzyme on pharmacokinetics and pharmacodynamics seems to be beneficial to obtain a prolonged pain relief in patients, as was observed in the quality of life in children during the first days after surgery (Murto et al. 2015). Nevertheless, pediatric poor metabolizers have an increased risk to obtain high plasma and tissue concentrations of the drug that requires a dose reduction to avoid adverse events like gastrointestinal and renal failure or other dose-related drug reactions of COX-2 inhibitor (Kirchheiner et al. 2003). For that reason, the US FDA requires to include this information in celecoxib label for pediatric patients.

9.4.3 Azathioprine, 6-Mercaptopurine (6-MP), and 6-Thioguanine (6-TG)

These three thiopurine medications are inactivated by thiopurine methyltransferase (*TPMT*) to obtain active thioguanine nucleotide (TGN) metabolites (Relling et al. 2011). Serious adverse effects are associated with toxic plasma concentration of thiopurines in poor metabolizer (Teusink et al. 2016).

In children with cancer, the identification of *TPMT* activity could be essential to reduce the risk of relapse or drug toxicity. Lennard et al. have demonstrated in the early 1990s that genetically determined *TPMT* activity may be a substantial regulator of the cytotoxic effect caused by 6-MP, which also can influence the outcome of therapy for childhood acute lymphoblastic leukemia (Lennard et al. 1987, 1990). Dose adjustment based on thiopurine methyltransferase (*TPMT*) (rs1800460 and rs1142345) genotype is now suggested (Relling et al. 2011). Variations in *TPMT* gene are prospectively studied in pediatric patients.

Before treating a child with thiopurine drugs, genotyping tests for TMPT should be performed to obtain information on clinical outcomes and the likelihood of adverse drug reactions (Relling et al. 2011).

9.4.4 Voriconazole

Voriconazole is a broad-spectrum antifungal agent with activity against *Aspergillus* and non-*Aspergillus* and many infections caused by *Candida* species (Teusink et al. 2016). Voriconazole is a second-generation azole antifungal and according to the latest consensus of the Infectious Diseases Society of America (IDSA) is the first-line treatment indicated in invasive aspergillosis.

In children, non-saturable elimination kinetics have been observed (Ramos-Martin et al. 2015). The CYP2C19 metabolizes this azole. Genetic polymorphisms such as CYP2C19*2, *3, and *17 have been reported to result in either increased (*17) or decreased (*2 and *3) metabolic activity, depending on the presence of concomitant treatments with inhibitory or inducer drugs (Bruggemann et al. 2009). Another vital factor in pharmacokinetic variability is age, with pediatric patients being fast metabolizers. The nonlinearity of voriconazole pediatric pharmacokinetics makes the current initial dose still under debate to ensure the target range is reached as quickly as possible (Walsh et al. 2004; Kadam and Van Den Anker 2016).

This high interindividual variability in voriconazole plasma concentration due to polymorphic CYP2C19 can even increase the total health cost. A delay until 29 days to obtain a therapeutic range of 1.0–5.5 µg/mL has been observed in pediatric patients with hematopoietic stem cell transplantation under prophylaxis with voriconazole (Teusink et al. 2016).

9.4.5 Methotrexate (MTX)

MTX is a drug mainly used in leukemia chemotherapy but also in autoimmune disorders. As a prodrug, MTX enters the cells mediated by different transporters and is polyglutamated to obtain cytotoxicity for pharmacological effect. It is well absorbed and reaches a rapid equilibrium of concentration but does not cross the blood-brain barrier in optimal concentrations, the reason why it needs to be administrated by intrathecal injection. The relationship between polymorphisms in genes encoding MTX transporters and MTX response is controversial. Elimination from plasma is mainly (80–90%) by renal excretion as unchanged molecule. In general, children have increased clearance of MTX than adults. As seen with other drugs, there is a general trend toward decreasing MTX clearance with increasing age until achieving adult's renal function values but with high variability in plasma concentration of MTX at any age (Crom 1994). Population pharmacokinetic studies have shown the relationship between polymorphisms and altered clearance (Wright et al. 2015). In pediatric patients with acute lymphoblastic leukemia, the single nucleotide polymorphisms (SNPs) in the organic anion-transporting polypeptide

SLCO1B1 were significantly associated in pharmacokinetic variability and toxicity (Liu et al. 2017), leading the focus on polymorphism in MTX transporters also related to changes in PK profile of MTX.

9.4.6   Tacrolimus (FK506)

Tacrolimus is an immunosuppressant calcineurin inhibitor that blocks calcineurin known to mediate the growth and differentiation of T cells of the immune system. Tacrolimus is used to reduce organ rejection before and after transplantation of solid organ and hematopoietic stem cell transplantation (Xie 2010). Plasma target concentrations and the subsequent dose depend on the transplanted organ like the heart, liver, or kidney and at the time of the evaluation of therapeutic drug concentrations. The interindividual pharmacokinetics is explained by changes in oral absorption, elimination, and metabolism. It was reported that, CYP3A5, P-glycoprotein (P-gp), and CYP3A5 play key roles in tacrolimus disposition (Xie 2010).

Depending on the CYP3A5 activity due to genetic variations, extensive and intermediate metabolizers generally have decreased dose-adjusted trough concentrations of tacrolimus as compared with those who are CYP3A5 non-expressers (poor metabolizers), possibly delaying achievement of target blood concentrations (Birdwell et al. 2015).

It has a narrow therapeutic range with highly variable pharmacokinetics, especially in pediatric patients, which makes it challenging to define standardized dosing. The developmental maturation and CYP3A5 polymorphism have significant impacts on tacrolimus disposition (Zhao et al. 2013).

Pharmacokinetic data in children are limited. Therapeutic drug monitoring still is the standard to adjust dose optimization to reduce the risk of organ rejection with subtherapeutic concentration or to avoid infection and other adverse events due to supratherapeutic plasma concentrations (Abaji and Krajinovic 2016; Zhao et al. 2009).

According to the pharmacokinetic studies in children and adolescents, the Clinical Pharmacogenetics Implementation Consortium (CPIC) guidelines suggest for children and adolescents with at least one CYP3A5*1 allele, a 1.5- to two-fold increase in dose followed by TDM (therapeutic drug monitoring) to obtain appropriate therapy (Birdwell et al. 2015). CYP3A5 polymorphism is the significant covariant to build a pharmacokinetic model in pediatric and adolescents (Zhao et al. 2009, 2013; Fukudo et al. 2006).

9.5   Conclusions and Future Directions

The information provided in this book chapter emphasizes the importance of considering the incorporation of pharmacogenomic information in the context of pediatric pharmacotherapy. Pharmacogenomics plays a significant role in understanding and quantifying the variability in pharmacokinetics and pharmacodynamics in pediatrics and is especially significant where maturational changes in pediatrics add another dimension to variability. It is evident from the examples discussed in this article that pharmacogenomic testing in pediatrics is slowly being integrated into the pediatric practice. However, the clinical significance of each pharmacogenomic test should be evaluated by prospective randomized trials to show improvement in patient outcomes. Furthermore, clinicians (physicians and other care providers) should be familiar with the terms used in pediatrics pharmacogenomics and their relationship to pharmacokinetics allowing them to make informed decisions for pediatric drug therapy. Nevertheless, to improve pediatric treatment outcomes, a specific pediatric pharmacogenomic database powered by bioinformatics tools which bank accumulative results and connect to international databases must be established.

References

Abaji R, Krajinovic M (2016) Current perspective on pediatric pharmacogenomics. Expert Opin Drug Metab Toxicol 12:363–365

Agunod M, Yamaguchi N, Lopez R et al (1969) Correlative study of hydrochloric acid, pepsin, and intrinsic factor secretion in newborns and infants. Am J Dig Dis 14:400–414

Anderson BJ, van Lingen RA, Hansen TG et al (2002) Acetaminophen developmental pharmacokinetics in premature neonates and infants: a pooled population analysis. Anesthesiology 96:1336–1345

Anderson GD, Lynn AM (2009) Optimizing pediatric dosing: a developmental pharmacologic approach. Pharmacotherapy 29:680–690

Berman W Jr, Whitman V, Marks KH, Friedman Z, Maisels MJ, Musselman J (1978) Inadvertent overadministration of digoxin to low-birth-weight infants. J Pediatr 92(6):1024–1025

Birdwell KA, Decker B, Barbarino JM et al (2015) Clinical Pharmacogenetics Implementation Consortium (CPIC) guidelines for CYP3A5 genotype and tacrolimus dosing. Clin Pharmacol Ther 98:19–24

Biss TT, Avery PJ, Brandao LR et al (2012) VKORC1 and CYP2C9 genotype and patient characteristics explain a large proportion of the variability in warfarin dose requirement among children. Blood 119:868–873

Blumer JL (1999) Off-label uses of drugs in children. Pediatrics 104:598–602

Brown AL, Lupo PJ, Okcu MF et al (2015) SOD2 genetic variant associated with treatment-related ototoxicity in cisplatin-treated pediatric medulloblastoma. Cancer Med 4:1679–1686

Brown RD, Campoli-Richards DM (1989) Antimicrobial therapy in neonates, infants and children. Clin Pharmacokinet 17:105–115

Bruggemann RJ, Alffenaar JW, Blijlevens NM et al (2009) Clinical relevance of the pharmacokinetic interactions of azole antifungal drugs with other coadministered agents. Clin Infect Dis 48:1441–1458

Center for Drug Evaluation and Research (CDER), Food and Drug Administration (1994) Guidance for industry: the content and format for pediatric use supplements

Constance JE, Campbell SC, Somani AA et al (2017) Pharmacokinetics, pharmacodynamics and pharmacogenetics associated with nonsteroidal anti-inflammatory drugs and opioids in pediatric cancer patients. Expert Opin Drug Metab Toxicol 13:715–724

Crews KR, Gaedigk A, Dunnenberger HM et al (2014) Clinical Pharmacogenetics Implementation Consortium guidelines for cytochrome P450 2D6 genotype and codeine therapy: 2014 update. Clin Pharmacol Ther 95:376–382

Crom WR (1994) Pharmacokinetics in the child. Environ Health Perspect 102:111–117

Cuzzolin L, Atzei A, Fanos V (2006) Off-label and unlicensed prescribing for newborns and children in different settings: a review of the literature and a consideration about drug safety. Expert Opin Drug Saf 5:703–718

Debotton N, Dahan A (2014) A mechanistic approach to understanding oral drug absorption in pediatrics: an overview of fundamentals. Drug Discov Today 19:1322–1336

Drew L (2016) Pharmacogenetics: the right drug for you. Nature 537(7619):S60–S62. https://doi.org/10.1038/537S60a

de Wildt SN, Kearns GL, Leeder JS et al (1999) Glucuronidation in humans. Pharmacogenetic and developmental aspects. Clin Pharmacokinet 36:439–452

Elens L, Capron A, van Schaik RH et al (2013) Impact of CYP3A4*22 allele on tacrolimus pharmacokinetics in early period after renal transplantation: toward updated genotype-based dosage guidelines. Ther Drug Monit 35:608–616

Engels MJ, Ciarkowski SL, Rood J et al (2016) Standardization of compounded oral liquids for pediatric patients in Michigan. Am J Health Syst Pharm 73:981–990

Evans WE, Relling MV, Petros WP et al (1989) Dextromethorphan and caffeine as probes for simultaneous determination of debrisoquin-oxidation and N-acetylation phenotypes in children. Clin Pharmacol Ther 45:568–573

Food and Drug Administration (2018) Table of pharmacogenomic biomarkers in drug labeling (2018). https://www.fda.gov/Drugs/ScienceResearch/ucm572698.htm

Frattarelli DA, Galinkin JL, Green TP et al (2014) Off-label use of drugs in children. Pediatrics 133:563–567

Fukudo M, Yano I, Masuda S et al (2006) Population pharmacokinetic and pharmacogenomic analysis of tacrolimus in pediatric living-donor liver transplant recipients. Clin Pharmacol Ther 80:331–345

Food and Drug Administration Modernization Act of 1997. (1997). https://www.govtrack.us/congress/bills/105/s830/text. Accessed 1 May 2018

Ganiere-Monteil C, Medard Y, Lejus C et al (2004) Phenotype and genotype for thiopurine methyltransferase activity in the French Caucasian population: impact of age. Eur J Clin Pharmacol 60:89–96

Giubergia V, Gravina L, Castanos C et al (2013) Influence of beta(2)-adrenergic receptor polymorphisms on asthma exacerbation in children with severe asthma regularly receiving salmeterol. Ann Allergy Asthma Immunol 110:156–160

Gershanik J, Boecler B, Ensley H et al (1982) The gasping syndrome and benzyl alcohol poisoning. New Engl J Med 307:1384–1388

Gore R, Chugh PK, Tripathi CD et al (2017) Pediatric off-label and unlicensed drug use and its implications. Curr Clin Pharmacol 12:18–25

Harries JT, Fraser AJ (1968) The acidity of the gastric contents of premature babies during the first fourteen days of life. Biol Neonat Neo-natal Stud 12(3):186–193

Hume R, Burchell A, Allan BB et al (1996) The ontogeny of key endoplasmic reticulum proteins in human embryonic and fetal red blood cells. Blood 87:762–770

Hume R, Coughtrie MW, Burchell B (1995) Differential localisation of UDP-glucuronosyltransferase in kidney during human embryonic and fetal development. Arch Toxicol 69:242–247

Johnson TN, Thomson M (2008) Intestinal metabolism and transport of drugs in children: the effects of age and disease. J Pediatr Gastroenterol Nutr 47:3–10

Kadam RS, Van Den Anker JN (2016) Pediatric clinical pharmacology of voriconazole: role of pharmacokinetic/pharmacodynamic modeling in pharmacotherapy. Clin Pharmacokinet 55:1031–1043

Kalra A, Goindi S (2014) Issues impacting therapeutic outcomes in pediatric patients: an overview. Curr Pediatr Rev 10:184–193

Kato Y, Ichida F, Saito K et al (2011) Effect of the VKORC1 genotype on warfarin dose requirements in Japanese pediatric patients. Drug Metab Pharmacokinet 26:295–299

Kearin M, Kelly JG, O'Malley K (1980) Digoxin "receptors" in neonates: an explanation of less sensitivity to digoxin than in adults. Clin Pharmacol Ther 28(3):346–349

Kearns GL, Abdel-Rahman SM, Alander SW et al (2003) Developmental pharmacology--drug disposition, action, and therapy in infants and children. N Engl J Med 349:1157–1167

Kearns GL (2015) Selecting the proper pediatric dose: it is more than size that matters. Clin Pharmacol Ther 98(3):238–240. https://doi.org/10.1002/cpt.168

Kirchheiner J, Stormer E, Meisel C et al (2003) Influence of CYP2C9 genetic polymorphisms on pharmacokinetics of celecoxib and its metabolites. Pharmacogenetics 13:473–480

Kirschner BS (1998) Safety of azathioprine and 6-mercaptopurine in pediatric patients with inflammatory bowel disease. Gastroenterology 115:813–821

Korbel L, George M, Kitzmiller J (2014) Clinically relevant pharmacogenomic testing in pediatric practice. Clin Pediatr (Phila) 53:831–838

Lam MS (2011) Extemporaneous compounding of oral liquid dosage formulations and alternative drug delivery methods for anticancer drugs. Pharmacotherapy 31(2):164–192. https://doi.org/10.1592/phco.31.2.164

Lavertu A, McInnes G, Daneshjou R et al (2018) Pharmacogenomics and big genomic data: from lab to clinic and back again. Hum Mol Genet 27:R72–R78

Lanvers-Kaminsky C, Sprowl JA, Malath I et al (2015) Human OCT2 variant c.808G>T confers protection effect against cisplatin-induced ototoxicity. Pharmacogenomics 16:323–332

Leeder JS, Kearns GL (1997) Pharmacogenetics in pediatrics. Implications for practice. Pediatr Clin North Am 44:55–77

Lennard L, Lilleyman JS, Van Loon J et al (1990) Genetic variation in response to 6-mercaptopurine for childhood acute lymphoblastic leukaemia. Lancet 336:225–229

Lennard L, Van Loon JA, Lilleyman JS et al (1987) Thiopurine pharmacogenetics in leukemia: correlation of erythrocyte thiopurine methyltransferase activity and 6-thioguanine nucleotide concentrations. Clin Pharmacol Ther 41:18–25

Lipworth BJ, Basu K, Donald HP et al (2013) Tailored second-line therapy in asthmatic children with the Arg(16) genotype. Clin Sci (Lond) 124:521–528. https://doi.org/10.1042/CS20120528

Liu SG, Gao C, Zhang RD, Zhao XX et al (2017) Polymorphisms in methotrexate transporters and their relationship to plasma methotrexate levels, toxicity of high-dose methotrexate, and outcome of pediatric acute lymphoblastic leukemia. Oncotarget 8:37761–37772

Maitland-van der Zee AH, Raaijmakers JA (2012) Variation at GLCCI1 and FCER2: one step closer to personalized asthma treatment. Pharmacogenomics 13:243–245

McLeod HL, Krynetski EY, Wilimas JA et al (1995) Higher activity of polymorphic thiopurine S-methyltransferase in erythrocytes from neonates compared to adults. Pharmacogenetics 5:281–286

Mlakar V, Huezo-Diaz Curtis P, Satyanarayana Uppugunduri CR, Krajinovic M, Ansari M (2016) Pharmacogenomics in pediatric oncology: review of gene-drug associations for clinical use. Int J Mol Sci 17(9). https://doi.org/10.3390/ijms17091502

Moreau C, Bajolle F, Siguret V et al (2012) Vitamin K antagonists in children with heart disease: height and VKORC1 genotype are the main determinants of the warfarin dose requirement. Blood 119:861–867

Moriyama T, Nishii R, Lin TN et al (2017a) The effects of inherited NUDT15 polymorphisms on thiopurine active metabolites in Japanese children with acute lymphoblastic leukemia. Pharmacogenet Genomics 27:236–239

Moriyama T, Yang YL, Nishii R et al (2017b) Novel variants in NUDT15 and thiopurine intolerance in children with acute lymphoblastic leukemia from diverse ancestry. Blood 130:1209–1212

Mukattash TL, Nuseir KQ, Jarab AS et al (2014) Sources of information used when prescribing for children, a survey of hospital based pediatricians. Curr Clin Pharmacol 9:395–398

Murto K, Lamontagne C, McFaul C et al (2015) Celecoxib pharmacogenetics and pediatric adenotonsillectomy: a double-blinded randomized controlled study. Can J Anaesth 62:785–797

Neville KA, Becker ML, Goldman JL et al (2011) Developmental pharmacogenomics. Paediatr Anaesth 21:255–265

Nowak-Gottl U, Dietrich K, Schaffranek D et al (2010) In pediatric patients, age has more impact on dosing of vitamin K antagonists than VKORC1 or CYP2C9 genotypes. Blood 116:6101–6105

Palmaro A, Bissuel R, Renaud N et al (2015) Off-label prescribing in pediatric outpatients. Pediatrics 135:49–58

Palmer CN, Lipworth BJ, Lee S et al (2006) Arginine-16 beta2 adrenoceptor genotype predisposes to exacerbations in young asthmatics taking regular salmeterol. Thorax 61:940–944

PharmGKB (2018) PharmGKB FAQs. https://www.pharmgkb.org/page/faqs#what-is-the-differ ence-between-pharmacogenetics-and-pharmacogenomics. Accessed 18 Apr 2018

Preventing errors relating to commonly used anticoagulants (2008). Sentinel Event Alert (41):1–4

Pussegoda K, Ross CJ, Visscher H, CPNDS Consortium et al (2013) Replication of *TPMT* and ABCC3 genetic variants highly associated with cisplatin-induced hearing loss in children. Clin Pharmacol Ther 94:243–251

Quiñones L, Roco A, Cayun JP et al (2017) Clinical applications of pharmacogenomics. Rev Med Chil 145:483–500

Ramos-Martin V, O'Connor O, Hope W (2015) Clinical pharmacology of antifungal agents in pediatrics: children are not small adults. Curr Opin Pharmacol 24:128–134

Relling MV, Gardner EE, Sandborn WJ et al (2011) Clinical Pharmacogenetics Implementation Consortium guidelines for thiopurine methyltransferase genotype and thiopurine dosing. Clin Pharmacol Ther 89:387–391

Roberts JK, Stockmann C, Constance JE et al (2014) Pharmacokinetics and pharmacodynamics of antibacterials, antifungals, and antivirals used most frequently in neonates and infants. Clin Pharmacokinet 53:581–610

Rood JM, Engels MJ, Ciarkowski SL et al (2014) Variability in compounding of oral liquids for pediatric patients: a patient safety concern. J Am Pharm Assoc: JAPhA 54:383–389

Ross CJ, Katzov-Eckert H, Dube MP, CPNDS Consortium et al (2009) Genetic variants in *TPMT* and COMT are associated with hearing loss in children receiving cisplatin chemotherapy. Nat Genet 41:1345–1349

Sachs AN, Avant D, Lee CS et al (2012) Pediatric information in drug product labeling. Jama 307:1914–1915

Schmiegelow K, Nielsen SN, Frandsen TL et al (2014) Mercaptopurine/methotrexate maintenance therapy of childhood acute lymphoblastic leukemia: clinical facts and fiction. J Pediatr Hematol Oncol 36:503–517

Sharma S, Ellis EC, Gramignoli R et al (2013) Hepatobiliary disposition of 17-OHPC and taurocholate in fetal human hepatocytes: a comparison with adult human hepatocytes. Drug Metab Dispos 41:296–304

Shaw K, Amstutz U, Hildebrand C et al (2014) VKORC1 and CYP2C9 genotypes are predictors of warfarin-related outcomes in children. Pediatr Blood Cancer 61:1055–1062

Stevens JC, Marsh SA, Zaya MJ et al (2008) Developmental changes in human liver CYP2D6 expression. Drug Metab Dispos 36:1587–1593

Stockmann C, Fassl B, Gaedigk R et al (2013) Fluticasone propionate pharmacogenetics: CYP3A4*22 polymorphism and pediatric asthma control. J Pediatr 162:1222–1227, 1227 e1221–1222

Stockmann C, Reilly CA, Fassl B et al (2015) Effect of CYP3A5*3 on asthma control among children treated with inhaled beclomethasone. J Allergy Clin Immunol 136:505–507

Table of Pharmacogenomic Biomarkers in Drug Labeling (2018) Center for Drug Evaluation and Research https://www.fda.gov/Drugs/ScienceResearch/ucm572698.htm

Teusink A, Vinks A, Zhang K et al (2016) Genotype-directed dosing leads to optimized voriconazole levels in pediatric patients receiving hematopoietic stem cell transplantation. Biol Blood Marrow Transplant 22:482–486

Turner S, Francis B, Vijverberg S et al (2016) Pharmacogenomics in childhood asthma C childhood asthma exacerbations and the Arg16 beta2-receptor polymorphism: a meta-analysis stratified by treatment. J Allergy Clin Immunol 138:107–113

Vear SI, Ayers GD, Van Driest SL et al (2014) The impact of age and CYP2C9 and VKORC1 variants on stable warfarin dose in the paediatric population. Br J Haematol 165:832–835

Walsh TJ, Karlsson MO, Driscoll T et al (2004) Pharmacokinetics and safety of intravenous voriconazole in children after single- or multiple-dose administration. Antimicrob Agents Chemother 48:2166–2172

Wright FA, Bebawy M, O'Brien TA (2015) An analysis of the therapeutic benefits of genotyping in pediatric hematopoietic stem cell transplantation. Future Oncol 11:833–851

Wehry AM, Ramsey L, Dulemba SE et al (2018) Pharmacogenomic testing in child and adolescent psychiatry: an evidence-based review. Curr Probl Pediatr Adolesc Health Care 48:40–49

Xie HG (2010) Personalized immunosuppressive therapy in pediatric heart transplantation: Progress, pitfalls and promises. Pharmacol Ther 126:146–158

Xu H, Robinson GW, Huang J et al (2015) Common variants in ACYP2 influence susceptibility to cisplatin-induced hearing loss. Nat Genet 47:263–266

Zhao W, Elie V, Roussey G et al (2009) Population pharmacokinetics and pharmacogenetics of tacrolimus in de novo pediatric kidney transplant recipients. Clin Pharmacol Ther 86:609–618

Zhao W, Fakhoury M, Baudouin V et al (2013) Population pharmacokinetics and pharmacogenetics of once daily prolonged-release formulation of tacrolimus in pediatric and adolescent kidney transplant recipients. Eur J Clin Pharmacol 69:189–195

Zuurhout MJ, Vijverberg SJ, Raaijmakers JA et al (2013) Arg16 ADRB2 genotype increases the risk of asthma exacerbation in children with a reported use of long-acting beta2-agonists: results of the PACMAN cohort. Pharmacogenomics 14:1965–1971

Bioavailability and Bioequivalence

10

Z. Gulsen Oner and James E. Polli

▶ **Objectives** After reading this chapter, a student will be able to:

- Describe terminology related to therapeutic equivalence
- Describe bioavailability/bioequivalence studies through a drug product life cycle
- Describe bioavailability/bioequivalence study types
- Describe a pharmacokinetic study design to assess in vivo bioequivalence
- Describe the Biopharmaceutics Classification System (BCS) and BCS-based biowaiver

Overview

A drug product may be placed on the market only after being reviewed and approved by the competent authority. In the United States, the Food and Drug Administration (FDA) ensures that safe and effective drugs are available by reviewing and approving New Drug Applications (NDAs), as well as Abbreviated New Drug Applications (ANDAs), or generic drug applications. Most types of NDAs typically include bioavailability (BA) and bioequivalence (BE) assessment. ANDAs typically include BE assessment.

(continued)

Z. G. Oner (✉)
Department of Pharmaceutical Sciences, University of Maryland, Baltimore, MD, USA

Turkish Medicines and Medical Devices Agency, Ankara, Turkey

J. E. Polli
Department of Pharmaceutical Sciences, University of Maryland, Baltimore, MD, USA
e-mail: jpolli@rx.umaryland.edu

© Springer Nature Switzerland AG 2018
A. Talevi, P. A. M. Quiroga (eds.), *ADME Processes in Pharmaceutical Sciences*,
https://doi.org/10.1007/978-3-319-99593-9_10

The FDA uses the Biologic License Application (BLA) pathway for innovator biological products and abbreviated licensure pathway for biosimilar biological products [denoted 351(k) pathway]. Biological products demonstrate comparability or biosimilarity. Biologics and biosimilars are not discussed here.

10.1   Introduction

A drug product may be placed on the market only after being reviewed and approved by the competent authority. In the United States, the Food and Drug Administration (FDA) ensures that safe and effective drugs are available by reviewing and approving New Drug Applications (NDAs), as well as Abbreviated New Drug Applications (ANDAs), or generic drug applications. Most types of NDAs typically include bioavailability (BA) and bioequivalence (BE) assessment. ANDAs typically include BE assessment.

The FDA uses the Biologic License Application (BLA) pathway for innovator biological products and abbreviated licensure pathway for biosimilar biological products [denoted 351(k) pathway]. Biological products demonstrate comparability or biosimilarity. Biologics and biosimilars are not discussed here.

For a drug product to exhibit a therapeutic effect, the active ingredient should be delivered to its site of action in an effective concentration and for a certain period of time. BA describes this phenomenon and emphasizes the rate and extent of absorption of the active ingredient or moiety to the site of action. The rate of absorption is often not simple to describe or measure, but it is sometimes desirable to increase or delay the absorption rate. The concentration of the active ingredient at the site of action is related to the concentration in the plasma, and therefore, it is desirable to measure the amount of drug absorbed.

Drug products having the same amounts of active ingredient may exhibit different therapeutic responses due to different drug plasma levels. To receive approval as a generic drug, an applicant must demonstrate that their proposed drug product is bioequivalent to the innovator drug (i.e., to the reference listed drug). BE is the absence of a significant difference in the rate and extent to which the active ingredient in two dosage forms becomes available at the site of drug action. Therefore, a generic drug product that is determined to be bioequivalent is expected to have equal safety and efficacy compared to its brand-name drug product.

▶ **Definitions** "Approved Drug Products with Therapeutic Equivalence Evaluations," commonly known as the "Orange Book," (U.S. Food & Drug Administration 2017a) is a source for drug products approved by the FDA, therapeutic equivalence (TE) evaluations, brand-name drug data, and patent and exclusivity information. Code of Federal Regulations Title 21 (21 CFR) Part 320 defines the BA and BE requirements in the United States. Below are some TE-related terms:

Pharmaceutical equivalents: Drug products that contain the same active ingredient (s), dosage form, route of administration, and amount of active ingredient while meeting the same standards (i.e., strength, quality, purity, and identity). They may differ in characteristics such as shape, scoring configuration, release mechanisms, packaging, excipients (including colors, flavors, preservatives), expiration time, and, within certain limits, labeling. Being pharmaceutical equivalents is a requirement to be therapeutic equivalents (e.g., to be a generic to the reference listed drug).

Pharmaceutical alternatives: Drug products that contain the identical therapeutic moiety, or its precursor, but not necessarily in the same amount or dosage form or as the same salt or ester (21CFR314.3). One example is tetracycline hydrochloride 250 mg capsules vs. tetracycline phosphate complex 250 mg capsules. Another example is quinidine sulfate 200 mg tablets vs. quinidine sulfate 200 mg capsules (U.S. Food & Drug Administration 2017a).

Therapeutic equivalents: Pharmaceutical equivalents for which bioequivalence has been demonstrated, and they can be expected to have the same clinical effect and safety profile when administered to patients under the conditions specified in the labeling.

FDA's criteria for therapeutically equivalent drug products are:

1. Approved as safe and effective
2. Pharmaceutical equivalents in that they (a) contain identical amounts of the same active drug ingredient in the same dosage form and route of administration and (b) meet compendial or other applicable standards of strength, quality, purity, and identity
3. Bioequivalent in that (a) they do not present a known or potential bioequivalence problem, and they meet an acceptable in vitro standard, or (b) if they do present such a known or potential problem, they are shown to meet an appropriate bioequivalence standard
4. Adequately labeled
5. Manufactured in compliance with Current Good Manufacturing Practice regulations

Bioequivalent drug products: Pharmaceutical equivalent or pharmaceutical alternative products that display comparable bioavailability when studied under similar experimental conditions (U.S. Food & Drug Administration 2017a).

Generic drug: A copy of a brand-name drug that one company makes that was developed by another company (U.S. Food & Drug Administration 2017b). Therefore, it has the following properties compared to its innovator drug:

• Contains the same active ingredients (inactive ingredients may vary)
• Is identical in strength, dosage form, and route of administration
• Has the same use indications

- Is bioequivalent
- Meets the same batch requirements for identity, strength, purity, and quality
- Is manufactured under the same strict standards of FDA's good manufacturing practice regulations required for innovator products (U.S. Food & Drug Administration 2017c)

The terms "generic drug" and "therapeutic equivalents" mean the same. "Generic drug" is usually used by healthcare professionals and patients, while "therapeutic equivalents" is more commonly used by regulatory scientists.

Reference listed drug: The listed drug identified by FDA as the drug product upon which an ANDA applicant relies on for seeking approval of their generic drug.

Reference standard: The drug product selected by FDA that an applicant seeking approval of an ANDA must use in conducting an in vivo bioequivalence study (21CFR314.3). This newer term generally is also the reference listed drug.

Bioavailability: The rate and extent to which the active ingredient or active moiety is absorbed from a drug product and becomes available at the site of drug action. Note that this definition in connection with BE testing emphasizes rate and extent (e.g., Cmax and AUC). This term "bioavailability" can also refer to a single number (see absolute bioavailability and relative bioavailability below).

Bioequivalence: The absence of a significant difference in the rate and extent to which the active ingredient/moiety in pharmaceutical equivalents/alternatives becomes available at the site of drug action when administered at the same molar dose under similar conditions in an appropriately designed study (21CFR314.3).

Generic substitution: The dispensing to a patient of a product classified as therapeutically equivalent.

Therapeutic substitution: The dispensing to a patient of not the prescribed drug product nor a therapeutic equivalent, but a drug product which is expected to yield the same therapeutic outcome. Therapeutic substitution generally occurs in closed health system (e.g., hospital) where healthcare professionals agree to the use of a more limited number of drugs than is commercially available, such as the use of one statin over all other statins. Therapeutic substitution has no conceptual overlap with therapeutic equivalents (i.e., generic drugs).

10.2 Bioavailability/Bioequivalence Studies Through a Drug Product Life Cycle

10.2.1 Investigational New Drug Period and New Drug Development

During a new drug's early preclinical development, the sponsor's primary goal is to determine if the product is safe for initial use in humans and if the compound exhibits pharmacological activity that justifies commercial development. These assessments are achieved by conducting animal pharmacology and toxicology studies (U.S. Food & Drug Administration 2015a). These assessments contribute to an Investigational

New Drug (IND) application. An IND is required to conduct human clinical trials on a new molecular entity. The IND period aims to lead to submission of an NDA.

After approval to begin human clinical trials, the primary goal is to determine if the product is safe and effective, which is achieved by clinical efficacy, safety, and pharmacokinetic studies. BA studies are one type of pharmacokinetic study. NDAs often include efforts to measuring in vivo BA. Several reasons can cause BA to be difficult to measure.

A typical NDA has four to six BE studies. BE documentation can be useful during the IND period to compare:

1. Early and late clinical trial formulations.
2. Formulations used in clinical trials and stability studies, if different.
3. Clinical trial formulations and to-be-marketed drug products, if different.
4. Product strength equivalence or proportionality (U.S. Food & Drug Administration 2014).

10.2.2 Generic Drug Development

During development of generic drugs, the sponsor's primary goal is to determine if the proposed product is bioequivalent to the innovator. However, clinical data to establish safety and efficacy is not required, since BE is taken as the major indicator of a proposed generic to be safe and effective, like the brand. BE is often in the form of plasma profiles of proposed generic and brand being similar, although in vitro or other studies can sometimes be applied in place of plasma profiles.

Similarly, any person submitting an ANDA must include evidence demonstrating that the drug product is bioequivalent to the RLD (or reference listed drug) or information that would permit FDA to waive the submission of in vivo BE evidence.

▶   **Important** NDA must undergo clinical trials to demonstrate safety and efficacy. Generic products must demonstrate bioequivalence with the innovator/reference product.

10.2.3 The Scale-Up and Post-approval Changes (SUPAC)

After a drug product is approved for marketing, the manufacturer may want or need certain changes in the finished product (e.g., manufacturing method or location, formulation composition) or in the active substance. BE (e.g., via human in vivo plasma profiles) is required for some manufacturing changes.

10.3 Bioavailability and Bioequivalence Studies

FDA has issued NDA BA and BE Draft Guidance (U.S. Food & Drug Administration 2014) and ANDA BE Draft Guidance (U.S. Food & Drug Administration 2013), in which study protocol details are outlined. BA depends on several factors including route of administration, dosage form, physicochemical properties of the drug(s) and excipient(s), and biologic factors.

10.3.1 Absolute and Relative Bioavailability Studies

BA may be absolute or relative:

Absolute bioavailability is the ratio of the BA of a given dosage form to that of the intravenous (IV) administration (100%) (e.g., oral solution vs. IV):

$$F = \frac{AUC_A/dose_A}{AUC_{IV}/dose_{IV}} \qquad (10.1)$$

where AUC_A and dose A are the area under the concentration-time curve and dose for the test drug product, respectively, and AUC_{IV} and dose IV refer to the area under the concentration-time curve and dose for the intravenous reference drug product. For example, $F = 40\%$ if AUC_A is 40% that of AUC_{IV}, after giving the same doses.

More generally, relative bioavailability is the ratio of the BA of a given dosage form to that of a form administered by the same or another non-intravenous route (e.g., tablet vs. oral solution):

$$F = \frac{AUC_A/dose_A}{AUC_B/dose_B} \qquad (10.2)$$

where AUC_A and dose A are the area under the concentration-time curve and dose for the test drug product, respectively, and AUC_B and dose B refer to the area under the concentration-time curve and dose for the reference drug product.

Absolute BA is often determined for a drug during the IND phase of drug development. One reason why absolute BA may not be measured is lack of availability of an IV formulation. An IV formulation requires effort to formulation, including sufficient drug solubility in such a formulation. Another reason is concern about drug safety after IV administration.

10.3.2 Mass Balance Studies

In the mass balance study, the drug is administered with a radioactive label in a metabolically stable position on the drug structure. The systemic exposure of drug and its metabolites, including drug and metabolites in urine and feces, is obtained and compared to the amount of total radioactive drug administered. The absolute BA can be measured through this approach, which involves identifying how much drug was excreted relative to how much drug was administered.

10.3.3 Bioequivalence Studies

A BE study is basically a comparative BA study designed to establish equivalence between test and reference products. Although BA and BE are related, BE comparisons normally rely on (1) a criterion, (2) a confidence interval for the criterion, and (3) a predetermined BE limit. BE comparisons could also be used in certain pharmaceutical product line extensions, such as additional strengths (e.g., introduce a 75 mg strength, between the existing 50 mg and 100 mg), new dosage forms [e.g., change from immediate-release (IR) to extended-release], and new routes of administration (e.g., oral to topical).

Approaches to determine BE generally follow approaches similar to those used for BA. The order of general preference to assess relative BA or BE is human pharmacokinetic (PK) studies, in vitro tests predictive of human in vivo BA (in vitro-in vivo correlation), pharmacodynamic (PD) studies, studies with clinical benefit endpoints, and other in vitro studies.

10.3.3.1 General Pharmacokinetic Study Design

The design of a BA/BE study depends on the objectives of the study, the ability to analyze the drug in biological fluids, the pharmacodynamics of the drug substance, the route of drug administration, and the nature of the drug and drug product. General requirements for the design and conduct of BA or BE studies are summarized below. Studies are usually evaluated by a single-dose, two-period, two-treatment, two-sequence, open-label, randomized crossover design, comparing equal doses of the test and reference products in fasted, adult, healthy volunteers. For example, a BE study may involve $n = 24$ subjects to compare generic to brand. There are two treatments (i.e., generic and brand). There are two sequences: TR and RT. TR that denotes test (or generic) is tested first in a subject, followed by reference (or brand) in that same subject. RT that denotes brand (or reference) is tested first. Each subject is randomized to be in sequence TR or sequence RT. In an open-label study, subjects (and immediate research staff) are not blinded but aware of what treatments are administered in each period. The study is two period, since treatments are tested over two occasions (i.e., once for generic and once for brand). This design is a non-replicate design, as each test and reference products is only evaluated in one period. By virtue of being a single-dose study, each period only involves drug administration once, rather than, for example, every 12 h for the entire period. A single-dose period can vary from several hours to several days, depending on how long it is necessary to measure drug levels in the blood to capture at least 80–90% of the area under the concentration-time curve, compared to if samples were collected for an infinite length of time. Long half-life drugs require longer sampling periods.

The study should be designed in such a way that the formulation effect can be distinguished from other effects. There are two ways of designing the study:

crossover and parallel. A crossover design is a study where each subject receives the test drug product and reference drug product separately, following a washout period between treatments. Compared to parallel design, the crossover design provides an advantage because it reduces variability caused by subject-specific factors, thereby increasing the ability to discern differences due to formulation. An adequate washout period should be scheduled between the two periods to separate each treatment before administration of the second dose (e.g., ≥ 5 half-lives of the moieties to be measured).

In a parallel design, subjects are divided randomly into groups, and each group receives only one treatment and is less powerful than crossover. The most common reason to employ a parallel design is long drug half-life, such that a crossover study with a washout period would be too long in duration to successfully complete (e.g., excessive number of subject dropouts due to loss to follow-up).

Single-dose PK studies recommended over steady-state (multiple dose) PK studies. Single-dose PK studies are generally more sensitive than steady-state studies in assessing rate of release of the drug substance from the drug product into the systemic circulation. Steady-state dose study is less sensitive in detecting differences in Cmax, since PK profiles from multiple dosing typically reflect not only the most recent dose but many of the most recent doses.

Meanwhile, steady-state PK studies involve subjects, usually patients, taking the drugs many times (e.g., every 12 h) over the study period. A main reason to use multiple-dose design is that the drug is too toxic for healthy volunteers, such that the BE study is only conducted in patients who can be expected to benefit from the drug, which generally necessitates multiple doses of the drug. Another reason for multiple dose is the ability to measure drug level after multiple dose, but not after single dose.

The BA or BE study should generally be conducted under fasting conditions. After an overnight fast of at least 10 h, a pre-dose blood sample is collected. Then, the test or reference products are administered with about 8 ounces (240 mL) of water. No food is allowed for at least 4 h post-dose. Blood samples are collected periodically after dosing, according to the protocol. PK sampling typically involves about 12–18 time points.

Food effect studies are usually conducted for new drugs and drug products during the IND period to assess the effects of food on the rate and extent of absorption of a drug when the drug product is administered shortly after a meal (under fed conditions), as compared to administration under fasting conditions. Fed studies are typically conducted using meals that provide the greatest effects on gastrointestinal (GI) physiology and systemic drug availability (e.g., high-fat meal).

Non-replicate crossover study designs are generally recommended for BA and BE studies of IR and modified-release dosage forms. Non-replicate means drug is only taken once (i.e., for only one period). Some BE studies involve replicate design, where at least one product is given twice (i.e., single doses are given in one period and then repeated in another period). Replicate crossover designs are used when evaluating BE of highly variable drugs or BE of narrow therapeutic index drugs. Replicate crossover designs are used to allow estimation of (1) with in-subject variance for the reference product, or for both the test and reference products, and

(2) the subject-by-formulation interaction variance component. This design accounts for the inter-occasion variability that may confound the interpretation of a BE study as compared to a non-replicate crossover approach.

Bioanalytical methods used for BA and BE studies should be accurate, precise, specific, sensitive, and reproducible, along with other properties.

10.3.3.2 Reference and Test Product

When in vivo BE testing is required, the FDA generally designates a single reference standard that ANDA applicants must use in in vivo BE testing. All generic versions must establish BE to that standard. Generally, FDA selects the RLD as the reference standard. However, in some instances (e.g., where the RLD has been withdrawn from sale and an ANDA is selected as the reference standard), the RLD and the reference standard may be different (21CFR314.3). The assayed drug content of the test product batch should not differ from the reference product by more than ±5%.

10.3.3.3 Subjects

In general, BA and BE studies should be conducted in healthy volunteers if the product can be safely administered to this population. Subjects should be 18 years of age or older, capable of giving informed consent, and together, represent the entire general population. Both male and female subjects should be enrolled in BA and BE studies unless there is a specific reason to exclude one sex. For example, oral contraceptives are only evaluated in female subjects because the indication is specific to females.

In addition, the total number of subjects in a study should be sufficient to provide adequate statistical power. The study should have 80% or 90% power to conclude BE between these two formulations.

According to EMA's BE guideline (European Medicines Agency 2010), at least 12 volunteers must be enrolled, where they should have a body mass index between 18.5 and 30 kg/m^2. Standardization of diet, fluid intake, exercise, and posture is recommended.

10.3.3.4 Sample Collection and Sampling Times

The active ingredient that is released from the dosage form or its active moiety should be measured in biological fluids in BA/BE studies. When appropriate, active metabolites should be collected. Under normal circumstances, blood, rather than urine or tissue, should be used. In most cases, drug or metabolites are measured in serum or plasma. However, in certain cases, such as when an assay of sufficient sensitivity cannot be developed for plasma, whole blood may be more appropriate for analysis.

Twelve to eighteen samples, including a pre-dose sample, should be collected per subject per dose. This sampling should continue for at least three or more terminal elimination half-lives of the drug to capture 80–90% of the relevant AUC. For multiple-dose studies, sampling should occur across the dose interval and include the beginning and the end of the interval. Three or more samples should be obtained during the terminal log-linear phase to obtain an accurate estimate of λ_z (i.e., elimination rate constant) from linear regression.

10.3.3.5 Subjects with Pre-dose Plasma Concentrations

A sufficiently long washout should result in zero or negligible drug in plasma for the second period, at $t = 0$. If the pre-dose concentration is $\leq 5\%$ of Cmax value in that subject, the subject's data without any adjustments can be included in all PK measurements and calculations. However, if the pre-dose value is $>5\%$ of Cmax, the subject should be dropped from all PK evaluations. The subject data should be reported and the subject should be included in safety evaluations.

10.3.3.6 Data Deletion Because of Vomiting

Occasionally, a subject vomits during a study. Data from subjects who experience emesis during the course of a study for IR products should be deleted from statistical analysis if vomiting occurs at or before two times median Tmax. For modified-release products, subjects who experience emesis at any time during the labeled dosing interval should not be included in PK analysis.

10.3.3.7 Evaluation and Methods to Document BA/BE: Use of Plasma Drug Concentration

BA and BE frequently rely on two main PK measures: area under the concentration-time curve (AUC) to assess the extent of systemic exposure and peak or maximum concentration (Cmax) to assess the rate of systemic absorption. $AUC_{(0-t)}$ denotes the area under the concentration-time curve from zero to the last measurable time point. $AUC_{(0-\infty)}$ denotes the area under the concentration-time curve extrapolated to infinity. If drug concentration is in terms of μg/mL, the units of AUC are μg/mL $\times$ h and generally represent the extent of drug exposure in the body. Tmax is the time to reach peak concentration and is also reported but is generally not used in decision-making since it has limitations due to sampling scheme and statistical properties.

Figure 10.1 illustrates the plasma drug concentrations over time following administration of a single oral dose drug. Once the minimum effective concentration (MEC) is reached, the pharmacological response begins. The concentration of the drug at the peak is Cmax, and the time required to reach Cmax is Tmax. At Cmax, in general, the rate of drug absorption is equal to the rate of drug elimination. In general, time that precedes Cmax (i.e., to the left of Cmax) represents the majority of the absorption phase. Meanwhile, time after Cmax largely represents the elimination phase. Within the therapeutic range, the desired pharmacological response is obtained (i.e., maximum efficacy and minimum toxicity). AUC expresses the total amount of drug exposure in the systemic circulation after drug administration. The duration of action is the time period that the drug concentration is greater than the MEC level. Minimum toxic concentration (or maximum therapeutic concentration (MTC)) is the drug concentration at which the drug becomes toxic. Therefore, the concentration range between MEC and MTC, the therapeutic window, gives the optimal drug efficacy while limiting toxicity.

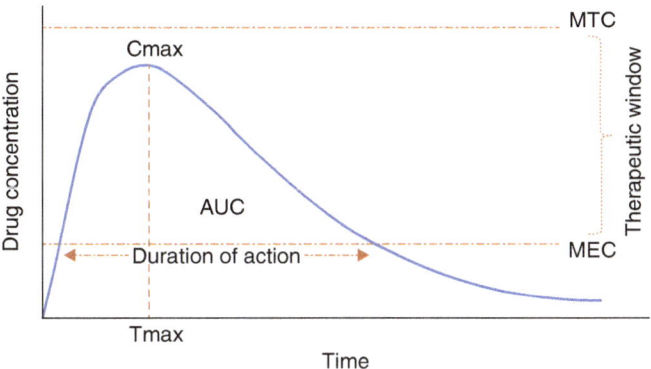

Fig. 10.1 Plasma concentration-time curve

For a drug with linear pharmacokinetics, the AUC and Cmax values increase or decrease proportionally with the dose. For a drug with nonlinear kinetics, the changes in AUC and Cmax values are not proportional to dose, and testing should be conducted at several different doses.

BE assessment is typically made based on Cmax and AUC. However, it is possible for two different profiles to have the same Cmax and AUC, but the profiles to be meaningfully different (e.g., differ in Tmax). Hence, it is possible that Cmax and AUC do not capture all important attributes of profile shape. Differences in the shape of the systemic concentration profile between the test and reference products could imply that the test product may not produce the same clinical response as the reference product. In such cases, additional data analysis (e.g., partial AUC_S), exposure-response evaluation, or clinical studies may be recommended to evaluate the BE of the two products. For example, the sleep medicine zolpidem extended-release tablet requires partial AUC0–1.5 and AUC1.5–t for evaluation of BE, under fasting conditions. These two metrics aim to reflect early drug exposure for sleep onset (i.e., 0–1.5 h) and later drug exposure for sleep maintenance (i.e., 1.5 h to the last measurable time point) (U.S. Food & Drug Administration 2011)).

10.3.3.8 Evaluation and Methods to Document BA/BE: Use of Urinary Drug Excretion

Urinary drug excretion studies are generally not encouraged. If serial measurements of the drug or its metabolites in plasma, serum, or blood cannot be accomplished, measurement of urinary excretion can be used to demonstrate BE. The concentration-time profile of the parent drug is more sensitive to changes in formulation performance than a metabolite. Cumulative urinary excretion (Ae) and its maximal rate (Rmax) are employed as extent and rate metrics. Ae(0–t) is cumulative urinary excretion of unchanged drug (mg). Rmax is maximal rate of urinary excretion (mg/h). Figure 10.2 illustrates urinary drug concentrations over time.

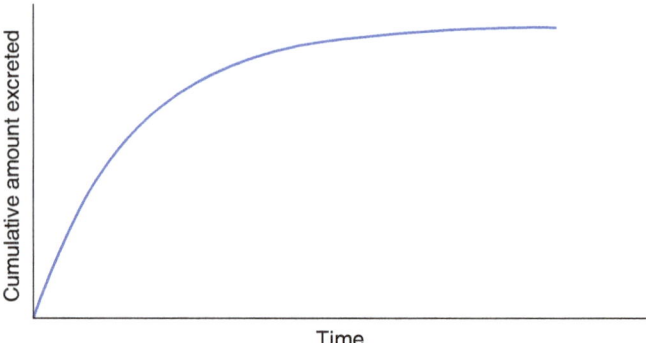

Fig. 10.2 Cumulative urinary excretion-time curve

10.3.4 Statistical Analysis and Acceptance Limits to Establish Bioequivalence

For AUC and Cmax, the 90% confidence interval for the ratio of the test and reference products should be contained within the acceptance interval of 80.00–125.00%. A statistical evaluation of Tmax is not required. However, if rapid release is claimed to be clinically relevant and of importance for onset of action or is related to adverse events, there should be no apparent difference in median Tmax and its variability between test and reference product.

For the demonstration of complete drug absorption, data from absolute bioavailability or mass balance studies could be used.

10.4 Bioequivalence Studies

10.4.1 The Basis for Determining Bioequivalence

BE is established if the in vivo BA of a test drug product does not differ significantly in rate and extent of drug absorption from that of the reference when administered under the same molar dose of the active moiety under similar experimental conditions, either as single or multiple dose. If two formulations show similar concentration-time profiles, they should exhibit similar therapeutic effects. The objective of the BE study is to measure and compare the formulation performance between two (or more) pharmaceutically equivalent drug products.

A drug product may be placed on the market in the United States after being approved by the FDA. The FDA uses the NDA pathway for innovators and the ANDA pathway for generics under the Federal Food, Drug, and Cosmetic Act. Table 10.1 summarizes the requirements imposed by the FDA for NDA and ANDA. Generics are generally not required to replicate the extensive clinical trials that have already been used in the development of the innovator, brand-name drug, since the safety and efficacy of the brand have already been well established. Instead, generic

Table 10.1 NDA and ANDA requirements

NDA	ANDA
Chemistry	Chemistry
Manufacturing	Manufacturing
Controls	Controls
Labeling	Labeling
Testing	Testing
Animal studies	
Clinical studies	Bioequivalence
Bioavailability	

drug products have to show they are pharmaceutical equivalent to and bioequivalent to innovator drug products, along with other manufacturing and labeling requirements.

There are several common scenarios for BE testing. In general, the BE testing occurs in the context of change, where the reference is the current or pre-change formulation and test is the post-change formulation. Innovator companies and generic companies both employ BE testing by virtue of various formulations or manufacturing changes. Scale-up and post-approval changes (SUPAC) are one context for BE testing. For example, perhaps soon after NDA approval, an innovator company may have the need to manufacture the product at a larger manufacturing scale or in a new location, such that the new "post-change" product needs to bioequivalent to the currently manufactured (i.e., currently approved) product. Generics also perform BE testing due to SUPAC. Generics also typically obtain ANDA approval through BE testing. Also, 505(b)(2) applications often rely on BE testing (e.g., approval of a new extended-release formulation based on BE to the approved IR formulation).

▶   **Important** Bioequivalence studies are performed during the development of generic products, which must prove bioequivalent to the innovator. When post-approval manufacturing changes are introduced, the reference will be the pre-change formulation.

10.4.2 Methods for Determining Bioequivalence

According to FDA's draft guidance for industry, "Bioavailability and Bioequivalence Studies Submitted in NDAs or INDs – General Considerations," in vivo and in vitro methods can be used to measure BA and establish BE. The following approaches are used to determine BE.

10.4.2.1 Pharmacokinetic (PK) Studies
BE frequently relies on pharmacokinetic endpoints such as Cmax and AUC, as explained above.

10.4.2.2 In Vitro Studies: Biopharmaceutics Classification System (BCS) and BCS-Based Biowaiver

21 CFR Part 320 addresses BA and BE requirements. One may request FDA to waive the requirement for the submission with evidence measuring the in vivo BA or demonstrating the in vivo BE of the drug product that is subject of an NDA, ANDA, or supplemental application. Some eligibility criteria:

1. The in vivo BA or BE of the drug product may be self-evident. Examples include parenteral solutions, ophthalmic or otic solutions, some inhalation products, and topical (skin) solutions, oral solutions, and nasal solutions.
2. BA may be measured, or BE may be demonstrated by in vitro in lieu of in vivo data, if the drug product is:
   • In a different strength to another product with demonstrated BE
   • Shown to meet an in vitro test that has been correlated with in vivo data
   • A reformulated product that is identical, except for a different color, flavor, or preservative that will not affect the BA
3. Protection of the public health (21CFR320.22)
4. A solid oral dosage form that meets the BCS criteria

Biopharmaceutics Classification System

The BCS is a scientific framework for classifying drug substances based on their aqueous solubility and intestinal permeability. According to the BCS, drug substances are classified as follows (Table 10.2).

When combined with in vitro dissolution characteristics of the drug product, the BCS takes into account three major factors: solubility, intestinal permeability, and dissolution.

▶ **Definitions** A drug substance is considered highly soluble when the highest strength is soluble in 250 mL or less of aqueous media over the pH range of 1–6.8. A drug substance is considered to be highly permeable when the extent of absorption in humans is determined to be 85% or more of an administered dose based on a mass balance determination (along with evidence showing stability of the drug in the GI tract) or in comparison to an intravenous reference dose.

An IR drug product is considered rapidly dissolving when 85% or more of the labeled amount of the drug substance dissolves within 30 min, using United States Pharmacopeia (USP) Apparatus I (basket method) at 100 rpm or Apparatus II

Table 10.2 Biopharmaceutics Classification System

BCS Class	Solubility	Permeability
1	High	High
2	Low	High
3	High	Low
4	Low	Low

(paddle method) at 50 rpm (or at 75 rpm when appropriately justified) in a volume of 500 mL or less in each of the following media:

1. 0.1 N HCl or simulated gastric fluid USP without enzymes.
2. A pH 4.5 buffer.
3. A pH 6.8 buffer or simulated intestinal fluid USP without enzymes.

An IR product is considered very rapidly dissolving when 85% or more of the labeled amount of the drug substance dissolves within 15 min using the above mentioned conditions.

The FDA is revising its guidance for industry on the waiver of in vivo bioavailability and bioequivalence studies for IR solid oral dosage forms based on BCS (U.S. Food & Drug Administration 2015b). These waivers are intended for (a) subsequent in vivo BA or BE studies of formulations after the initial establishment of the in vivo BA of IR dosage forms during the IND period and (b) in vivo BE studies of IR dosage forms in ANDAs.

This guidance is applicable for BA/BE waivers (biowaivers) based on BCS, for BCS class 1 and class 3 IR solid oral dosage forms.

For BCS class 1 drug products, the following should be demonstrated to obtain a biowaiver:

• The drug substance is highly soluble.
• The drug substance is highly permeable.
• The drug product (test and reference) is rapidly dissolving (85% within 30 min).
• The product does not contain any excipients that will affect the rate or extent of absorption of the drug.

For BCS class 3 drug products, the following should be demonstrated to obtain a biowaiver:

• The drug substance is highly soluble.
• The drug product (test and reference) is very rapidly dissolving (>85% within 15 min).
• The test product formulation is qualitatively the same and quantitatively very similar, e.g., falls within SUPAC IR level 1 and 2 changes, in composition to the reference.

The concept is applicable to IR, solid pharmaceutical products for oral administration and systemic action having the same pharmaceutical form. BCS-based biowaivers are not applicable for narrow therapeutic range drugs and products designed to be absorbed in the oral cavity.

The scientific basis of BCS considers how drug is absorbed. Observed in vivo differences in the rate and extent of absorption of a drug from two pharmaceutically equivalent solid oral products may be due to differences in drug dissolution in vivo.

However, when the in vivo dissolution of an IR solid oral dosage form is rapid or very rapid in relation to gastric emptying and the drug has high solubility, the rate and extent of drug absorption are unlikely to be dependent on drug dissolution and/or GI transit time. Under such circumstances, demonstration of in vivo BA or BE may not be necessary for drug products containing class 1 and class 3 drug substances, as long as the inactive ingredients used in the dosage form do not significantly affect absorption of the active ingredients.

▶   **Important** Biowaivers are applicable to IR, solid pharmaceutical
        products for oral administration and systemic action having the same
        dosage form. BCS-based biowaivers are not applicable for narrow thera-
        peutic range drugs and products designed to be absorbed in the oral
        cavity.

10.4.2.3 Other In Vitro Studies

Dissolution/Drug-Release Testing
In vitro dissolution studies have several uses, including contributing to waiver of in vivo BE for lower strengths of an IR solid oral dosage form. Most products have several strengths (e.g., 10 mg, 25 mg, 75 mg, and 100 mg tablets). In vivo BE is typically demonstrated using the highest strength (e.g., 100 mg tablet), and waivers of in vivo studies may be granted on lower strengths (e.g., 10 mg, 25 mg, and 75 mg tablets), based on in vitro dissolution testing. However, two requirements are that the multiple strengths of IR products show linear kinetics and that the lower strengths are proportionately similar in composition to the highest strength product (e.g., the 50 mg tablet uses half of the same powder as the 100 mg tablet). The approval of the lower strengths is based on dissolution profile comparisons between these lower strengths and the strength of the batch used in the BE study (i.e., typically the highest strength). The dissolution profile of the lower strength must be similar to the dissolution profile of the highest strength.

For IR formulations, comparison at 15 min is essential. Where more than 85% of the drug is dissolved within 15 min, dissolution profiles may be accepted as similar. That is, comparison is not necessary for very rapidly dissolving drug products (i.e., when over 85% is dissolved at 15 min). When less than 85% is dissolved at 15 min, further mathematical evaluation should be used to compare profiles, often using the f_2 similarity factor. An f_2 value >50 suggests that two dissolution profiles are similar (U.S. Food & Drug Administration 1997).

10.4.2.4 Pharmacodynamic Studies

Pharmacodynamics (PD) concerns drug action or drug effect (e.g., reduction in blood pressure). PD studies should consider the relationship of drug concentrations to drug effect. PD studies are not recommended for orally administered drug products when the drug is absorbed into systemic circulation; rather, a PK approach can be used to assess systemic exposure and evaluate BA or BE. PK endpoints are preferred when they are the most accurate, sensitive, and reproducible approach. However, in instances where a PK endpoint is not possible, a well-justified PD endpoint can be used to demonstrate BA or BE.

10.4.3 Statistical Approaches for Bioequivalence

The only applied method to test BE is average bioequivalence (ABE), which focuses on the comparison of population averages response of a test and reference product, not on the variances. Meanwhile, population and individual BE approaches are alternative potential approaches to compare both averages and variances of the two (or more) formulations. The population BE approach assesses total variability of the measure in the population. The individual BE approach assesses within-subject variability for the test and reference products, as well as the subject-by-formulation interaction (U.S. Food & Drug Administration 2001).

10.4.4 Supra-bioavailability

If the test product shows larger bioavailability than reference (i.e., more than the upper limit of 125.00%), than the reference product, the test product cannot be considered as therapeutic equivalent. The new product may be considered as a new medicinal product.

10.5 Conclusion

BE studies are frequently used through the pharmaceutical industry. BE has acceptable approaches to determine the safety and efficacy of drug products, which accelerate the approval process and reduce the cost of product development.

References

21CFR314.3. Accessed 28 Aug 2017. https://www.ecfr.gov/cgi-bin/text-idx?SID=d1173eb1e0b5c915 1b05905ffe183de3&mc=true&node=se21.5.314_13&rgn=div8.

21CFR320.22. Accessed 28 Aug 2017. https://www.ecfr.gov/cgi-bin/text-idx?SID=d1173eb1e0b5c9 151b05905ffe183de3&mc=true&node=se21.5.320_122&rgn=div8.

European Medicines Agency (2010) Guideline on the investigation of bioequivalence. Accessed 28 Aug 2017. http://www.ema.europa.eu/docs/en_GB/document_library/Scientific_guideline/ 2010/01/WC500070039.pdf.

U.S. Food & Drug Administration (1997) Guidance for industry, dissolution testing of immediate release solid oral dosage forms. Accessed 28 Aug 2017. https://www.fda.gov/downloads/drugs/ guidances/ucm070237.pdf.

U.S. Food & Drug Administration (2001) Guidance for industry, statistical approaches to establishing bioequivalence. Accessed 28 Aug 2017. https://www.fda.gov/downloads/Drugs/GuidanceComplianceRegulatoryInformation/Guidances/UCM070244.pdf.

U.S. Food & Drug Administration (2011) Guidance on zolpidem. Accessed 28 Aug 2017. https://www.fda.gov/downloads/Drugs/GuidanceComplianceRegulatoryInformation/Guidances/UCM175029.pdf.

U.S. Food & Drug Administration (2013) Guidance for industry, bioequivalence studies with pharmacokinetic endpoints for drugs submitted under an ANDA. Accessed 28 Aug 2017. https://www.fda.gov/downloads/Drugs/GuidanceComplianceRegulatoryInformation/Guidances/UCM377465.pdf.

U.S. Food & Drug Administration (2014) Guidance for industry, bioavailability and bioequivalence studies submitted in NDAs or INDs – general considerations. Accessed 28 Aug 2017. https://www.fda.gov/downloads/Drugs/GuidanceComplianceRegulatoryInformation/Guidances/UCM389370.pdf.

U.S. Food & Drug Administration (2015a) Drug development and review definitions. Accessed 28 Aug 2017. https://www.fda.gov/drugs/developmentapprovalprocess/howdrugsaredeveloped andapproved/approvalapplications/investigationalnewdrugindapplication/ucm176522.htm.

U.S. Food & Drug Administration (2015b) Guidance for industry, waiver of in vivo bioavailability and bioequivalence studies for immediate-release solid oral dosage forms based on a biopharmaceutics classification system. Accessed 28 Aug 2017. https://www.fda.gov/downloads/Drugs/GuidanceComplianceRegulatoryInformation/Guidances/UCM070246.pdf.

U.S. Food & Drug Administration (2017a) Approved drug products with therapeutic equivalence evaluations. 37th edn. Accessed 28 Aug 2017. https://www.fda.gov/downloads/Drugs/DevelopmentApprovalProcess/UCM071436.pdf.

U.S. Food & Drug Administration (2017b) What is the approval process for generic drugs? Accessed 28 Aug 2017. https://www.fda.gov/Drugs/ResourcesForYou/Consumers/BuyingUsingMedicine Safely/UnderstandingGenericDrugs/ucm506040.htm.

U.S. Food & Drug Administration (2017c) Generic drugs: questions & answers. Accessed 28 Aug 2017. https://www.fda.gov/drugs/resourcesforyou/consumers/questionsanswers/ucm100100.htm.

Absorption, Distribution, Metabolism, and Excretion of Biopharmaceutical Drug Products

11

Molly Graveno and Robert E. Stratford, Jr.

11.1 Introduction

▶ **Definition** The term "biopharmaceuticals" refers to a class of pharmaceutical products, also frequently referred to as "biotherapeutics," that are derived from living sources. These sources can be of microbial, fungal, plant, animal, or human origin. Peptide and protein drugs are the most common type of product in this drug class; accordingly, emphasis will be on the absorption, distribution, metabolism, and excretion (ADME) properties of these agents. However, biopharmaceuticals can also be comprised of sugars (heparin being a notable example), nucleic acids (DNA or RNA), or even whole cells (cell- or gene-based therapy). The ADME properties of these nonprotein biopharmaceuticals will be briefly covered.

▶   **Important** Recombinant DNA (rDNA) and hybridoma technologies are the most common means for manufacture of peptide and protein therapeutics. rDNA technology involves the insertion of the gene coding the desired peptide or protein into a host cell, typically of bacterial or animal origin, which is then cultured on a large scale to produce the drug substance. Human insulin was the first biopharmaceutical to be sold, marketed in 1982 by Eli Lilly and Company for the treatment of diabetes mellitus (Goeddel et al. 1979); it is produced using rDNA technology. Monoclonal antibody drug production commonly uses hybridomas, cells derived from the fusion of a lymphocyte B cell (that produces the

M. Graveno (✉)
Duquesne University School of Pharmacy (MG), Pittsburgh, PA, USA
e-mail: gravenom@duq.edu

R. E. Stratford, Jr.
Clinical Pharmacology, Indiana University School of Medicine (RES), Indianapolis, IN, USA
e-mail: robstrat@iu.edu

© Springer Nature Switzerland AG 2018   241
A. Talevi, P. A. M. Quiroga (eds.), *ADME Processes in Pharmaceutical Sciences*,
https://doi.org/10.1007/978-3-319-99593-9_11

antibody) with a mouse myeloma cell. The resulting hybridoma is stable for large-scale cell culture and production of the monoclonal antibody. More recently, monoclonal antibody production via rDNA technology has also been used.

Since the launch of human insulin, numerous peptide and protein drugs are sold worldwide. As of the end of 2016, there were 380 marketed peptide and protein drugs in the United States (Sedelmeier and Sedelmeier 2017). Their impact on the pharmaceutical market has substantially grown over these years such that six of the top 10 ranked drugs in terms of sales in the United States are biopharmaceuticals (Usmani et al. 2017), including the number one selling drug, Humira® (adalimumab), a monoclonal antibody used for the treatment of various autoimmune disorders. Tables 11.1, 11.2, 11.3, and 11.4 provide a summary of several top-selling peptide and protein drugs routinely used in various therapeutic classes. Also included are some of the key ADME properties of these drugs.

Because peptide and protein biopharmaceuticals are comprised of naturally occurring amino acids linked by amide bonds, they are relatively safe and well tolerated compared to traditional small nonprotein drugs. However, owing to their size, which can range from 10 to more than 1000 amino acids, and therefore molecular weights ranging from roughly 1500 to 150,000 Da, their ADME properties are quite different from traditional small drug molecules. Given the expanding use of peptide and protein drugs in pharmacotherapy, understanding their ADME properties is important for professionals involved in the continuum spanning the manufacture to ultimate point-of-care delivery of these agents to the patient.

This chapter will first address absorption of the peptide and protein class of biopharmaceuticals and then cover their distribution, metabolism, and excretion characteristics. In the following, the important role that absorption rate modification can play in controlling plasma concentration versus time courses of these agents will be described. Specifically, alteration of insulin structure to modify its solution structure and solubility to achieve either rapid onset of action or basal control of serum glucose in the treatment of diabetes mellitus following subcutaneous administration will be discussed. The student will learn the unique drug disposition properties of protein biotherapeutics from this comprehensive presentation of the ADME properties of this major and growing drug class. To support this goal, the impact of size and susceptibility to degradation by proteinases on their rate and extent of absorption, distribution, and excretion will be covered.

The chapter will conclude with a discussion of the unique ADME properties of heparin and its low-molecular-weight derivatives and the challenges associated with characterizing the drug disposition of two novel gene-based drugs (introduced to the US market in the second half of 2017): tisagenlecleucel (Kymriah™), for the treatment of B-cell precursor acute lymphoblastic leukemia, and axicabtagene ciloleucel (Yescarta™), for the treatment of large B-cell lymphoma.

Protein biopharmaceuticals are also used as vaccines and in a diagnostic capacity; however, these two applications will not be discussed because of the minimal impact

Table 11.1 Examples of peptide and protein drugs used in cancer treatment

Generic name	Molecular weight (Da)	Route of administration	Pharmacokinetic properties			
			Bioavailability (%)	Volume of distribution	Clearance	Half-life
Bevacizumab	149,000	Intravenous	100	0.05 L/kg	0.2 L/day	20 days
Rituximab	143,860	Intravenous	100	3.1 L	0.34 L/day	3–5 weeks
Trastuzumab	145,532	Intravenous	100	0.04 L/kg	Not available	28.5 days
Filgrastim	18,800	Intravenous or subcutaneous	SC is 60–70	0.15 L/kg	0.5–0.7 mL/min/kg	3.5 h

1. Bevacizumab. Lexi-Drugs. Lexi-Comp Online. Hudson, OH: Lexi-Comp; 2017. Accessed December 17, 2017
2. Bevacizumab. DrugBank. https://www.drugbank.ca/drugs/DB00112. Published December 10, 2017. Accessed December 17, 2017
3. Rituximab. Lexi-Drugs. Lexi-Comp Online. Hudson, OH: Lexi-Comp; 2017. Accessed December 17, 2017
4. Rituximab. DrugBank. https://www.drugbank.ca/drugs/DB00073. Published December 10, 2017. Accessed December 17, 2017
5. Trastuzumab. Lexi-Drugs. Lexi-Comp Online. Hudson, OH: Lexi-Comp; 2017. Accessed December 17, 2017
6. Trastuzumab. DrugBank. https://www.drugbank.ca/drugs/DB00072. Published December 10, 2017. Accessed December 17, 2017
7. Filgrastim. Lexi-Drugs. Lexi-Comp Online. Hudson, OH: Lexi-Comp; 2017. Accessed December 17, 2017
8. Filgrastim. DrugBank. https://www.drugbank.ca/drugs/DB00099. Published December 10, 2017. Accessed December 17, 2017

244 M. Graveno and R. E. Stratford Jr.

Table 11.2 Examples of peptide and protein drugs used in the treatment of inflammatory diseases

Generic name	Molecular weight (Da)	Route of administration	Pharmacokinetic properties			
			Bioavailability (%)	Volume of distribution	Clearance	Half-life
Adalimumab	144,190	Subcutaneous	64	5–6 L	12 mL/h	10–20 days
Golimumab	146,943	Subcutaneous	53	0.15 L/kg	4.9–6.7 mL/kg/day	2 weeks
Certolizumab pegol	91,000	Subcutaneous	76–86	6–8 L	9.21–21 mL/h	14 days
Infliximab	144,190	Intravenous	100	3–6 L	5–20 mL/h	7–12 days
Belimumab	147,000	Intravenous or subcutaneous	SC is 74	5 L	215 mL/day	18–19 days
Abatacept	92,300	Intravenous or subcutaneous	SC is 78.6	0.07 L/kg	0.22–0.4 mL/h/kg	17 days
Etanercept	51,235	Subcutaneous	60	Not available	160 mL/h	102 h

1. Adalimumab. Lexi-Drugs. Lexi-Comp Online. Hudson, OH: Lexi-Comp; 2017. Accessed December 6, 2017
2. Adalimumab. Clinical Pharmacology [database online]. Tampa, FL: Gold Standard; 2017. http://www.clinicalpharmacology.com. Accessed December 6, 2017
3. Golimumab. Lexi-Drugs. Lexi-Comp Online. Hudson, OH: Lexi-Comp; 2017. Accessed December 17, 2017
4. Golimumab. DrugBank. https://www.drugbank.ca/drugs/DB06674. Published December 10, 2017. Accessed December 17, 2017
5. Certolizumab Pegol. Lexi-Drugs. Lexi-Comp Online. Hudson, OH: Lexi-Comp; 2017. Accessed December 17, 2017
6. Certolizumab Pegol. DrugBank. https://www.drugbank.ca/drugs/DB08904. Published December 10, 2017. Accessed December 17, 2017
7. Infliximab. Lexi-Drugs. Lexi-Comp Online. Hudson, OH: Lexi-Comp; 2017. Accessed December 17, 2017
8. Infliximab. DrugBank. https://www.drugbank.ca/drugs/DB00065. Published December 10, 2017. Accessed December 17, 2017
9. Belimumab. Lexi-Drugs. Lexi-Comp Online. Hudson, OH: Lexi-Comp; 2017. Accessed December 17, 2017
10. Belimumab. DrugBank. https://www.drugbank.ca/drugs/DB08879. Published December 10, 2017. Accessed December 17, 2017
11. Abatacept. Lexi-Drugs. Lexi-Comp Online. Hudson, OH: Lexi-Comp; 2017. Accessed December 17, 2017
12. Abatacept. DrugBank. https://www.drugbank.ca/drugs/DB01281. Published December 10, 2017. Accessed December 17, 2017
13. Efficacy and Safety Study of abatacept to Treat Lupus Nephritis - Full Text View. ClinicalTrials.gov. https://clinicaltrials.gov/ct2/show/NCT01714817. Published August 17, 2017. Accessed December 17, 2017
14. Etanercept. Lexi-Drugs. Lexi-Comp Online. Hudson, OH: Lexi-Comp; 2017. Accessed December 17, 2017
15. Etanercept. DrugBank. https://www.drugbank.ca/drugs/DB00005. Published December 10, 2017. Accessed December 17, 2017

Table 11.3 Examples of peptide and protein drugs used in hematology

Generic name	Molecular weight (Da)	Route of administration	Pharmacokinetic properties			
			Bioavailability (%)	Volume of distribution	Clearance	Half-life of effect
Bivalirudin (anticoagulant)	2180	Intravenous	100	0.2 L/kg	3.4 mL/min/kg, reduced if there is kidney impairment	25 min, 57 min if creatinine clearance <30
Thrombolytic agents						
Alteplase	59,042	Intravenous	100	Approximates plasma volume	572 mL/min, primarily metabolized by the liver	5 min
Tenecteplase	58,591	Intravenous	100	Approximates plasma volume	99–119 mL/min	20–24 min
Anistreplase	59,042	Intravenous	100	Approximates plasma volume	Not available	90 min
Antianemic agents						
Darbepoetin alfa	18,296	Intravenous or subcutaneous	SC is 37	0.05 L/kg	1.6 mL/h/kg	70–74 h
Epoetin alfa	18,396	Intravenous or subcutaneous	SC is 36	9 L	9.2–23.6 mL/h/kg	16–67 h in cancer, 4–13 h in chronic kidney disease

1. Bivalrudin. Lexi-Drugs. Lexi-Comp Online. Hudson, OH: Lexi-Comp; 2017. Accessed December 17, 2017
2. Bivalrudin. DrugBank. https://www.drugbank.ca/drugs/DB00006. Published December 10, 2017. Accessed December 17, 2017
3. Alteplase. Lexi-Drugs. Lexi-Comp Online. Hudson, OH: Lexi-Comp; 2017. Accessed December 17, 2017
4. Alteplase. DrugBank. https://www.drugbank.ca/drugs/DB00009. Published December 10, 2017. Accessed December 17, 2017
5. Tenecteplase. Lexi-Drugs. Lexi-Comp Online. Hudson, OH: Lexi-Comp; 2017. Accessed December 17, 2017
6. Tenecteplase. DrugBank. https://www.drugbank.ca/drugs/DB00029. Published December 10, 2017. Accessed December 17, 2017
7. Anistreplase. DrugBank. https://www.drugbank.ca/drugs/DB00029. Published on November 06, 2017. Accessed January 12, 2018
8. Darbepoetin alpha. Lexi-Drugs. Lexi-Comp Online. Hudson, OH: Lexi-Comp; 2017. Accessed December 17, 2017
9. Darbepoetin alpha. DrugBank. https://www.drugbank.ca/drugs/DB00012. Published December 10, 2017. Accessed December 17, 2017
10. Epoetin alpha. Lexi-Drugs. Lexi-Comp Online. Hudson, OH: Lexi-Comp; 2017. Accessed December 17, 2017
11. Erythropoietin. DrugBank. https://www.drugbank.ca/drugs/DB00016. Published December 10, 2017. Accessed December 17, 2017

Table 11.4 Peptide and protein drugs used in the treatment of miscellaneous conditions

Generic name (indication)	Molecular weight (Da)	Route of administration	Pharmacokinetic properties			
			Bioavailability (%)	Volume of distribution	Clearance	Half-life
Daptomycin (skin and soft tissue infections)	1621	Intravenous	100	0.1 L/kg	8.3–9 mL/h/kg	8–9 h
Somatotropin (growth hormone disorders and prevention of wasting in HIV-AIDS)	22,129	Subcutaneous	81	Not available	2.3 mL/min/kg	Not available
Teriparatide (osteoporosis)	4118	Subcutaneous	95	0.12 L/kg	62–94 L/h	1 h
Leuprolide (precocious puberty, endometriosis, prostate cancer, uterine leiomyomata)	1209	Subcutaneous	94	27 L	8.34 L/h	3 h

1. Afibercept. Lexi-Drugs. Lexi-Comp Online. Hudson, OH: Lexi-Comp; 2017. Accessed December 17, 2017
2. Afibercept. DrugBank. https://www.drugbank.ca/drugs/DB08885. Published December 10, 2017. Accessed December 17, 2017
3. Daptomycin. Lexi-Drugs. Lexi-Comp Online. Hudson, OH: Lexi-Comp; 2017. Accessed December 17, 2017
4. Daptomycin. DrugBank. https://www.drugbank.ca/drugs/DB00080. Published December 10, 2017. Accessed December 17, 2017
5. Somatotropin. DrugBank. https://www.drugbank.ca/drugs/DB00052. Published January 07, 2018. Accessed January 12, 2018
6. Teriparatide. Lexi-Drugs. Lexi-Comp Online. Hudson, OH: Lexi-Comp; 2017. Accessed December 17, 2017
7. Teriparatide. DrugBank. https://www.drugbank.ca/drugs/DB06285. Published December 10, 2017. Accessed December 17, 2017
8. Leuprolide. Lexi-Drugs. Lexi-Comp Online. Hudson, OH: Lexi-Comp; 2017. Accessed December 17, 2017
9. Leuprolide. DrugBank. https://www.drugbank.ca/drugs/DB00007. Published December 10, 2017. Accessed December 17, 2017
10. Anistreplase. DrugBank. https://www.drugbank.ca/drugs/DB00029. Published November 06, 2017. Accessed January 8, 2018

of ADME properties on their safety and efficacy. The reader is referred to Leader et al. for a discussion and examples of these two additional uses (Leader et al. 2008).

11.2   Absorption of Protein Drugs

▶ **Definition** As a general guideline, the term "peptide" refers to proteins with approximately 50 or fewer amino acids. Insulin, having 51 amino acids (30 in the A-chain and 21 in the B-chain), can thus be referred to as a "peptide" or a "protein." For the sake of clarity in presentation, "protein" will be used in this chapter to describe the ADME properties of this class of drugs.

Largely because of its convenience, the oral route of administration is the most common means used to achieve reliable systemic absorption of small drug molecules. Unfortunately, this convenience does not cross over to protein drugs. Inability to achieve reliable oral bioavailability of protein drugs is due to (1) their degradation in the gastrointestinal tract by the same proteinases responsible for the digestion of protein components of meals and (2) their size (even if they were resistant to proteolytic degradation), being too large to achieve appreciable diffusion across the protective epithelial lining of the small and large intestine. The immuno-suppressant drug, cyclosporine, is a notable exception. This 11-amino acid peptide (molecular weight = 1202.63) has an oral bioavailability >30% when administered orally in a formulation that forms a microemulsion when it is exposed to water (Drewe et al. 1992). Its sufficiently extensive and reliable oral absorption is attributed to its cyclized sequence (as opposed to linear sequence) and that some of its amino acids are modified such that their peptide bonds are resistant to gastrointestinal proteases.

Alternative non-oral and non-parenteral administration of protein drugs is also limited, there being a few noteworthy instances: intranasal administration of salmon calcitonin for the treatment of postmenopausal osteoporosis, intranasal administration of desmopressin for antidiuretic hormone replacement therapy, and oral inhalation administration of human insulin for the treatment of diabetes mellitus. Salmon calcitonin has a bioavailability <10% compared to intramuscular administration, but this extent of absorption is considered sufficiently reproducible to support its clinical use by this route (Lee et al. 1994). Desmopressin is a modified form of the natural arginine vasopressin peptide. Replacing L-arginine with D-arginine makes it resistant to hydrolysis. With respect to insulin, the Food and Drug Administration (FDA) approved Afrezza® for the treatment of diabetes in June 2014. Relative to short-acting insulin analogs administered subcutaneously, inhalation of the native insulin used in this drug product has a faster onset of action by this route (Boss et al. 2012) and is considered an ultrarapid-acting form of insulin (Heinemann and Muchmore 2012).

▶ **Important** Intravenous (IV) and subcutaneous (SC) administrations are
the two most common routes to achieve systemic therapy with protein
drugs. According to a published database of FDA-approved protein
drugs (Usmani et al. 2017), 44% and 33% of such drugs are administered
by the IV and SC routes, respectively. The intramuscular (IM) route
accounts for only 14%, with approximately half of the agents
administered by this route used as vaccines, which are not for systemic
absorption. From the perspective of patient convenience, the SC route is
preferred relative to IV administration. However, achieving consistent
bioavailability is not always possible for some proteins, which can be
problematic for drugs with narrow therapeutic indexes. Frequency of
administration is another factor influencing IV versus SC administration,
in which the SC route is preferred for drugs with a half-life requiring daily
or more frequent administration.

As shown in Tables 11.1, 11.2, 11.3, and 11.4, bioavailability following SC
administration of the various examples of protein drugs given varies from approximately 30% to more than 90%. The range observed is representative of this class of
drugs given by this route (Usmani et al. 2017). Importantly, there is no simple
relationship between bioavailability and protein molecular weight. This may seem
surprising given the importance of molecular size in determining drug transport
across cell barriers, including the capillary endothelial barrier between SC interstitial
fluid and the systemic circulation. However, absorption through lymphatic
capillaries, which are more permeable than blood capillaries (Kagan 2014), is a
parallel route that exists in SC tissue. Thus, the fraction of drug absorbed into the
lymph increases with protein size, with drugs smaller than 5000 Da thought to be
primarily absorbed directly into the blood (Supersaxo et al. 1990). Figure 11.1
provides a schematic representation of the two pathways for absorption of protein
drugs following SC dosing.

Absorption of proteins into lymph leads to systemic absorption via the lymphatic
circulation due to eventual emptying of collected lymph into the bloodstream
through the right lymphatic duct (associated with the right subclavian vein) or the
thoracic duct (associated with the left subclavian vein). While absorption across
blood capillaries occurs by passive diffusion between adjacent endothelial cells,
absorption into the lymph capillaries occurs by convective flow of interstitial fluid
(Richter and Jacobsen 2014). Hydrostatic pressure moves plasma water into the
interstitial space; subsequently, this pressure plus osmotic pressure that exists due to
impermeable solute concentration differences across capillary and lymph relative to
interstitial fluid results in vectorial movement of water and dissolved protein drug
with it, into the lymph (Richter and Jacobsen 2014).

Although the extent of absorption does not correlate with molecular weight, the
rate of absorption does fall off as size increases (Wu et al. 2012). This trend is
thought to be the result of the fraction of a dose absorbed into lymph increasing as
drug size increases (Supersaxo et al. 1990). Convective movement of drug in the

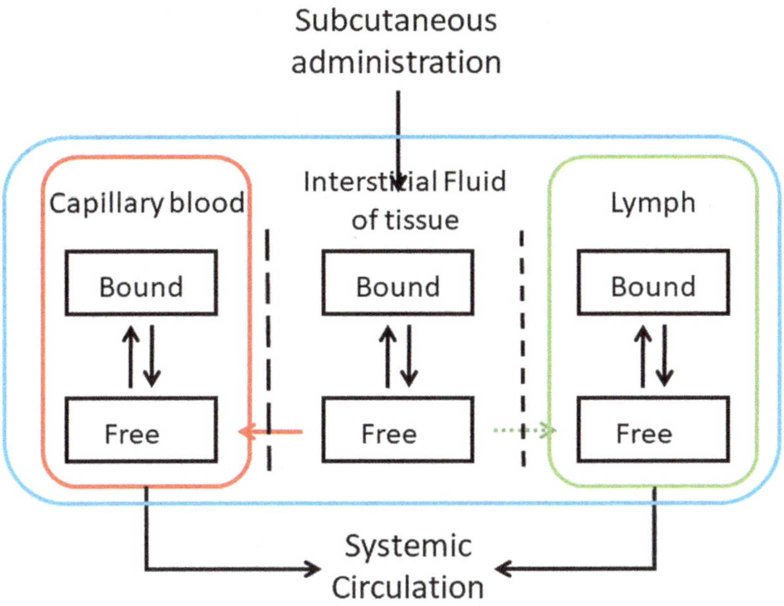

Fig. 11.1 Schematic diagram of subcutaneous region of a tissue (bound by solid blue line) showing routes of protein drug absorption across blood and lymphatic capillary barriers (denoted by dashed lines) with eventual delivery of both pathways to the general systemic circulation. The different arrow types for drug absorption into blood and lymph reflect the different mechanisms of drug absorption across these two barriers, being largely by passive diffusion between cells for blood and convective flow for lymph. The two-way arrows between free form (drug not bound to endogenous carrier proteins) and bound form represent the reversible nature of drug binding and emphasize that only the unbound (free) form can be absorbed across the blood and lymph and capillary barriers

interstitial space is size dependent because of the presence of collagen and glycosaminoglycans in the interstitial space that impede protein drug diffusion and convection (Richter and Jacobsen 2014). Thus, typical peak blood concentrations of monoclonal antibody drugs, which have molecular weights >150,000 Da, can take several days (Richter and Jacobsen 2014). For example, the reported time to maximum plasma concentration (T_{max}), a parameter that depends on rate of absorption, is 5.5 days for adalimumab, which has a molecular weight of 148,000 Da (Daily Med, https://dailymed.nlm.nih.gov/dailymed/). This contrasts with a peak time of the "regular" formulation of human insulin (molecular weight of 5808 Da) of 1.5–3.5 h (Daily Med, https://dailymed.nlm.nih.gov/dailymed/).

In addition to such factors as molecular weight (as discussed above), or the important role that physicochemical property modification on absorption rate can have (as will be exemplified for insulin), several other factors can have an impact on the rate and extent of absorption of protein drugs following SC administration (Richter et al. 2012; Richter and Jacobsen 2014). Electrostatic forces can influence the rate of absorption (Reddy et al. 2006). Subcutaneous space bears a net negative charge; thus proteins that are net positively charged at physiologic pH (determined by their isoelectric point, pI, relative to the pH of the SC fluid, 7.4) will electrostatically interact with SC matrix to slow their movement in this space (Mach et al. 2011). The site of administration can also influence the rate of absorption. Regional differences in blood and lymph flow are the attributed cause of location differences (Porter and Charman 2000). For example, absorption of insulin (Ter Braak et al. 1996) and human growth hormone (Beshyah et al. 1991) was faster following injection in the abdomen relative to the thigh. Host factors, such as age, disease, heat, massage, and movement, can also influence both the rate and extent of absorption (Richter et al. 2012; Richter and Jacobsen 2014). Catabolism at the site of administration can occur for both antibody and non-antibody drugs (Wang et al. 2008; Richter et al. 2012). For non-antibody drugs, injection site metabolism can occur in the interstitial fluid and/or the lymph that drains the site (Richter and Jacobsen 2014). For monoclonal antibodies, the presence of neonatal Fc receptor (FcRn) in cells at the injection site influences bioavailability (Richter and Jacobsen 2014). The influence of this receptor with respect to both the distribution and elimination of monoclonal antibody drugs will be described in more detail later. In brief, binding of an antibody to FcRn expressed in cells at the site of administration can circumvent local catabolism resulting in improved bioavailability (Richter and Jacobsen 2014; Liu 2018).

Several monoclonal antibodies in cancer chemotherapy require much larger doses and volumes than practical for SC administration (Jackisch et al. 2014; Salar et al. 2014; Shpilberg and Jackisch 2013). For example, a typical maintenance dose of Herceptin® (trastuzumab) for use as a single agent in the treatment of breast cancer is over 300 mg once a week. With a solubility <100 mg/mL, which is typical for these drugs (Liu 2018), over 3 mL is required. This volume is too large to be administered subcutaneously, as this route has a limit of 2 mL in order to avoid patient discomfort (Hunter 2008); thus, the drug must be administered by IV infusion on a weekly basis. Clearly, SC administration would be more practical for patients, and less expensive, as they could self-administer from home. In order to circumvent this volume limitation, a novel approach involving co-administration of the monoclonal antibody with a recombinant source of hyaluronidase has been developed. This enzyme transiently degrades the glycosaminoglycan hyaluronan in the SC space to allow an expansion of void volume. The result is that up to 10 mL can be administered comfortably. The approach has been applied to trastuzumab, tocilizumab, and rituximab, all of which have been approved for sale in Europe and the latter two for sale in the United States for tumor therapy (Cui et al. 2017). Expansion to indications beyond cancer has occurred recently, suggesting this approach will gain broader application for this subclass of protein drugs, namely, the FDA-approved tocilizumab for the treatment of rheumatoid arthritis in 2013 and HyQvia immune globulin for the treatment of primary immunodeficiency in 2016.

11.3   Distribution of Protein Drugs

Distribution of this class of drugs is limited largely to extracellular water (plasma plus interstitial space of tissues and organs). This is because of the relatively large size and hydrophilic nature of protein drugs, both of which make it difficult for them to cross the vascular endothelial cells that comprise the capillaries and subsequently move from interstitial water to intracellular water by crossing the membranes of cells that comprise an organ or tissue. Figure 11.2 provides a schematic summary of the principal water compartments of the body, shown at the level of an individual tissue or organ.

Typical distribution volumes of protein drugs range from that of the plasma (0.04 L/kg or approximately 3 L in a 70 kg person) to total extracellular water (0.23 L/kg or approximately 15 L in a 70 kg person). The example drugs listed in Tables 11.1, 11.2, 11.3, and 11.4 fall within this range, leuprolide being the only possible exception, having an average volume of distribution just beyond the upper limit. As with the process of absorption, cyclosporine is a clear exception in the protein drug class, having a volume of distribution ranging from 3 to 5 L/kg. This

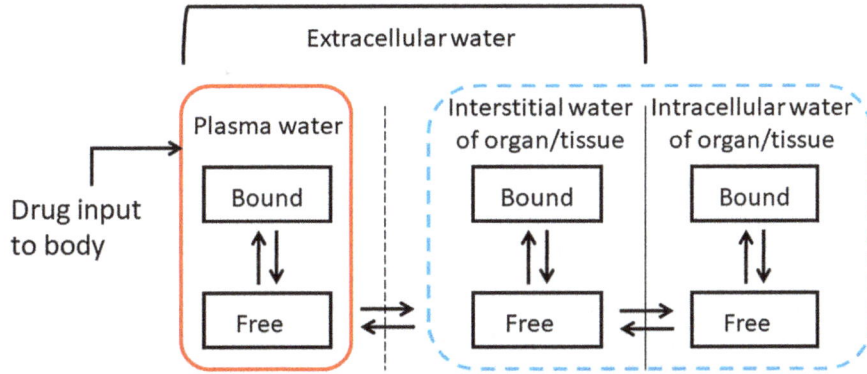

Water Volume	Plasma	Interstitial	Intracellular
L/kg	0.04	0.17	0.40
L (70 kg body weight)	3	12	28

Fig. 11.2 Schematic diagram representing drug distribution at the level of an individual tissue or organ following drug absorption into the systemic circulation ("drug input"). The three principal water compartments of the body are plasma, interstitial water, and cell water. Extracellular water is the sum of plasma and interstitial water. The blue dashed line denotes a given tissue or organ. The vertical dashed line represents the capillary barrier that separates plasma water from interstitial water, and the solid vertical line represents the cell membrane barrier separating interstitial water from intracellular water. Water volumes associated with the three compartments are also shown

outlier is attributed to the lipophilic nature of the drug, as measured by its log octanol/water partition coefficient of approximately 3, and that it bears no net charge at pH 7.4. Several N-methylated amino acids, which increase its lipophilicity, undoubtedly contribute to its unique properties.

Figure 11.2 also depicts the reversible nature of drug distribution and that only the unbound form of the drug can cross capillary and cellular barriers. While small drug molecules commonly achieve distributional equilibrium of unbound drug between the three water compartments in minutes, such is not the case for proteins. As stated, their size, which is the cause of their constrained extent of distribution, also prolongs their rate of distribution from plasma to interstitial water. In general, movement of proteins across capillary barriers is slower than their removal from tissue by lymph drainage and/or local tissue metabolism (catabolism); therefore, unbound concentrations in interstitial fluid are commonly lower than plasma and with perhaps no attainment of distributional equilibrium. Direct measurements and physiological-based pharmacokinetic modeling estimates of tissue-to-plasma ratios of monoclonal antibodies place these in the range of <0.1 to roughly 0.5 (Shah and Betts 2013; Lobo et al. 2004). These ratios roughly correlate with capillary structure. Tissues such as the brain, muscle, lung, and skin that possess continuous non-fenestrated capillaries had the lowest ratios, whereas tissues such as renal glomeruli and intestine, possessing fenestrated continuous capillaries, were intermediate, and liver, spleen, and bone marrow, tissues having discontinuous endothelium, had the highest ratios.

Unlike most small drugs that display linear pharmacokinetics, which manifests as constant PK parameters (clearance, volume of distribution, and half-life) over several orders of magnitude of exposure, observation of concentration-dependent (nonlinear) pharmacokinetics is common for many monoclonal antibody drugs. For several products in this subclass of protein drugs, increasing dose reduces their volume of distribution. Saturation of binding to their target antigen is the attributed cause of this nonlinearity. Target-mediated drug distribution (TMDD) is a phenomenon common to monoclonal antibody drugs (Mager 2006); it is used to describe this saturable distribution, as well as the saturable elimination of these drugs (to be discussed later). With respect to distribution, at lower doses high-affinity target binding maximizes distribution outside of the plasma; however, as dose is increased, target binding can saturate. This saturation manifests as a dose-dependent decline in distribution volume and can eventually reach a plateau indicative of complete saturation of antigen binding. The distribution volumes given in Tables 11.1, 11.2, 11.3, and 11.4 for monoclonal antibodies reflect this minimal value, which commonly occurs at clinically relevant doses of these drugs.

11.4   Metabolism and Excretion of Protein Drugs

Protein drugs are removed from the body by extrarenal hydrolysis of their amide bonds (metabolism) and/or glomerular filtration followed by hydrolysis in the proximal tubule (excretion). Collectively, these two pathways account for the elimination of protein drugs. Elimination half-life of peptides and smaller proteins (2000–30,000 Da) can be limited to a few minutes if they are comprised of L-amino acids, since this would make them susceptible to both elimination pathways. This is the case for endogenous peptide hormones, such as GLP-1, gonadotropin-releasing hormone, somatostatin, and vasopressin, to name a few, all of which have elimination half-lives ≤30 min. The presence of protein hydrolases (peptidases and proteinases) in blood can even result in systemic clearances of endogenous peptide hormones that are faster than cardiac output. This rapid clearance provides exquisite control of pharmacologic effects.

Several strategies have been used to prolong the half-life of these endogenous hormones in order to make them viable for use in replacement or augmentation therapy (Di 2015). Modification of absorption rate is an approach that will be described in detail later. Briefly, this approach transfers the rate-limiting step controlling the time course of drug in the plasma from elimination to absorption. Aside from this approach, strategies to reduce elimination have been pursued. One approach involves blocking the N- or C-terminus to hydrolysis by aminopeptidases or carboxypeptidases, respectively. An example is tesamorelin, which is an analog of growth hormone-releasing hormone (GHRH) in which the N-terminal tyrosine is protected from aminopeptidase-mediated hydrolysis. An increase in half-life from <10 min to >30 min is the outcome of this approach. Replacement of L-amino acids with D-amino acids is another strategy that reduces susceptibility to hydrolysis. Desmopressin is an example of this approach. Replacement of L-arginine in the eighth position of the endogenous undecapeptide (vasopressin) with the D-amino acid increases the half-life from 10–35 min to 3.7 h (Agerso et al. 2004).

Use of modified amino acids and cyclization are two additional approaches that confer resistance to hydrolysis. These were previously mentioned as responsible for the unique absorption and distribution properties of cyclosporine, but they also contribute to its resistance to hydrolysis post-absorption such that the drug has an elimination half-life of approximately 8.4 h. Selective substitutions of amino acids in the native hormone is a related strategy used to confer resistance to hydrolysis. For example, endogenous GLP-1 has a half-life of approximately 2 min (Drucker and Nauck 2006); this is due to removal of the N-terminal dipeptide by dipeptidyl peptidase-4. Selective substitution of the alanine at position 2 with a glycine (such as in exenatide and lixisenatide) or α-amino butyric acid (semaglutide) dramatically increases half-life, as shown in Table 11.5. Figure 11.3 provides a comparison of the amino sequences of these peptides relative to endogenous GLP-1.

Conjugation to polymers can also slow elimination. Attachment of C-16 or C-18 fatty acids via acylation at the lysine position 26 of GLP-1 (liraglutide and semaglutide, respectively) promotes reversible binding to albumin. This plasma protein binding reduces the glomerular filtration of these peptide-fatty acid constructs, as protein-bound drug is too large for glomerular filtration. Conjugation with polyethylene glycol (PEG) has also been used to reduce glomerular filtration by converting a protein susceptible to filtration to one that is poorly filtered because of size (>60,000 Da). Certolizumab pegol (Table 11.2) is an example of such a drug. In this drug, a humanized antibody Fab fragment specific for tumor necrosis factor

Table 11.5 Glucagon-like peptide-1 (GLP-1) receptor agonist analog drug products

Generic name	Description	T_{max}	Bioavailability (%)	Half-life
Albiglutide	Analog that is dipeptidyl peptidase-4 resistant. A dimer of the analog is attached to human albumin	3–5 days	Not known	4–7 days
Dulaglutide	Analog that is dipeptidyl peptidase-4 resistant and covalently linked to the Fc heavy chain of human IgF4	1–3 days	Not known	5 days
Exenatide	Analog found in the saliva of the Gila monster lizard	2 h	Not known	2–3 h
Liraglutide	Analog in which palmitic acid is attached through a glutamic acid spacer on a lysine residue at position 26	8–12 h	55	13 h
Lixisenatide	Analog of GLP-1	1–3.5 h	Not known	3 h
Semaglutide	Analog in which stearic acid is attached through a spacer on lysine residue at position 26. Analog is also resistant to dipeptidyl peptidase-4	1–3 days	89	1 week

1. Albiglutide. DrugBank. https://www.drugbank.ca/drugs/DB09043. Published January 7, 2018. Accessed January 8, 2018
2. Dulaglutide. DrugBank. https://www.drugbank.ca/drugs/DB09045. Published January 7, 2018. Accessed January 8, 2018
3. Exenatide. DrugBank. https://www.drugbank.ca/drugs/DB01276. Published January 7, 2018. Accessed January 8, 2018
4. Liraglutide. DrugBank. https://www.drugbank.ca/drugs/DB06655. Published January 7, 2018. Accessed January 8, 2018
5. Lixisenatide. DrugBank. https://www.drugbank.ca/drugs/DB09265. Published January 7, 2018. Accessed January 8, 2018
6. Semaglutide. DrugBank. https://www.drugbank.ca/drugs/DB13928. Published December 22, 2017. Accessed January 8, 2018

alpha (TNF-α) is attached to a PEG polymer of approximately 40,000 Da. The final molecular weight is approximately 91,000 Da, a size that makes the conjugate too large for efficient filtration. Possibly, PEG also blocks access of proteinases to the Fab protein, so that both reduced filtration and hydrolysis contribute to the drug's 14-day half-life.

Covalent attachment to macromolecules, such as albumin or the Fc domain of IgG, is yet another strategy used to reduce the clearance of protein drugs. Application of this approach has resulted in GLP-1 agonist constructs that have half-lives of several days. As described in Table 11.5 and shown in Fig. 11.3, albiglutide consists of albumin covalently attached to a dipeptidyl peptidase-4-resistant GLP-1 agonist dimer. In the case of dulaglutide, two GLP-1 dipeptidyl peptidase-4-resistant agonists are attached to the Fc fragment of an IgG. Both conjugation approaches effectively remove glomerular filtration as an elimination pathway. Abatacept and etanercept are two additional examples, whereby attachment of the Fc domain of IgG is used to reduce hydrolysis and practically remove the glomerular filtration

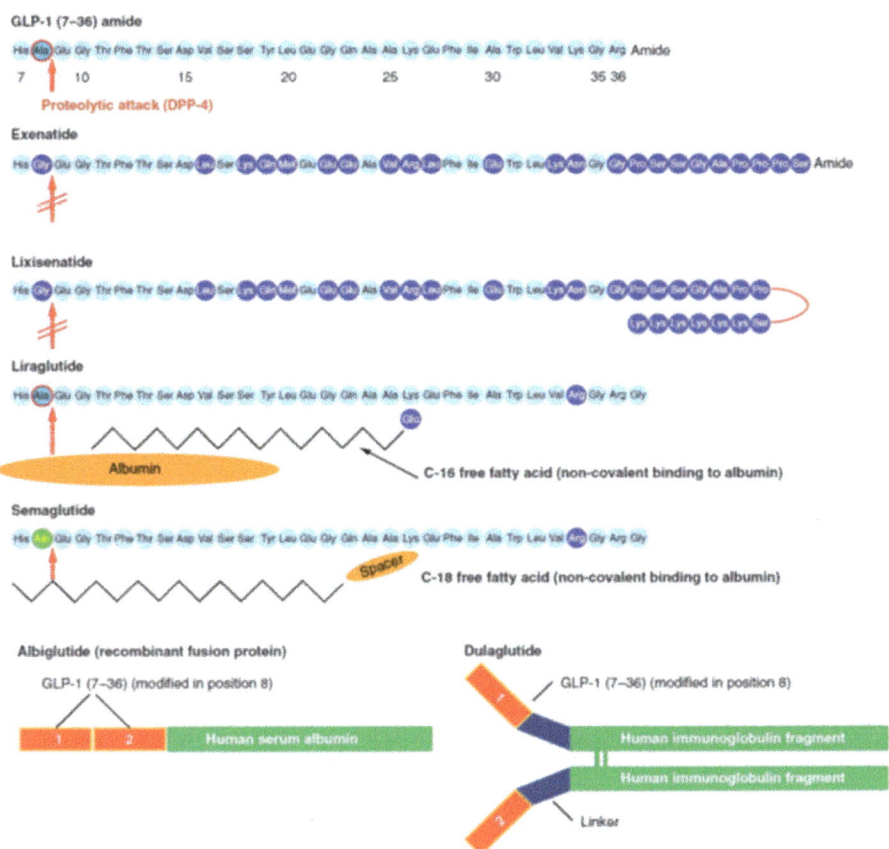

Fig. 11.3 Amino acid sequences of native GLP-1 (7–38) and GLP-1 agonist analogs. (Figure is used by permission from Nauck 2016, published by John Wiley & Sons, Ltd.)

clearance pathway of the pharmacologically active protein. As shown in Table 11.2, these so-called "fusion proteins" have half-lives of several days. Eventual hydrolysis is the clearance mechanism for elimination of these examples of proteins too large to be filtered by the kidney. The location of this hydrolysis is intracellular following endocytosis by cells throughout the body, including their target tissues, and by circulating phagocytic cells (Keizer et al. 2010). Intracellular proteinases, present largely in the lysosomes, degrade such drugs. Research on the elimination of monoclonal antibody drugs, having molecular weights >150,000 Da, has contributed significantly to our understanding of this elimination process. Results from this research, described in the next few paragraphs, have identified a combination of drug-related and physiologic factors that influence elimination of such drugs.

Tables 11.1 and 11.2 provide examples of the long half-life of monoclonal antibody drugs following IV administration, a route that provides a more direct measure of elimination kinetics than SC administration by omitting the absorption

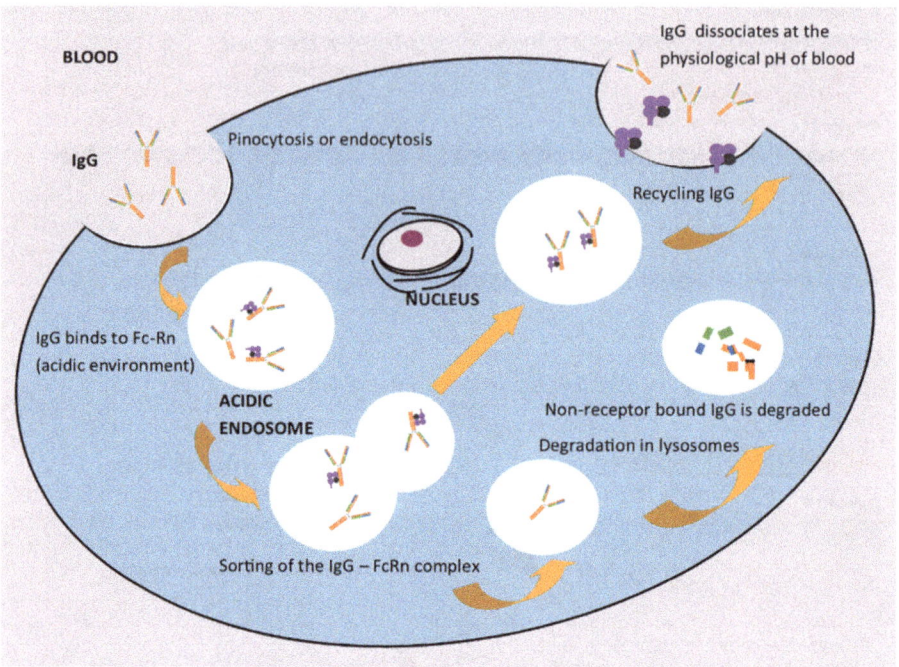

Fig. 11.4 Diagram of monoclonal antibody recycling in blood via the neonatal Fc receptor (FcRn) mechanism present in capillary endothelial cells and monocytes

kinetics. From these we discern that the eventual hydrolytic metabolism of large proteins is slow. An elaborate recycling process is the principal cause of the slow elimination. Figure 11.4 *shows an overview of this process.* Endocytosis of antibodies by capillary endothelial cells and monocytes in the bloodstream is followed by specific binding to the neonatal Fc receptor (FcRn) located in acidic endosomes (Liu 2018). This binding is pH dependent, having its highest affinity around pH 6 and lowest near pH 7.4 (Ghetie and Ward 1997; Roopenian and Akilesh 2007). This pH-dependent antibody binding to the FcRn receptor spares the antibody from degradation in lysosomes. Through an intracellular sorting process, bound antibody is recycled back to the cell surface where the complex encounters the higher extracellular pH of approximately 7.4 to release the antibody back into the bloodstream or into interstitial space. The saturable nature of the FcRn-IgG binding ensures that as IgG concentrations become too large, excess IgG, including the antibody, is sorted to lysosomes for degradation by the proteinases contained in these subcellular structures. As mentioned in the preceding paragraph, this recycling process also applies to albumin and Fc domain fusion protein drugs.

The FcRn receptor binds to the Fc portion of the IgG in a region located between the C_{H2} and C_{H3} domains (Fig. 11.5). Three conserved methionine residues have been found to be important to the affinity of this binding (Liu 2018); in fact, oxidation of these during manufacturing can result in products with reduced half-

Fig. 11.5 Diagram of an IgG antibody showing key structural features. Fab (variable) domain provides specificity for antigen binding, while the Fc (constant) domain is important for antibody disposition in the body. V refers to variable, C refers to constant, H refers to heavy, and L refers to light. Glycosylation of the antibody at the C_{H2} region of the Fc domain is also shown. (Diagram is used by permission from Liu 2018, published by Springer Nature)

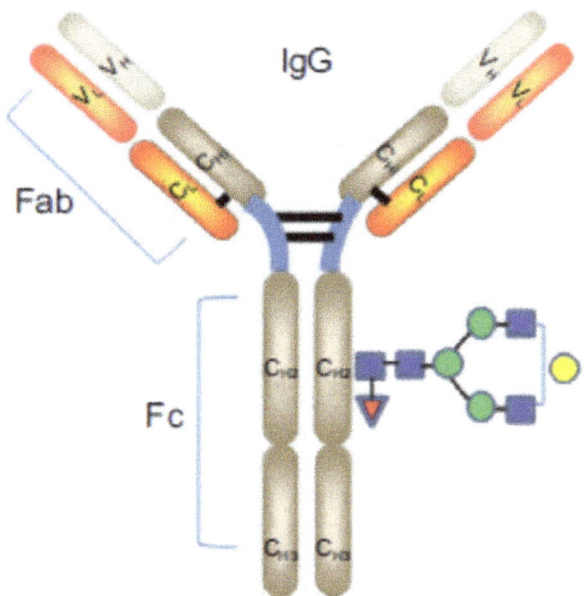

life (Alt et al. 2016). Participation of the Fab region to FcRn binding can also occur and influence half-life. Ustekinumab is used in the treatment of various autoimmune disorders. A variant of this drug, briakinumab, has an almost identical Fc fragment and similar binding at pH 6; however, a positive charge region in the Fv portion of briakinumab renders stronger binding to FcRn at higher pH. This reduces extracellular release of the antibody, subjecting more of it to intracellular metabolism, thus shortening its half-life relative to ustekinumab (Schoch et al. 2015).

A hallmark of monoclonal antibody drugs is their target specificity. This highly selective drug-target binding can result in saturable distribution and elimination, which, as described earlier, has pharmacokinetic implications. Saturation of these two processes is collectively referred to as TMDD (Dirks and Meibohm 2010). With respect to elimination, binding of monoclonal antibodies to their soluble or membrane targets (enzymes, receptors, or transporters) results in endocytosis of the complex with subsequent intracellular catabolism of the drug and target. This process is saturable because of the finite number of targets. The potential for saturation is responsible in part for the saturable, nonlinear kinetics of monoclonal antibody drugs (Ait-Oudhia et al. 2017). The number of targets available, their rate of synthesis and catabolism, and the dose of the monoclonal antibody are factors that influence the contribution of TMDD to the pharmacokinetics of these drugs and the potential for observation of nonlinearity (Ait-Oudhia et al. 2017). Pharmacokinetic models that incorporate TMDD have been developed to describe and predict its contribution to monoclonal antibody exposure and efficacy (Ait-Oudhia et al. 2017).

Development of antibody-drug conjugates (ADCs) is a proactive application of the exceptional target selectivity possible with monoclonal antibodies. The approach involves attachment of a small drug molecule to a monoclonal antibody via linker chemistry that is sufficiently labile in order to release the small drug near its target. The approach can be very useful in cancer chemotherapy to maximize therapeutic benefit by targeting a cytotoxic agent to cancer cells and limiting adverse effects on healthy cells. Two ADCs currently approved by the FDA for treatment of various cancers are brentuximab and ado-trastuzumab emtansine. Several more are in clinical development for a broad range of cancers. Kraynov et al. and Vezina et al. provide excellent overviews of the ADME challenges associated with this novel biopharmaceutical drug class (Kraynov et al. 2016; Vezina et al. 2017). As with nonconjugated monoclonal antibody drugs, TMDD plays an important role in their disposition.

Two additional phenomena that can influence the elimination of monoclonal antibodies are pI (the isoelectric point, that is, the pH at which the antibody bears no net electrical charge) and the production of anti-drug antibodies (ADAs) to the antibody drug. Most monoclonal antibodies have a pI in the range of 7–9 (Bumbaca et al. 2012). Based largely upon nonclinical studies and selective engineering of antibodies to achieve a range of pIs, clearance increased and half-life decreased as pI increased (Igawa et al. 2010). Preferential binding of more alkaline (higher pI) antibodies to the negative charges on cell surfaces, which then presumably enhances endocytosis and intracellular catabolism, was the attributed cause of the trend (Liu 2018). The majority of monoclonal antibody drugs are immunogenic, meaning that they produce an antibody response, as manifested by the production of ADAs (Krishna and Nadler 2016). In fact, generally speaking, for the class of protein biopharmaceuticals, because of their size, complex structure, and variation in structure, the latter arising from posttranslational modifications (e.g., glycosylation), ADAs are quite common (Liu 2018). The effect of ADAs on monoclonal antibodies is to increase their clearance (which leads to a reduced half-life), with a consequent loss of efficacy (Chirmule et al. 2012). The timing and intensity of the ADA response are drug and patient specific (Krishna and Nadler 2016). Importantly, there is no evidence that fully human antibodies are less immunogenic than humanized ones (Liu 2018). In fact, in a recent multinational study, 31.2% of patients treated with adalimumab, a fully human antibody, had detectable ADAs (Moots et al. 2017). These patients also had lower trough adalimumab concentrations that correlated with poorer clinical outcomes. Another recent multicenter study with this drug reported a similar result (Pecoraro et al. 2017). Overall, there remains much research to understand the potential for development of ADAs and the ability to predict their clinical impact on monoclonal antibody therapy and generally protein drug-based therapy (Smith et al. 2016).

As shown in Fig. 11.4, monoclonal antibodies contain a carbohydrate region. Posttranslational addition of the carbohydrate is referred to as glycosylation. The size of the carbohydrate and its monomeric composition vary. Not surprisingly, the impact of glycosylation on the distribution and elimination of monoclonal antibody drugs and Fc-fusion protein drugs varies (Tibbitts et al. 2016). Monoclonal antibodies with terminal mannose residues have more rapid clearance (Goetze et al. 2011; Liu et al. 2011). This is attributed to receptor-mediated endocytosis into hepatocytes and Kupffer cells elicited by binding to the mannose receptor (Mi et al. 2014). Glycosylation seems to have a more substantial impact on the clearance of Fc-fusion proteins, specifically in reducing their clearance relative to the aglycosylated variant (Liu 2015).

> ► **Important** In summary, enzyme-mediated hydrolysis of protein drugs, collectively referred to as metabolism or catabolism, is ultimately responsible for their elimination from the body. What differs among these biotherapeutics is where this hydrolysis occurs: either in the kidney post-glomerular filtration or in tissues throughout the body and in the circulation by phagocytic cells. There is a strong size dependency associated with pathway selection: for drugs <60,000 Da, renal filtration followed by hydrolysis represents a significant component of their elimination, whereas for drugs >60,000 Da, which are too large to be filtered, elimination is largely intracellular by lysosomal proteinases following endocytosis. The latter process is much slower. Indeed, such awareness has led to approaches that strive to switch from the former to the latter. These include acylation with fatty acids in order to enhance albumin binding, or outright conjugation with albumin or other polymers (e.g., PEG), or the Fc fragment (so-called fusion proteins) to increase molecular weight in order to obviate glomerular filtration.

11.5 Physicochemical Property Modification to Alter Drug Pharmacokinetics: Application to Subcutaneous Insulin

As previously indicated, protein drugs with molecular weights <30,000 Da can have short half-lives (<30 min) due to efficient renal elimination (filtration followed by catabolism) and extrarenal catabolism. This short half-life is a disadvantage if sustained levels of the drug are needed, as the drug would have to be administered several times a day. Requirement for administration by injection makes the rapid elimination problem even more cumbersome and runs counter to encouraging patient compliance. Slowing of absorption rate can be a proactive way to address this short half-life challenge. Modification of the physicochemical properties of insulin for the treatment of diabetes mellitus provides an excellent example of accomplishing this. However, before discussing this approach further, it is important first to describe the objective of exogenous insulin therapy in managing diabetes.

To realize the objective of 24-h glucose regulation with exogenous insulin, the goal is to mimic physiologic insulin secretion by the pancreas in healthy individuals. There are two main phases to this release: immediate postprandial release to control meal-derived increase in serum glucose and basal insulin release to achieve between meal, including overnight, glucose homeostasis (Home 2015). Control of serum glucose between meals has been achieved through modifications of insulin structure that slow its rate of absorption. Moreover, modification of insulin structure to actually increase rate of absorption to achieve the postprandial goal has also been accomplished. Table 11.6 categorizes by rate of onset of effect the different approaches currently available to patients having this chronic disease, with the rapid-acting analogs shown first and the basal insulin analogs shown last. The drug substances listed are specific to human insulin, or chemical modifications thereof, and do not include drug product mixtures, such as 70/30 insulin regular/insulin NPH.

Both endogenous and exogenous insulins have a serum half-life of only 4–5 min (Home 2012). Degradation by insulin-degrading enzyme in hepatocytes and filtration by the kidney with subsequent metabolism are primarily responsible for this short half-life (Duckworth et al. 1998). For all the products listed in Table 11.6, the terminal half-life of the drug substance is longer than 4–5 min. This fact indicates that the rate of absorption, rather than the rate of elimination, controls systemic exposure over time. This type of pharmacokinetic profile in which the process of absorption is slower than that for elimination is termed "flip-flop" kinetics, and the post-peak serum profile represents the kinetics of absorption rather than the kinetics of elimination. Figure 11.6 provides a graphical depiction of this phenomenon. The post-peak slopes showing the decline of insulin concentration for the exogenous SC administered insulins have much longer half-lives than the 4–5-min half-life of insulin once it is in the blood.

The first human insulin marketed was regular insulin (referred to as "insulin regular" in Table 11.6) in 1982. As with endogenous insulin stored in beta cells of the pancreas, in the presence of zinc, insulin monomers have a propensity to self-associate into hexamers. These hexamers are in equilibrium with insulin dimers and monomers. The equilibrium is constantly shifting toward dimer and monomer due to dilution in interstitial fluids following administration and subsequent predominant monomer absorption. This process takes time and is responsible for the absorption profile of regular insulin (short-acting) shown in Fig. 11.6. Through various spectroscopic techniques, X-ray crystallography, and molecular modeling approaches, modifications to the insulin molecule favoring the monomeric form at concentrations relevant to therapeutic doses were developed. Figure 11.7 shows the amino acid sequences of these so-called rapid-acting analogs in relation to the native human insulin used in the "insulin regular" formulation. A representative concentration-time profile of a rapid-acting analog is shown in Fig. 11.6 in comparison to regular insulin. It has a shorter time to peak and a faster decline than regular insulin, both of which are preferable for controlling postprandial glucose.

Understanding the need for basal insulin led to the development of neutral protamine Hagedorn (NPH) beef- and pork-sourced insulins in the 1940s and was quickly adapted to human insulin when it became available in the 1980s (Owens 2011). Principally, insulin solubility is substantially reduced in the presence of protamine; thus the drug product is a suspension. Unfortunately, the need for resuspension prior to dosing increases variability in dose and ensuing response (Binder 1969). This realization, coupled with the therapeutic need for basal insulin, stimulated similar research efforts used for discovery of the rapid-acting analogs and

Table 11.6 Descriptive formulation and pharmacodynamic effects on glucose of human insulin and human insulin analog drug products administered subcutaneously

Generic name	Description	Onset (min)	Peak (h)	Duration (h)
Rapid-acting insulin analogs				
Insulin lispro	Two amino acids in the B-chain of human insulin are switched: B28 proline and B29 lysine. The switch to B28 lysine, B29 proline, stabilizes the monomer form to increase the rate of absorption and onset of action compared to regular insulin	15–30	0.5–2.5	3–6.5
Insulin aspart	Aspartic acid replaces proline in the B29 position of human insulin. This modification stabilizes the monomer form to increase the rate of absorption and onset of action compared to regular insulin	10–20	1–3	3–5
Insulin glulisine	Lysine replaces arginine in human insulin at B3, and glutamate replaces lysine at B29. These substitutions stabilize the monomer form to increase the rate of absorption and onset of action compared to regular insulin	10–15	1–1.5	3–5
Short-acting insulin				
Insulin regular	Identical to human insulin; peptide hormone composed of two amino acid chains (A and B) covalently linked by disulfide bonds. The peptide forms a hexamer at the formulated concentration	30–60	1–5	6–10
Intermediate-acting insulin				
Insulin isophane	Insulin is bound to protamine, which reduces solubility to create a suspension product. Injection of this suspension slows down the rate of absorption and onset of action relative to regular insulin	60–120	6–14	16–24
Basal insulin analogs				
Insulin glargine	Soluble insulin analog with glycine instead of asparagine at A21 and two arginines at the C-terminus. This protein is soluble at pH of 4 but forms microprecipitates at physiologic pH upon administration, thus decreasing absorption rate and onset of action relative to regular insulin	66	Not applicable	>24
Insulin detemir	Soluble insulin analog with a fatty acid attached to the lysine at B29. This modification increases self-association, resulting in decreased absorption rate and onset of action relative to regular insulin	48–120	Not applicable	>24
Insulin degludec	Soluble insulin analog with a fatty acid attached to the lysine at B29. This modification results in the formation of multiple hexamers at the injection site, thus decreasing absorption rate and onset of action relative to regular insulin	60–120	Not applicable	>36

Descriptions and data used for this table are from DailyMed: https://dailymed.nlm.nih.gov/dailymed/index.cfm

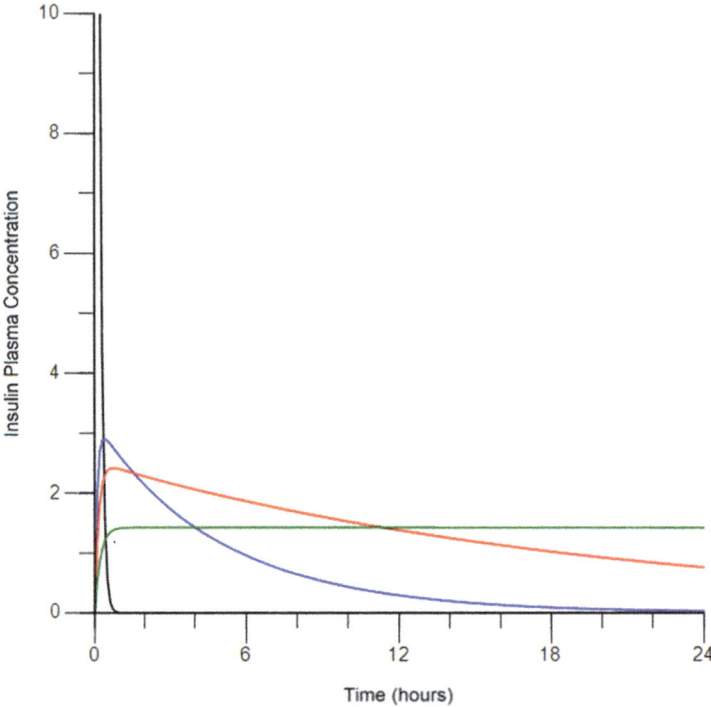

Fig. 11.6 Pharmacokinetic profiles of human insulin and insulin analogs. Elimination profile of human insulin from plasma is shown as the black solid line. Profiles following subcutaneous administration of regular insulin (red), a rapid-acting insulin analog (blue), and a basal insulin analog (green) are also shown

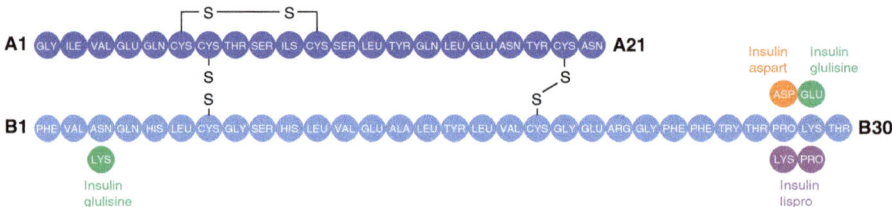

Fig. 11.7 Amino acid sequences of human insulin and rapid-acting analogs: insulin lispro, insulin aspart, and insulin glulisine. (Figure is used by permission from Home 2012, John Wiley & Sons, Inc.)

ultimately led to the discovery of basal insulins that were soluble drug products. Figure 11.8 shows the amino acid sequences of the three products currently available largely worldwide. Table 11.6 provides a brief description of how they slow the rate of absorption, thus resulting in slower onset of action, delayed peak effect, and longer duration of action. Figure 11.6 also shows a concentration-time profile of a basal insulin.

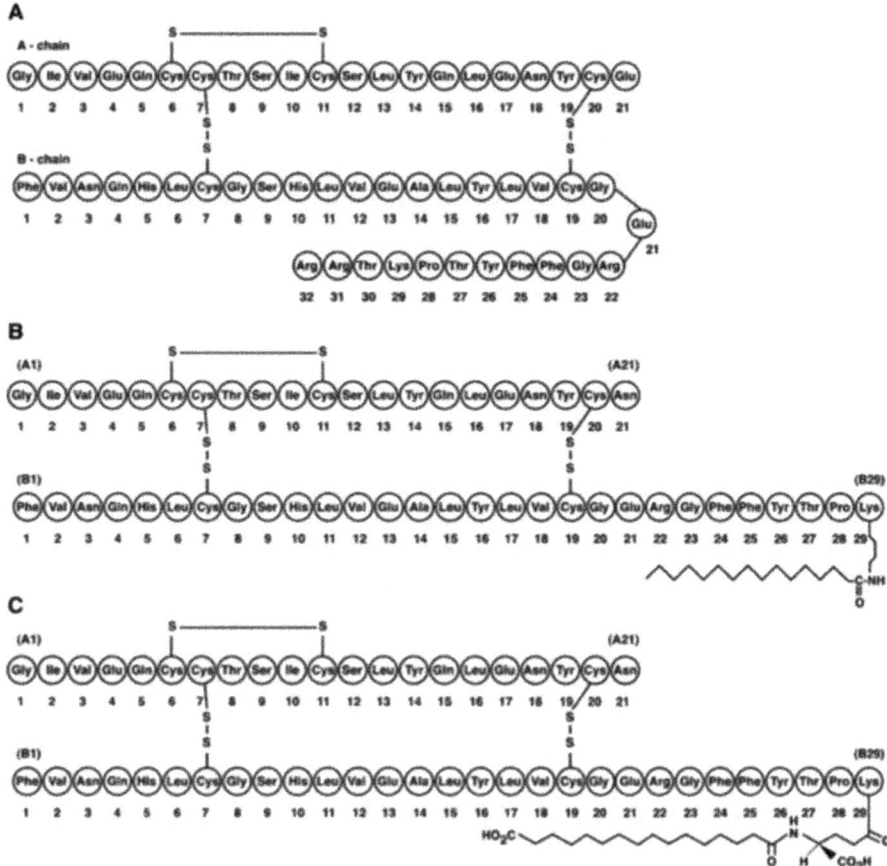

Fig. 11.8 Amino acid sequences of basal insulin analogs: (**a**) insulin glargine, (**b**) insulin detemir, (**c**) insulin degludec. (Figure is used by permission from Owens 2011, Mary Ann Liebert, Inc.)

▶ **Important** In summary, this overview of physicochemical property modification through manipulation of insulin structure is an excellent example of how knowledge of the ADME properties of protein-based drugs can be used to improve therapy. Although proteins are large and complex molecules, posing unique challenges relative to small molecules, these same properties provide multiple opportunities to create therapeutically superior molecules relative to the native protein.

11.6   ADME Properties of Heparin and Low-Molecular-Weight Heparins

Heparin is a mucopolysaccharide consisting of various sugars present in varying proportions. Several of these sugars are negatively charged due to the presence of sulfate or carboxylic acid; thus heparin is largely an anionic polymer. It is also a glycosaminoglycan because of the presence of amino sugars. IV infusion or SC administered exogenous heparin, largely extracted from mucosal tissues (porcine intestine and bovine lung) (Linhardt and Gunay 1999), inhibits clot formation or the progression of clot growth by inactivating several enzymes in the clotting cascade. This inhibition is largely indirect, requiring obligatory binding to antithrombin III, which then results in rapid activation of the latter to a potent inhibitor of several clotting factors (Hirsh and Raschke 2004). Its mainstay use as a rapid-acting anticoagulant has continued for approximately 80 years.

Heparin used clinically is more precisely referred to as unfractionated heparin (UFH). This is because it is a heterogeneous polysaccharide with a molecular weight range of 3000–30,000 Da, with a mean molecular weight of 15,000 (approximately 45 monosaccharide units). Not surprisingly, its ADME properties are also heterogeneous (Hirsh and Raschke 2004). SC administration is largely limited to prophylaxis of clot formation in pregnancy, since warfarin is contraindicated in such patients and the bioavailability is limited (Bara et al. 1985). Administered chronically by this route, peak levels are on average 3 h following administration and onset 1–2 h, too slow for acute treatment of thrombosis. Its slow, variable, and limited absorption are attributed to its size and heterogeneous character (Hirsh and Raschke 2004). The extent of its distribution is limited to the systemic circulation, with a reported distribution volume of 0.04–0.06 L/kg (Bjornsson et al. 1982). The size of UFH limits its ability to exit the systemic circulation across capillary endothelial cell barriers. Elimination of heparin is dose-dependent. This nonlinear elimination is a consequence of saturable binding to endothelial cells (Bjornsson et al. 1982; De Swart et al. 1982). Following endocytosis by these cells, heparin molecules are depolymerized. Thus the half-life of heparin decreases as dose increases, starting at 30 min following an IV dose of 25 units/kg to 60 min following an IV dose of 100 units/kg to 150 min following an IV dose of 400 units/kg (Hirsh and Raschke 2004). Glomerular filtration is a minor pathway and, because of its size dependency, decreases in importance as polymer size increases, thus contributing to the heterogeneous clearance of UFH (De Swart et al. 1982; Hirsh and Raschke 2004).

Realization that the ability of UFH to inhibit various clotting factors was related to molecular size led to clinical studies to develop selective factor Xa inhibitors in the 1980s (Hirsh and Raschke 2004). This work led to the introduction of low-molecular-weight heparins (LMWHs) into clinical practice. These molecules are approximately one-third the size of UFH, with a range of 4500–6500 Da. The average size and size distribution of the final product depend on the method used in preparation, being either chemical or enzymatic depolymerization of UFH. The selective inactivation of factor Xa improves the safety and efficacy of heparin-based therapy. As it turns out, LMWHs also have superior ADME properties, which have led to their principal use in the prophylaxis of clot formation on an outpatient basis. Their excellent bioavailability (approximately 90%) following SC administration certainly supports such use. Elimination of LMWHs is largely by glomerular filtration. This process is not dose-dependent and, therefore, contributes to their superior pharmacokinetics relative to UFH. However, because they are still relatively large molecules, their filtration efficiency is sensitive to size. As the different manufacturers of LMWHs use different UFH depolymerization and subsequent fractionation procedures, the molecular weight distribution of these products, and hence their half-life, is product specific. Overall, their half-lives are longer than UFH, ranging by largely by-product source from 3 to 4.5 h.

11.7 ADME Properties of Gene-Based Biopharmaceuticals

Tisagenlecleucel (Kymriah™) and axicabtagene ciloleucel (Yescarta™) are cell-based, gene-based biotherapeutics recently approved by the FDA. Both products consist of CD-19-directed T cells reintroduced by IV infusion following genetic modification of the T cells originating from the patient.

Tisagenlecleucel is prepared from a patient's T cells by lentiviral-mediated insertion of a gene coding for an anti-CD19 chimeric antigen receptor (CAR), the receptor portion of the construct derived from mouse DNA (Mueller et al. 2017). Following transduction, the patient's transformed cells are expanded in culture, washed, and formulated into a cryopreserved cell suspension. Axicabtagene is prepared similarly, except that an inactivated retrovirus is the vector used to introduce the CAR gene into a patient's T cells (Roberts et al. 2018). Ultimately, both products kill CD-19 expressing normal and cancer cells. Since both products are cell-based therapies, characterization of their ADME properties depends on measurement of CAR cells in blood. Quantitative polymerase chain reaction (qPCR) of CAR DNA provides the measure CAR cells. This measure is normalized to total DNA in a sample for the expression of transformed cell concentration. There is no "absorption" component to the pharmacokinetics since cells are administered intravenously. ADME properties in blood are based on descriptive measures of the normalized CAR DNA concentration versus time profile. These measures are maximum observed normalized CAR DNA (Cmax), associated time to maximum (T_{max}), area under the concentration (AUC) versus time curve, and half-life of the terminal phase of the normalized CAR DNA measure.

For both products, a positive relationship between the magnitude of AUC and Cmax, with treatment response, has been observed (Mueller et al. 2017; Roberts et al. 2018). For tisagenlecleucel, persistence of transformed cells in blood also positively correlates with treatment response (Mueller et al. 2017). For example, the median half-life of tisagenlecleucel in responders was 14 days versus 1.9 days in non-responders (Mueller et al. 2017). The association of treatment response with persistence of transformed cells following axicabtagene administration is inconclusive (Roberts et al. 2018).

Following IV infusion of tisagenlecleucel, which takes <10 min, levels of transformed cells (based on qPCR of CAR DNA) decline during the first week. In responders, these levels then increase to reach Cmax in approximately 2 weeks. This increase reflects expansion of the transformed cell population in the patient, and its magnitude positively correlates with clinical response (Mueller et al. 2017). The decline observed during the first week following infusion has been attributed to redistribution of cells. In this regard, transformed cells were detected in bone marrow and cerebrospinal fluid, indicating that cells can distribute outside of the systemic circulation. Decline in these extra-systemic levels was proportional to blood concentration. Research to understand the mechanism and extent of this redistribution and its relationship to safety and efficacy of tisagenlecleucel is needed (Mueller et al. 2017).

▶ **Important** Use of genetically modified autologous cell therapy to deliver DNA to patients is a novel field. The limited success with tisagenlecleucel and axicabtagene stimulates continued investment to improve safety and efficacy and apply this approach to other cancers and diseases. Understanding the ADME properties of these approaches will be a necessary component in their ultimate clinical acceptance.

11.8   Conclusions

Protein-based biopharmaceuticals (protein biotherapeutics) represent a growing class of drugs used to treat a variety of diseases. Relative to small drugs, they have virtually no oral bioavailability, so that their route of elimination is largely constrained to IV and SC administration. As a class, the range of molecular weights is broad: 1000–150,000 Da. Peptides are considered small proteins (<5000 Da), thus constituting the lower end of the range, and monoclonal antibodies represent the upper end of the range.

Surprisingly, SC bioavailability of these drugs is not related to their size. Contribution of lymphatic absorption to bioavailability of the larger proteins maintains bioavailability to a sufficient level to support reproducible performance. Given that lymphatic absorption is a slower process than absorption of drug directly into the blood via the capillaries, there is a relationship between the rate of absorption (represented by the time to peak blood levels, T_{max}) and molecular weight.

As with absorption, the rate of distribution of these agents is slow, sometimes taking hours to reach equilibrium between the plasma and water spaces outside of the systemic circulation. The extent of their distribution (expressed as their volume of distribution) is largely limited to the systemic circulation but can also include some interstitial space distribution, the extent of which is roughly related to size. Volume of distribution of monoclonal antibody drugs and Fc domain fusion protein drugs can be dose-dependent due to saturation of target binding.

For the monoclonal antibody and Fc domain fusion protein subclass of biotherapeutic proteins, nonlinear (dose-dependent) elimination is observed, decreasing to a limiting value of clearance and associated half-life as dose is increased. Like saturation of the extent of their distribution, saturation of target binding is the attributed cause to this dose-dependent behavior. Contribution of target binding to distribution and elimination of these drugs to their post-absorption pharmacokinetics has been termed TMDD.

The important role that modification of physicochemical properties, achieved through structural modifications, can play in altering the time course of drug concentration, particularly in controlling the frequency of administration of drugs with a short elimination half-life, was discussed. Insulin, having a half-life of 4–5 min, was used to demonstrate this approach, which leverages the slower process of absorption relative to elimination to achieve different patterns of glucose control that better mimic endogenous control.

Finally, the ADME properties on nonprotein biotherapeutics were presented. Heparin properties, a carbohydrate biopharmaceutical, and the superior ADME properties of the LMWHs, namely, obviation of nonlinear elimination, were discussed. The drug disposition properties (distribution and elimination) of two novel cell and gene-based biopharmaceuticals for the treatment of blood-borne cancers were presented. Emphasis was placed on the positive relationship between measures of exposure in blood and their clinical efficacy.

References

Agerso H, Seiding Larsen L, Riis A et al (2004) Pharmacokinetics and renal excretion of desmopressin after intravenous administration to healthy subjects and renally impaired patients. Br J Clin Pharmacol 58:352–358

Ait-Oudhia S, Ovacik MA, Mager DE (2017) Systems pharmacology and enhanced pharmacodynamic models for understanding antibody-based drug action and toxicity. MAbs 9:15–28

Alt N, Zhang TY, Motchnik P et al (2016) Determination of critical quality attributes for monoclonal antibodies using quality by design principles. Biologicals 44:291–305

Bara L, Billaud E, Gramond G et al (1985) Comparative pharmacokinetics of a low molecular weight heparin (PK 10 169) and unfractionated heparin after intravenous and subcutaneous administration. Thromb Res 39:631–636

Beshyah SA, Anyaoku V, Niththyananthan R et al (1991) The effect of subcutaneous injection site on absorption of human growth hormone: abdomen versus thigh. Clin Endocrinol 35:409–412

Binder C (1969) Absorption of injected insulin. A clinical-pharmacological study. Acta Pharmacol Toxicol (Copenh) 27:1–84

Bjornsson TD, Wolfram KM, Kitchell BB (1982) Heparin kinetics determined by three assay methods. Clin Pharmacol Ther 31:104–113

Boss AH, Petrucci R, Lorber D (2012) Coverage of prandial insulin requirements by means of an ultra-rapid-acting inhaled insulin. J Diabetes Sci Technol 6:773–779

Bumbaca D, Boswell CA, Fielder PJ et al (2012) Physiochemical and biochemical factors influencing the pharmacokinetics of antibody therapeutics. AAPS J 14:554–558

Chirmule N, Jawa V, Meibohm B (2012) Immunogenicity to therapeutic proteins: impact on PK/PD and efficacy. AAPS J 14:296–302

Cui Y, Cui P, Chen B et al (2017) Monoclonal antibodies: formulations of marketed products and recent advances in novel delivery system. Drug Dev Ind Pharm 43:519–530

De Swart CA, Numeyer B, Roelofs JM et al (1982) Kinetics of intravenously administered heparin in normal humans. Blood 60:1251–1258

Di L (2015) Strategic approaches to optimizing peptide ADME properties. AAPS J 17:134–143

Dirks NL, Meibohm B (2010) Population pharmacokinetics of therapeutic monoclonal antibodies. Clin Pharmacokinet 49:633–659

Drewe J, Meier R, Vonderscher J et al (1992) Enhancement of the oral absorption of cyclosporin in man. Br J Clin Pharmacol 34:60–64

Drucker DJ, Nauck MA (2006) The incretin system: glucagon-like peptide-1 receptor agonists and dipeptidyl peptidase-4 inhibitors in type 2 diabetes. Lancet 368:1696–1705

Duckworth WC, Bennett RG, Hamel FG (1998) Insulin degradation: progress and potential. Endocr Rev 19:608–624

Ghetie V, Ward ES (1997) FcRn: the MHC class I-related receptor that is more than an IgG transporter. Immunol Today 18:592–598

Goeddel DV, Kleid DG, Bolivar F et al (1979) Expression in Escherichia coli of chemically synthesized genes for human insulin. Proc Natl Acad Sci U S A 76:106–110

Goetze AM, Liu YD, Zhang Z et al (2011) High-mannose glycans on the Fc region of therapeutic IgG antibodies increase serum clearance in humans. Glycobiology 21:949–959

Heinemann L, Muchmore DB (2012) Ultrafast-acting insulins: state of the art. J Diabetes Sci Technol 6:728–742

Hirsh J, Raschke R (2004) Heparin and low-molecular-weight heparin: the seventh ACCP conference on antithrombotic and thrombolytic therapy. Chest 126:188S–203S

Home PD (2012) The pharmacokinetics and pharmacodynamics of rapid-acting insulin analogues and their clinical consequences. Diabetes Obes Metab 14:780–788

Home PD (2015) Plasma insulin profiles after subcutaneous injection: how close can we get to physiology in people with diabetes? Diabetes Obes Metab 17:1011–1020

Hunter J (2008) Subcutaneous injection technique. Nurs Stand 22:41–44

Igawa T, Tsunoda H, Tachibana T et al (2010) Reduced elimination of IgG antibodies by engineering the variable region. Protein Eng Des Sel 23:385–392

Jackisch C, Muller V, Maintz C et al (2014) Subcutaneous administration of monoclonal antibodies in oncology. Geburtshilfe Frauenheilkd 74:343–349

Kagan L (2014) Pharmacokinetic modeling of the subcutaneous absorption of therapeutic proteins. Drug Metab Dispos 42:1890–1905

Keizer RJ, Huitema AD, Schellens JH et al (2010) Clinical pharmacokinetics of therapeutic monoclonal antibodies. Clin Pharmacokinet 49:493–507

Kraynov E, Kamath AV, Walles M et al (2016) Current approaches for absorption, distribution, metabolism, and excretion characterization of antibody-drug conjugates: an industry white paper. Drug Metab Dispos 44:617–623

Krishna M, Nadler SG (2016) Immunogenicity to biotherapeutics – the role of anti-drug immune complexes. Front Immunol 7:21

Leader B, Baca QJ, Golan DE (2008) Protein therapeutics: a summary and pharmacological classification. Nat Rev Drug Discov 7:21–39

Lee WA, Ennis RD, Longenecker JP (1994) The bioavailability of intranasal salmon calcitonin in healthy volunteers with and without a permeation enhancer. Pharm Res 11:747–750

Linhardt RJ, Gunay NS (1999) Production and chemical processing of low molecular weight heparins. Semin Thromb Hemost 25:5–16

Liu L (2015) Antibody glycosylation and its impact on the pharmacokinetics and pharmacodynamics of monoclonal antibodies and Fc-fusion proteins. J Pharm Sci 104:1866–1884

Liu L (2018) Pharmacokinetics of monoclonal antibodies and Fc-fusion proteins. Protein Cell 9:15–32

Liu L, Stadheim A, Hamuro L et al (2011) Pharmacokinetics of IgG1 monoclonal antibodies produced in humanized Pichia pastoris with specific glycoforms: a comparative study with CHO produced materials. Biologicals 39:205–210

Lobo ED, Hansen RJ, Balthasar JP (2004) Antibody pharmacokinetics and pharmacodynamics. J Pharm Sci 93:2645–2668

Mach H, Gregory SM, Mackiewicz A et al (2011) Electrostatic interactions of monoclonal antibodies with subcutaneous tissue. Ther Deliv 2:727–736

Mager DE (2006) Target-mediated drug disposition and dynamics. Biochem Pharmacol 72:1–10

Mi Y, Lin A, Fiete D et al (2014) Modulation of mannose and asialoglycoprotein receptor expression determines glycoprotein hormone half-life at critical points in the reproductive cycle. J Biol Chem 289:12157–12167

Moots RJ, Xavier RM, Mok CC et al (2017) The impact of anti-drug antibodies on drug concentrations and clinical outcomes in rheumatoid arthritis patients treated with adalimumab, etanercept, or infliximab: results from a multinational, real-world clinical practice, non-interventional study. PLoS One 12:e0175207

Mueller KT, Maude SL, Porter DL et al (2017) Cellular kinetics of CTL019 in relapsed/refractory B-cell acute lymphoblastic leukemia and chronic lymphocytic leukemia. Blood 130:2317–2325

Nauck M (2016) Incretin therapies: highlighting common features and differences in the modes of action of glucagon-like peptide-1 receptor agonists and dipeptidyl peptidase-4 inhibitors. Diabetes Obes Metab 18:203–216

Owens DR (2011) Insulin preparations with prolonged effect. Diabetes Technol Ther 13:S5–S14

Pecoraro V, De Santis E, Melegari A et al (2017) The impact of immunogenicity of TNFalpha inhibitors in autoimmune inflammatory disease. A systematic review and meta-analysis. Autoimmun Rev 16:564–575

Porter CJ, Charman SA (2000) Lymphatic transport of proteins after subcutaneous administration. J Pharm Sci 89:297–310

Reddy ST, Berk DA, Jain RK et al (2006) A sensitive in vivo model for quantifying interstitial convective transport of injected macromolecules and nanoparticles. J Appl Physiol 101:1162–1169

Richter WF, Jacobsen B (2014) Subcutaneous absorption of biotherapeutics: knowns and unknowns. Drug Metab Dispos 42:1881–1889

Richter WF, Bhansali SG, Morris ME (2012) Mechanistic determinants of biotherapeutics absorption following SC administration. AAPS J 14:559–570

Roberts ZJ, Better M, Bot A et al (2018) Axicabtagene ciloleucel, a first-in-class CART T cell therapy for aggressive NHL. Leuk Lymphoma 59(8):1785–1796

Roopenian DC, Akilesh S (2007) FcRn: the neonatal Fc receptor comes of age. Nat Rev Immunol 7:715–725

Salar A, Avivi I, Bittner B et al (2014) Comparison of subcutaneous versus intravenous administration of rituximab as maintenance treatment for follicular lymphoma: results from a two-stage, phase IB study. J Clin Oncol 32:1782–1791

Schoch A, Kettenbergher H, Mundigl O et al (2015) Charge-mediated influence of the antibody variable domain on FcRn-dependent pharmacokinetics. Proc Natl Acad Sci U S A 112:5997–6002

Sedelmeier G, Sedelmeier J (2017) Top 200 drugs by worldwide sales 2016. Chimia (Aarau) 71:730

Shah DK, Betts AM (2013) Antibody biodistribution coefficients: inferring tissue concentrations of monoclonal antibodies based on the plasma concentrations in several preclinical species and human. MAbs 5:297–305

Shpilberg O, Jackisch C (2013) Subcutaneous administration of rituximab (MabThera) and trastuzumab (Herceptin) using hyaluronidase. Br J Cancer 109:1556–1561

Smith A, Manoli H, Jaw S (2016) Unraveling the effect of immunogenicity on the PK/PD, efficacy, and safety of therapeutic proteins. J Immunol Res 2016:9: ID 2342187

Supersaxo A, Hein WR, Steffen H (1990) Effect of molecular weight on the lymphatic absorption of water-soluble compounds following subcutaneous administration. Pharm Res 7:167–169

Ter Braak EW, Woodworth JR, Bianchi R et al (1996) Injection site effects on the pharmacokinetics and glucodynamics of insulin lispro and regular insulin. Diabetes Care 19:1437–1440

Tibbitts J, Canter D, Graff R et al (2016) Key factors influencing ADME properties of therapeutic proteins: a need for ADME characterization in drug discovery and development. MAbs 8:229–245

Usmani SS, Bedi G, Samuel JS et al (2017) THPdb: database of FDA-approved peptide and protein therapeutics. PLoS One 12:e0181748

Vezina HE, Cotreau M, Han TH et al (2017) Antibody-drug conjugates as cancer therapeutics: past, present, and future. J Clin Pharmacol 57:S11–S25

Wang W, Wang EQ, Balthasar JP (2008) Monoclonal antibody pharmacokinetics and pharmaco-dynamics. Clin Pharmacol Ther 84:548–558

Wu F, Bhansali SG, Law WC et al (2012) Fluorescence imaging of the lymph node uptake of proteins in mice after subcutaneous injection: molecular weight dependence. Pharm Res 29:1843–1853

Drug-Drug and Food-Drug Interactions of Pharmacokinetic Nature

12

Pietro Fagiolino, Marta Vázquez, Manuel Ibarra, Cecilia Maldonado, and Rosa Eiraldi

12.1 Introduction

▶ **Definition** Pharmacokinetic drug-drug interactions are those changes in the action of the individual over a drug (interacted) caused by the presence of other drug in the body (interacting).

Pharmacokinetic food-drug interactions are those changes in the action of the individual over a drug (interacted) caused by the presence of food in the body (interacting).

Nowadays, drug-drug (DD) interactions are a common problem since more drugs are concurrently prescribed to patients, mainly if they are aged. Sometimes serious or even fatal adverse events can happen either related with toxic or ineffective outcomes. Because of such polypharmacy, there would be greater possibility that foods interact with some of the medicines received by patients.

▶ **Important** The risk of DD interactions increases with the number of drugs simultaneously taken by a patient.

The potential for DD interactions is being now considered in the benefit-risk evaluation of medicines when a new drug product is under development, before asking for marketing authorization (EMA 2012).

Pharmacodynamic interactions are related to the opposite or similar clinical response that two molecules could have in the individual. It might happen between two drugs that have opposite or similar effects mediated by different mechanisms of

P. Fagiolino (✉) · M. Vázquez · M. Ibarra · C. Maldonado · R. Eiraldi
Pharmaceutical Sciences Department, Faculty of Chemistry, Universidad de La República, Montevideo, Uruguay
e-mail: pfagioli@fq.edu.uy

© Springer Nature Switzerland AG 2018
A. Talevi, P. A. M. Quiroga (eds.), *ADME Processes in Pharmaceutical Sciences*,
https://doi.org/10.1007/978-3-319-99593-9_12

actions. However, in most of the cases, they are the consequence of different affinities that such molecules have to the same molecular target. If this were the case, their pharmacokinetics would be relevant to determine the amount of each specimen that attains the action site.

More fascinating, and intriguing, could be the interaction between a drug and a substrate with no pharmacodynamic action. This could happen with racemates, being one of the stereoisomers the active principle and the other one the non-active species that antagonizes the former at the level of the molecular target. It has been reported that R-warfarin (the inactive isomer, R-WF) would compete with S-warfarin (S-WF) for its receptor (Xue et al. 2017). Therefore, any change in the amount of R-WF at the biophase might alter the intensity of bleeding even though no changes in S-WF concentration had happened.

Racemates represent a challenge since most of the time, they introduce a non-required molecule to the individual, which is structurally very close to the active stereoisomer and, hence, prone to interact with it. Because of this, some of the benefits they entail, such as their low cost, might be overridden by the cost entrained to solve those clinically relevant interactions.

On the other hand, some essential nutrients or vitamins or endogenous substances compete pharmacodynamically with the drug; therefore, its clinical response might change with high risk for the patient when the pharmacokinetics of interactors are altered. Subjects receiving long-term WF therapy are sensitive to fluctuating levels of dietary vitamin K (Hirsh et al. 2003). Drugs may influence the pharmacodynamics of WF by inhibiting absorption or increasing clearance of vitamin K. Adverse events caused by the co-administration of ezetimibe and WF were rescued by oral vitamin K supplementation, suggesting that the DD interaction effects observed were the consequence of ezetimibe-mediated vitamin K malabsorption (Takada et al. 2015).

12.2   Plasma Protein Binding

The risk of clinically relevant interactions via displacement from plasma protein binding is low, even for highly bound drugs. This is because free drug plasma concentration is the only moiety responsible for its body disposition. Figure 12.1 and Eqs. 12.1, 12.2, 12.3, 12.4, 12.5, 12.6, 12.7 and 12.8 will demonstrate this statement. Drug comes into blood (compartment 1) as a free entity, and then, it displays different equilibriums with plasma proteins (PP) and blood cells (BC), among other sites. Let's assume that the drug is being administered through a zero-order rate process.

Fig. 12.1 Tricompartmental model with intravascular zero-order drug input and elimination from an extravascular compartment (2). Blood cells (BC) and plasma protein (PP) play the roles of drug reservoirs in the blood compartment (1)

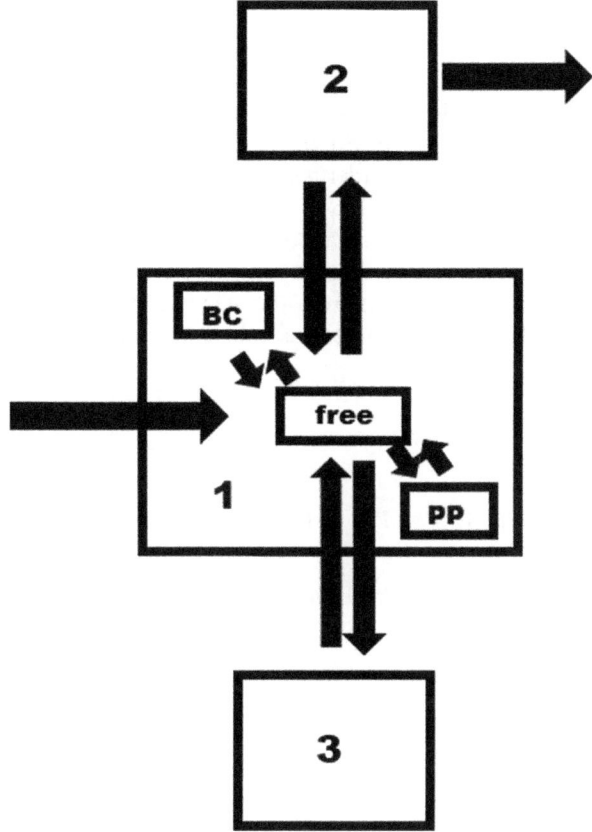

Differential equations describing the model shown in Fig. 12.1 are:

$$\frac{dX_{\text{free}}}{dt} = k_0 - k_{\text{free-PP}} * X_{\text{free}} + k_{\text{PP-free}} * X_{\text{PP}} - k_{\text{free-BC}} * X_{\text{free}}$$

$$+ k_{\text{BC-free}} * X_{\text{BC}} - k_{\text{free-2}} * X_{\text{free}} + k_{2\text{-free}} * X_2 - k_{\text{free-3}} * X_{\text{free}}$$
$$+ k_{3\text{-free}} * X_3 \tag{12.1}$$

$$\frac{dX_{\text{PP}}}{dt} = k_{\text{free-PP}} * X_{\text{free}} - k_{\text{PP-free}} * X_{\text{PP}} \tag{12.2}$$

$$\frac{dX_{\text{BC}}}{dt} = k_{\text{free-BC}} * X_{\text{free}} - k_{\text{BC-free}} * X_{\text{BC}} \tag{12.3}$$

$$\frac{dX_2}{dt} = k_{\text{free-2}} * X_{\text{free}} - k_{2\text{-free}} * X_2 - k_{20} * X_2 \tag{12.4}$$

$$\frac{dX_3}{dt} = k_{\text{free}-3} * X_{\text{free}} - k_{3-\text{free}} * X_3 \tag{12.5}$$

These equations represent the mass balance at each compartment or sub-compartment. In Eq. 12.1, except for the zero-order input rate (k_0), the other kinetics are of first order, since the transfer rates are proportional to the amount of drug in the site (X_i). Subscript of the first-order rate constants, k_{ij}, denotes the sense of the mass transfer, from compartment "i" to compartment "j".

Once the equilibrium is attained, all differentials will become null, and the fundamental equation from the steady state is verified ($k_0 = k_{20} * X_{2\text{ss}}$). Subsequently, drug concentrations in compartment 2 and other different sites can be obtained:

$$C_{2\text{ss}} = \frac{k_0}{(k_{20} * V_2)} \tag{12.6}$$

From Eqs. 12.4 and 12.6, free drug concentration in plasma:

$$C_{1\text{freess}} = \frac{k_0}{\frac{k_{\text{free}-2} * k_{20}}{(k_{2-\text{free}} + k_{20})} * V_{1\text{free}}} = \frac{k_0}{k_{20} * V_2} \tag{12.7}$$

From Eqs. 12.4 and 12.5:

$$C_{3\text{ss}} = \frac{k_{\text{free}-3} * k_0}{\frac{k_{3-\text{free}} * k_{\text{free}-2} * k_{20}}{(k_{2-\text{free}} + k_{20})} * V_3} \tag{12.8}$$

As it can be seen from Eqs. 12.6, 12.7 and 12.8, no affectation of the free plasma concentration and the concentrations in compartments 2 and 3 is expected following a change in binding to either plasma proteins ($k_{\text{free} - \text{PP}}$ or $k_{\text{PP} - \text{free}}$) or blood cells ($k_{\text{free} - \text{BC}}$ or $k_{\text{BC} - \text{free}}$). Free plasma clearance remains constant even though $V_{1\text{free}}$ increases or decreases, as a consequence of an increased or a decreased protein/blood cell binding, due to the reciprocate change that happened in $k_{\text{free} - x}$ (Fagiolino 2004). The same changes occur on both $k_{\text{free} - 3}$ and $k_{\text{free} - 2}$, and then $C_{3\text{ss}}$ remains unchanged. In other words, no changes in either free plasma or extravascular drug levels could be envisaged if drug binding at the intravascular space were modified. Only a brief disturb would be seen after such changes, but once the new equilibrium is reached, the steady-state drug concentrations will remain unaltered. Nevertheless, the total plasma drug concentration changes dramatically and definitively because of the particular relationship that exists between free and total drug concentrations:

$$C_{1\text{free ss}} = f * C_{1\text{total ss}} \quad \text{being } f \text{ the free fraction of drug} \tag{12.9}$$

This phenomenon applies for both linear and nonlinear pharmacokinetic systems because drug input rate (k_0) does not change, and such sudden supplement of drug rapidly disappears from the system.

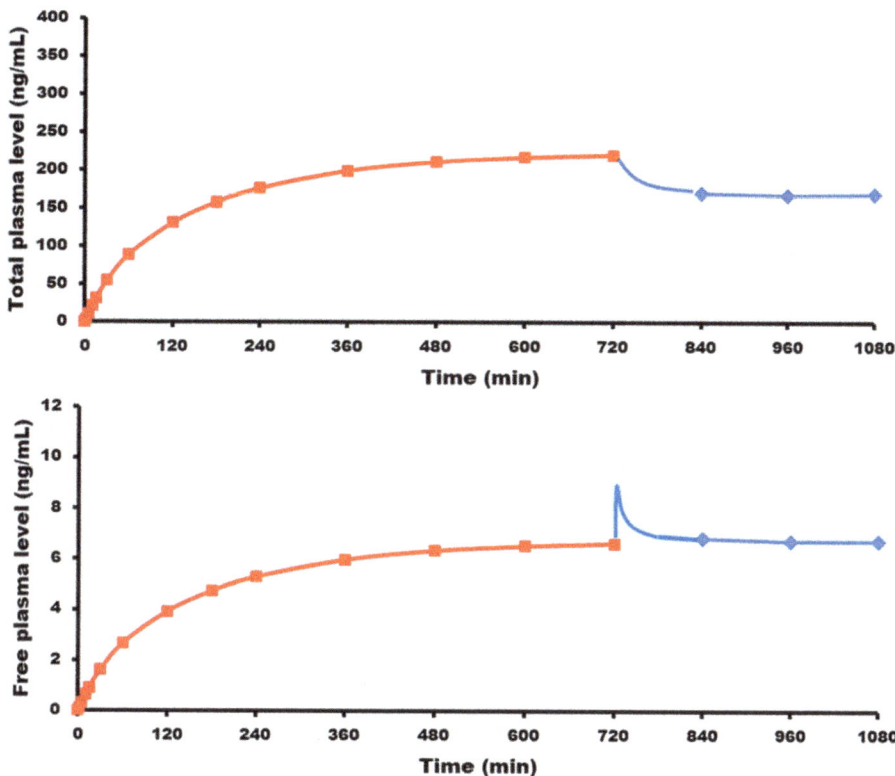

Fig. 12.2 Midazolam infusion given at a rate of 4 mg/h. Once the steady state is attained (red line), a sudden change in its protein binding from 97% to 96% happens (blue line). While total plasma drug concentration falls to a three quarter of its previous level (upper panel), the free one increases for a short period up to retrieve the previous level (lower panel)

Figure 12.2 shows a hypothetical situation with midazolam, which is given at a rate of 4 mg/h, and suddenly, a change in its protein binding from 97% to 96% happens. A transient increase in its free concentration can be observed, while a permanent decreased total plasma level becomes installed. This should not justify any increase in the rate of administration since free levels would not have been assessed. Unfortunately, free drug concentrations are not commonly monitored in the clinical setting, and therefore bad decisions could be taken.

As the action sites are normally placed outside the intravascular space, a transient increase in the intensity of drug effect, if it happens, would be foreseen.

▶   **Important** DD interactions are usually insignificant from a clinical perspective, unless the drug subjected to displacement has high fraction bound, narrow therapeutic window, small volume of distribution, and diminished clearance because, even for a short time, the change in drug concentration might be enough to produce severe adverse effects. Nevertheless, once the interacting drug remains as a therapeutic agent, dosage of the interacted drug should not be changed.

12.2.1 Cyclosporine-Statins: A Case Where the Bound Moiety Is Relevant

Cyclosporine (CYA) is a powerful immunosuppressive drug used in transplantation medicine and to treat autoimmune diseases (Freeman 1991). Dyslipidemia is frequent in solid-organ transplant recipients primarily as a result of immunosuppressive treatment. In order to reduce cardiovascular risk in these patients, lipid-lowering drugs have become widely used especially hydroxymethylglutaryl-CoA (HMG-CoA) reductase inhibitors (statins). This reductase catalyzes the conversion of HMG-CoA to mevalonate, the rate-limiting step in de novo cholesterol synthesis. Competitive inhibition of this enzyme by the statins decreases hepatocyte cholesterol synthesis. The associated reduction in intracellular cholesterol concentration induces low-density lipoprotein (LDL) receptor expression on the hepatocyte cell surface, which results in increased extraction of LDL-cholesterol from the blood and decreased circulating LDL-cholesterol concentrations (Schachter 2004).

In plasma, CYA binds primarily to lipoproteins, including high-density, low-density, and very-low-density lipoprotein, and, to a lesser extent, albumin. The unbound fraction of CYA in plasma is approximately 2%. It has been shown that both the pharmacokinetic and pharmacodynamic properties of CYA are related to its binding characteristics in plasma (Akhlaghi and Trull 2002).

CYA is mainly metabolized in the liver by CYP3A4 enzyme. As stated before free drug plasma concentration is the only moiety responsible for body disposition. However, in the case of CYA, its disposition depends not only on physicochemical characteristics but also on plasma carriers such as lipoproteins. The complex CYA-LDL can reach the intracellular space via LDL membrane receptor. When combined with statins, an increase in clearance of CYA can be observed (Asberg 2003) as statins reduce plasma LDL increasing the free fraction of CYA. Besides, due to the upregulation of LDL receptors in the liver, CYA biotransformation increases. Both factors are then the cause of the decrease of both free and total CYA plasma concentrations, and thereafter its level at the action site also diminishes. A case was reported (Eiraldi et al. 2004) in which the combination of both drugs resulted in an acute rejection episode.

This case emphasizes the role that membrane transporter has on drug disposition, not only carrying the free drug, as it will be dealt with in the next section, but also its bound moiety. For this exceptional case, Fig. 12.1 should add a transport arrow driving mass from the PP moiety toward the compartment 2.

12.3 Membrane Transport

Drug diffusion into tissues mostly takes place passively; however, pharmacokinetics is now challenged by the growing importance of membrane transporters, a relatively new and potentially relevant factor in drug distribution and elimination. As it is shown in Eq. 12.7, drug elimination could be affected if some other substances altered $k_{free - 2}$ or k_{20}. If the drug quits the system by excretion, k_{20} will be the constant rate for an efflux transport, being $k_{free - 2}$ the one for an influx transport.

Different hypotheses have been developed for the well-known amiodarone-digoxin (AMD-DGX) interaction. The one by which AMD inhibits the efflux carrier P-glycoprotein (Pgp), and thereafter both intestinal and renal clearance of DGX decreases, is the most predominant (Wessler et al. 2013). This is supported by the fact that quinidine (QND) interacts with DGX using the same mechanism of Pgp inhibition (Warner et al. 1985). However, the difference between QND and AMD is that the latter reduces DGX distribution into the liver and kidneys, and then, influx carriers should also be involved apart from the Pgp transporter. New evidences show AMD inhibition of organic anion transporters OATP (Lambert et al. 1989; Kodawara et al. 2002; Funakoshi et al. 2005; Roth et al. 2012).

AMD produces an important and lasting increase in DGX concentration because not only the drug exerts inhibition of transporters but also its metabolite desethylamiodarone (Sibl et al. 2017) does. Due to the competitive antagonism between AMD and DGX for the influx carrier, DGX might also inhibit the hepatic elimination of AMD (Eiraldi 1997), giving then a further clinical implication in the sense of enhancing the risk of such association.

Transporters can also be induced. Both rifampicin (RFP) and carbamazepine (CBZ) induce Pgp by means of a transcriptional mechanism. It takes the membrane around 48 h to overexpress Pgp (Maldonado et al. 2011). DGX increased its clearance and reduced its bioavailability after a daily oral dose of 600 mg of RFP for 23 days (Greiner et al. 1999). The effect was more pronounced when DGX was given by the oral route in comparison to intravenous dosing. Since induction is concentration-dependent, RFP would have produced a higher overexpression of Pgp at the intestine because local concentration must have been much higher than the one attained systemically and then renal clearance of DGX might have not increased at the same extent than the intestinal one. Besides, oral bioavailability reduction was another factor that could have contributed to the higher interaction oral-RFP-oral-DGX since it was not present after intravenous dose of DGX.

Another important aspect to take into account in the RFP-DGX interaction is RFP systemic residence time. Due to its short half-life, RFP cannot achieve systemic inductive concentration for a long time, and then, its induction becomes circumscribed at the splanchnic region. Because of the longer half-life CBZ has, its systemic inductive effect might persist all day long in chronic treatments, as it was recently shown through an increased saliva DGX concentrations following a 10-day chronic dosing of 200 mg of CBZ every 12 h (unpublished data). Sustained plasma concentration of Pgp inducer maintains the stimulus for overexpressing Pgp at the salivary gland more efficiently. This was demonstrated for a 100 mg of phenytoin

(PHT) every 12 h in comparison with a dosage regime of 600 mg every 72 h (Alvariza et al. 2014).

12.3.1 Carnitine-Valproic Acid as a Substrate for Drug Transportation

Valproic acid (VPA), a branched short-chain fatty acid, has multiple mechanisms of action. It increases the concentration of the neurotransmitter GABA in the human brain by inhibiting GABA transaminase and increasing GABA synthesis and release (Losher 1999), but it also blocks the voltage-gated sodium channels and T-type calcium channels (Ghodke-Puranik et al. 2013). It is extensively metabolized in the liver by three routes: glucuronidation (accounting for 40% of dose), β-oxidation in the mitochondria facilitated by carnitine (40% of dose), and ω-oxidation (considered a minor route, approximately 10%). The latter is responsible for the formation of a toxic metabolite (4-en-VPA) (Ghodke-Puranik et al. 2013; Siemes et al. 1993).

VPA crosses the membrane of liver mitochondria via the facilitation of L-carnitine. This pathway consists of several steps until valproylcarnitine is formed and in the mitochondrial matrix turns into valproyl-CoA, which is able to get into the β-oxidation process. L-carnitine is reabsorbed by means of the organic cation transporter 2 (OCTN2), a saturable influx carrier located in the apical membrane of renal tubular cells. Valproylcarnitine inhibits the OCTN2 transporter, thereby decreasing L-carnitine and acylcarnitine tubular reabsorption (Lheureux et al. 2005). Because of this, chronic VPA therapy or acute VPA overdose induces carnitine depletion (Maldonado et al. 2016) and then increases the β-oxidation route, yielding higher concentration of 4-en-VPA (Vázquez et al. 2014).

Toxic metabolites of VPA inhibit carbamoyl phosphate synthetase (CPS), which catalyzes the conversion of ammonia to carbamoyl phosphate in the first step of the urea cycle (Mehndiratta et al. 2008). Moreover, impairment of β-oxidation produces acetyl-CoA depletion leading to a decreased synthesis of N-acetyl glutamic acid, an allosteric activator of CPS. All this could result in incorrect ammonium elimination through urea cycle providing the basis for the development of hyperammonemia that could lead to encephalopathy, seizures, and other neurological dysfunctions.

A strategy to reverse hyperammonemia is not to discontinue VPA administration but to use VPA extended-release formulations avoiding peak-trough oscillations or L-carnitine as a supplement (Vázquez et al. 2014; Maldonado et al. 2017).

A higher risk factor for hyperammonemia is the combination of VPA with other antiepileptic medications, particularly phenobarbital, PHT, and CBZ. The mechanisms of action are thought to be related to an increase in the production of toxic VPA metabolites because all of them are inducers of the ω-oxidation route (Carr and Shrewsbury 2007). Topiramate, a weak carbonic anhydrase inhibitor, can increase ammonia levels, as a result of urine alkalinization, synergizing VPA effect (Hamer et al. 2000; Vázquez et al. 2013).

VPA is extensively bound to plasma protein (89–93%), mainly albumin. Increases in VPA concentration may result in the saturation of the protein binding sites (Bowdle et al. 1979). This fact explains the reported nonlinear pharmacokinetics of VPA (Gugler and von Unruh 1980), in which dosage increases result in less than proportional increases in serum concentrations. In other words, the mean free fraction increases with dose. However, as it was predicted by Eq. 12.7, free drug concentrations show a linear kinetics after the equilibrium is reached.

Interestingly, the fact that VPA itself or in combination with other antiepileptic drugs can cause carnitine deficiency can serve to better explain VPA nonlinear kinetics. A decrease in carnitine level alters β-oxidation pathway, decreasing VPA clearance, resulting in increased VPA concentrations (total and free concentrations). Each time the dose of VPA increases (k_0 in Eq. 12.7), its clearance decreases (because of k_{20} decrease). Therefore, a mixed nonlinear kinetics can be envisaged for VPA.

After adding carnitine, not only a decrease in ammonia levels was found but also a decrease in VPA and 4-en-VPA concentrations. This perhaps is indicating that L-carnitine supplementation descends ammonia levels because it reestablishes VPA altered kinetics (Vázquez et al. 2014; Maldonado et al. 2017) and in this case only the nonlinear protein binding kinetics operates.

12.4 Metabolism

Drug-drug interaction at the level of enzymes is the most referenced mechanism. Here below, some examples will be given in order to highlight the mechanisms of such interactions.

12.4.1 Carbamazepine-Phenytoin: A Mixture Between Enzymes and Transporters

An interesting finding was the dual effect observed in the interaction between CBZ and PHT (Zielinski et al. 1985; Zielinski and Haidukewych 1987). In some cases, PHT concentration increased, but in others, its levels decreased when CBZ dose increased. In order to understand the possible mechanism of such interaction, it should be better to go deeper about the way PHT displays its nonlinear pharmacokinetics. Figure 12.3 shows the tricompartmental model seen in Fig. 12.1 but where compartment 2 was subdivided according to the different organs that eliminate PHT. As it can be seen, the hepatic (2-H) is the main route of elimination (the liver has the widest arrow getting the drug out of the system), followed by the intestinal (2-I) and renal (2-R) ones. Blood compartment exchanges mass with sub-compartments 2 with different preferences depending on the expression of transporter at their respective interphase (arrows with wide borders denote the presence of membrane transport). A more detailed mathematical deduction has already been published (Fagiolino et al. 2011), but Fig. 12.3 illustrates the most relevant conclusions.

As PHT overexpresses efflux transporters, it moves away from the liver and goes toward the intestine and renal tissues. Therefore, hepatic clearance diminishes

Fig. 12.3 Tricompartmental model where the second compartment was subdivided accordingly with the organs that participate in the elimination of PHT (H, hepatic; I, intestinal; R, renal). The width of violet arrows indicates the intensity of drug elimination. Arrows with orange borderlines refer to process of transfer assisted by membrane transporters. Due to the action of overexpressed transporters, PHT accumulates dissimilarly among the compartments

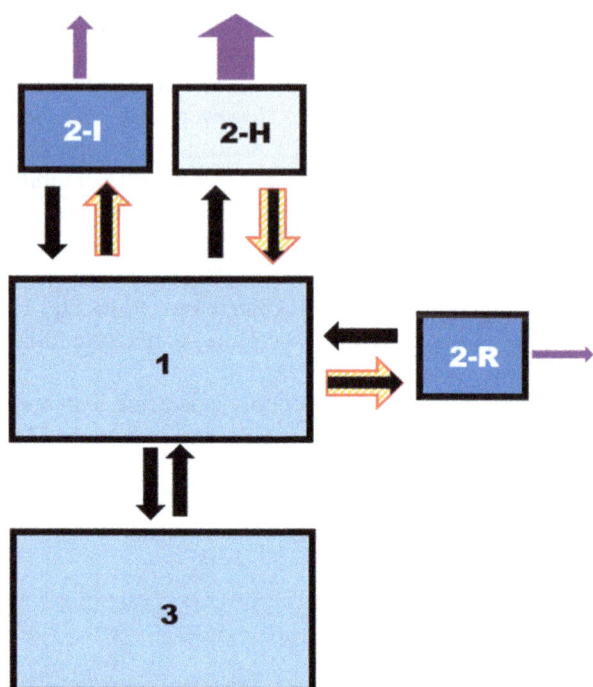

meanwhile intestinal and renal clearance might increase. Despite the fact that PHT is able to induce its own enzymes, this effect is not enough to reverse the reduction of total clearance. This is because the higher number of molecules driven to the intestine override the renewed capacity of the enzymes here expressed. In order to maintain this inductive status, drug input should be kept constant. If a higher dose of PHT is given, a higher inductive status will be attained. However, the intestinal metabolism will remain unaltered because this supplement of PHT saturates the overexpressed enzymes. Consequently, each time the dose of PHT increases, its clearance decreases.

The situation is different when PHT dose remains constant and CBZ is added to the treatment. Now, CBZ plays the role of an inducer agent. When the dose of CBZ increases, both transporter and enzyme will be overexpressed, but the activity of intestinal enzymes will not be counterbalanced because CBZ does not compete with PHT for them. Therefore, the total clearance of PHT might be reduced or increased, depending on the intensity of CBZ induction (Fagiolino et al. 2014).

12.4.2 Rifampicin: Another Example of Enzyme and Transporter Induction

The previous examples presented a common scenario in epilepsy treatment when monotherapy does not work and two anticonvulsants are needed.

Tuberculosis is common among HIV patients and allografts recipients due to their immune system frailty (WHO 2012; Aguado et al. 2009). Both conditions are per se treated using polypharmacy; as a consequence, the probability of drug interactions arises. The commonest treatment for the disease entails among others the use of RFP, whose potential drug-drug interactions have been studied in many works (Greiner et al. 1999; Mahatthanatrakul et al. 2007) as it is considered the prototype for enzyme and transporter inducer but has also placed the drug at a difficult position in the clinical setting.

As it was stated before, this antibiotic exerts its mechanism as an inducer by a nuclear mechanism that entails the activation of pregnane X receptor (PXR), which functions as a ligand-activated transcription factor (Chen and Raymond 2006; Yamashita et al. 2013). When a PXR ligand binds to PXR, it in turn activates transcription of CYP enzymes and several other genes such as those coding aldehyde dehydrogenases, UDP-glucuronosyltransferases (UGTs), sulfonyltransferases, glutathione-S-transferases (GSTs), and Pgp. This explains the high potential for drug-drug interactions RFP has. Regarding CYP isoenzymes, RFP induces mainly CYP3A4, displaying weak induction over other isoenzymes such as CYP2A6, CYP2Cs, and CYP2B6 (Palleria et al. 2013).

HIV treatment comprises a wide variety of drugs, several of them are metabolized by CYP enzymes (Malaty and Kuper 1999; Usach et al. 2013; Vourvahis and Kashuba 2007; Ramanathan et al. 2011), and some such as protease inhibitors (PIs) are Pgp substrates as well (Kim et al. 1998). To show the different interaction profiles the same inducer may have, three different drugs will be discussed: lopinavir (LPV), enzyme and transporter substrate; efavirenz (EFV), enzyme substrate and inducer; and nevirapine (NVP), enzyme substrate.

LPV exposure may decrease by up to 75% in the presence of RFP (Bertz et al. 2000). The extent of this phenomenon can be explained by the induction of both enzymes and transporters in the intestine. At the enterocyte, Pgp is expressed in the apical membrane and works coordinately with intracellular enzymes, acting as a barricade against xenobiotics. The extrusion of the drug performed by Pgp is a way of improving enzymatic efficiency at the enterocyte, especially for CYP3A4 which is the isoenzyme with the highest content. Shortly after oral administration, drug concentration in the enterocyte is high; thus most enzymes are working at the highest possible rate; therefore, the metabolic system starts working at zero-order rate. Drug entrance to the portal vein is not counterbalanced by its presystemic elimination, and then, its bioavailability results high. By enhancing the expression of enzymes and efflux transporters, inducers like RFP lead to diminish oral drug bioavailability. Figure 12.4 aids the readers to understand this mechanism. An increase in the number of enzymes enhances the maximum rate of metabolism (blue arrow). On the other hand, an increase of Pgp reduces the amount of drug molecules in the enterocyte (orange arrow) due to their extrusion by pump action. Therefore, in both cases, the influx through the basal membrane and the presystemic elimination at the enterocyte becomes competitive, and thus the passage of unaltered molecules to the portal vein diminishes.

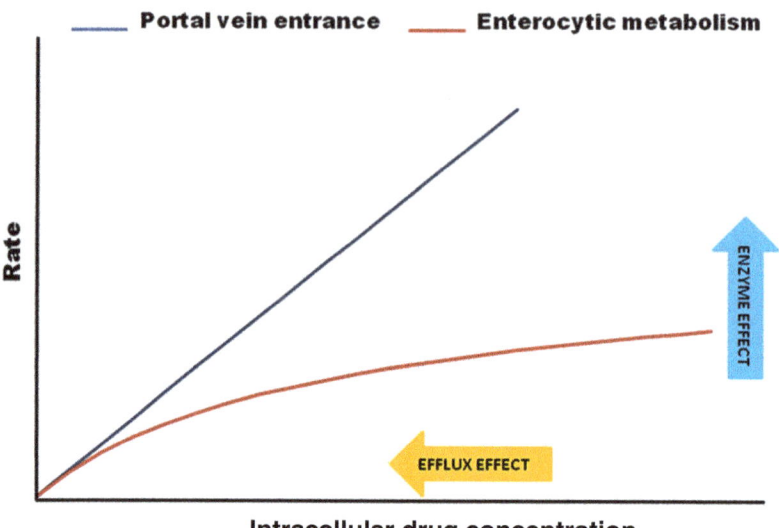

Fig. 12.4 The graph represents the effect both efflux transporter and enzyme inductions have over the competition between drug entrance to portal circulation and metabolism at the enterocyte. Under normal conditions, the rate for both processes may differ significantly. Induction of transporters diminishes drug concentration within the cells (orange arrow), while the induction of enzymes enhances the metabolic rate (blue arrow)

The knowledge of this coordinated system was used as the basis for the inhibition of the system to reversely enhance body exposure to PIs. First PIs needed great doses to be effective due to very low bioavailability. The discovery of ritonavir (RTV) inhibition of CYP3A4 and Pgp allowed the formulation of a pharmaceutical form that changed the patients' quality of life to a great extent, not only for the decrease in the number of pills taken but also for diminishing adverse drug reactions (Boffito et al. 2004). The most direct explanation is the inhibition of CYP3A4; however, the inhibition of the transporter enhances absorption and at the same time declines metabolism rate by exposing enzymes to higher concentrations that are unable to efficiently eliminate.

Findings regarding the interaction between PIs and RFP changed the course of treatment. Nowadays this combination is discouraged by many guidelines based on some works that found adverse events after PI dose modification (Nijland et al. 2008). Nevertheless, evidence may result controversial, as some authors have found effective and safe combinations of RFP and PIs (La Porte et al. 2004).

The importance of transporter activity in controlling a drug's fate can be easily understood on a slightly different scenario, the addition of RFP to a treatment including drugs that are metabolized but are not Pgp substrates; this would be the case of non-nucleoside reverse transcriptase inhibitors (NNRTIs), NVP and EFV.

Both drugs are CYP3A4 substrates mainly followed by CYP2B6, but the pharmacokinetic results of the interaction with RFP differ slightly. On the one hand, some studies have drawn the conclusion that there is no need for dose adjustment in EFV treatments, whereas patients on NVP may need to increase dose. To understand where the difference may lie, we need to go over each drug pharmacokinetics.

Apart from being CYP3A4 substrate (Usach et al. 2013), EFV induces its own metabolism by this isoenzyme. Multiple doses of 200–400 mg during 10 days produced 22–42% lower concentrations than expected and a half-life of 40–55 h, shorter than the 52–76 h reported after a single dose (Usach et al. 2013).

In the case of EFV, minor or no changes in concentrations have been reported with the addition of RFP; apparently EFV concentrations seem to be more dependent on subject's genotype (Ramachandran et al. 2009). What could be fairly stated in this case is that EFV previous autoinduction leads to a system less vulnerable to an enzyme inducer such as RFP. Therefore, its pharmacokinetics may remain unchanged despite the inducer addition. Most likely, however, the previous induction performed by EFV should have increased the amount of efflux transporter at the apical membrane of enterocytes and at the hepatobiliary membrane of hepatocytes. Consequently, substrates of Pgp, like RFP (Hosagrahara et al. 2013), would not be able to exert its enzymatic inductive effect either in the intestine or in the liver because of the lower concentrations of RFP in both tissues caused by Pgp extrusion.

In the case of NVP, elimination depends upon mainly CYP3A4 followed by CYP2B6, being responsible for hydroxyl derivatives (Riska et al. 1999). Although metabolism autoinduction has been reported, resulting in a decrease in the terminal-phase half-life of NVP from approximately 45 h (single dose) to approximately 25–30 h (multiple dosing with 200–400 mg/day) (Mugabo et al. 2011), it should be taken into consideration that this study was carried out in preterm infants, so the evolution in clearance may be explained by the natural maturation of eliminating pathways in the liver. More recent data show that metabolic ratio may change unevenly. An increase in metabolic ratios for CYP2B6 was reported, whereas metabolic ratio decreased for metabolites produced by CYP3A4 isoenzyme. This study was carried out in two different populations: single oral dose of 200 mg NVP in 10 HIV-infected African-Americans and at steady state 200 mg twice daily in 10 HIV-infected Cambodians (Fan-Havard et al. 2013).

Under this evidence, NVP metabolism autoinduction seems to be at least controversial. More reasonable is to think that NVP fictitiously reduces its half-life when passing from single- to multiple-dose regimens. Due to the multiple secondary peaks NVP shows after single-dose administration, its longer terminal pharmacokinetic phase is confounded as having a higher half-life than it really has (Ibarra et al. 2014). After multiple-dose administration is not possible to see its longer terminal phase, hence a higher decline in drug plasma concentrations is reported. The finding that RPF addition to treatments reduces NVP concentration significantly, leading to the need for dose adjustment after tuberculosis treatment initiation (Cohen and Meintjes 2010), supports its distinctive behavior with respect to EFV.

The same reasoning as for PIs may be applied to calcineurin inhibitors, tacrolimus (TAC) and CYA. Both drugs share not only a similar mechanism of action but also their metabolism profile, being substrate of CYP3A and Pgp (Amundsen et al. 2012; Fricker et al. 1996; Saeki et al. 1993). Again, the incorporation of RFP to the treatment has proved to be hazardous for patients due to the plummeting of the immunosuppressant concentrations (Naylor and Robichaud 2013; Valk-Swinkels et al. 2013).

The interaction of drugs that are substrate of enzymes and/or transporters with RFP may pose patients at higher risks for treatment failures, but a thorough study is needed with the objective of evaluating the real significance in clinical practice.

12.4.3 Valproic Acid-Lamotrigine

Lamotrigine (LTG) is an important anticonvulsant which has a wide range of activity against both partial and generalized onset seizures (Fitton and Goa 1995), but its use is associated with a significant incidence of idiosyncratic drug reactions, mainly when it is administered with VPA (Schlienger et al. 1998).

The pharmacokinetic profile of LTG demonstrates good absorption after oral administration, a linear relation between dose and plasma concentrations, 55% protein binding, and an elimination half-life of 25–30 h under monotherapy (Fitton and Goa 1995) and 60 h approximately when combined with VPA (Yuen et al. 1992).

LTG is primarily metabolized by UGTs leading to the formation of two major metabolites: the N-2 glucuronide and the N-5 glucuronide. A minor metabolism pathway is the formation of the N-2 methyl and the N-2-oxide-LTG (Doig and Clare 1991). Because of this level of dependency on glucuronidation for LTG inactivation and elimination, anything that inhibits UGT enzymes will affect LTG levels. VPA is a known inhibitor of UDP-glucuronosyltransferases (Lu and Uetrecht 2007), and LTG clearance decreases if both drugs are administered concomitantly (Lalic et al. 2009). The o-dichlorophenyl moiety is a potential site for bioactivation of the drug to an arene oxide in the absence of a major competing pathway such as N-glucuronidation (Maggs et al. 2000). This reactive metabolite is believed to be an essential component in the development of hypersensitivity reactions as it interacts covalently (hapten) with cellular macromolecules (Naisbitt 2004). Microsomal epoxide hydrolase is responsible for the detoxification of the arenoxide; however such enzyme is inhibited by VPA (Kerr et al. 1989) resulting in increased covalent adduct formation. This could lead to the frequent adverse effects, mainly rash and other cutaneous reactions, reported in the literature (Faught et al. 1999).

12.4.4 Azole Antifungals-Cyclosporine

Calcineurin inhibitors such as CYA have been the mainstay of immunosuppression for the last decades and nowadays still remain as relevant drugs in the setting of solid-organ transplantation and autoimmune diseases.

Increased emphasis is now placed on the role of Pgp in the low and variable oral bioavailability of CYA (Lown et al. 1997). Mainly CYP3A4, located in the liver and the intestine, is responsible for the biotransformation of CYA. This enzyme together with Pgp is involved in the absorption and presystemic elimination of CYA. CYA seems not only to be substrate of CYP3A and Pgp but has also been described as inhibitor of both the enzyme and the efflux transporter (Amundsen et al. 2012).

Immunosuppressive drugs put down the immune system, so individuals under these medications are at an increased risk for infections. Systemic fungal infections are a major threat for immunocompromised patients, and antifungal therapy is commonly administered. Among the most widely prescribed antifungal drugs are the azoles. Ketoconazole (KCZ) was the first drug used for many fungal infections. It was announced (FDA 2013) that clinicians should no longer prescribe KCZ tablets as a first-line therapy for any fungal infection, because of the risk for severe liver injury, adrenal insufficiency, and adverse drug interactions, and therefore, it has been replaced by other azoles such as itraconazole (ICZ), fluconazole (FCZ), and voriconazole (VCZ). All azoles are substrates for and inhibitors of cytochrome P450 (CYP450) isoenzymes, as well as inhibitors of Pgp. KCZ is the most potent inhibitor of CYP3A4, followed by ICZ and VCZ (approximately equipotent) and then FCZ. As with CYP enzymes, azoles differ in their affinity for, as well as their ability to inhibit, Pgp (Wang et al. 2002). So, the clinical relevance of interactions may vary.

Many studies were carried out regarding the interaction between KCZ and CYA, showing an important increase in blood CYA concentrations (Ferguson et al. 1982; Daneshmend 1982).

FCZ may be a substrate for active transport by Pgp, with very slow passive diffusion across the membrane, thus impeding reentry and interaction with the substrate binding site (Eytan et al. 1996; Ferte 2000). Indeed, the FCZ molecule is significantly shorter, smaller, and more hydrophilic than the other antifungals considered herein. Some authors found that oral and intravenous immunosuppressive therapies do not appear to demonstrate the same degree of interaction with FCZ. While given orally CYA concentrations increased 2–3-fold in that time period after starting FCZ (Canafax et al. 1991). However, no clinically significant increases in CYA levels, given intravenously, were observed with administration of high-dose FCZ 400 mg/day (Osowski et al. 1996). This lack of interaction most likely highlights the role that FCZ inhibition has on gut metabolism of CYA, enhancing its bioavailability and reducing its clearance. Then, monitoring of blood levels is required when oral doses of CYA are used.

Some authors (Leather et al. 2006) conducted a study of the interaction between intravenous ICZ and intravenous CYA, and a mean increase of 80% in CYA concentration was observed. In a double-blind, randomized, placebo-controlled, crossover trial in seven renal transplant recipients, a clinically significant increase in CYA exposure (AUC increased 1.7-fold) was observed during concomitant administration of VCZ (Romero et al. 2002).

For KCZ, ICZ, and VCZ, systemic as well as presystemic inhibition is operating. The data suggest that dose reductions of CYA, independently on the administration route used, are necessary when therapy with these antifungals is initiated and that subsequent close monitoring of blood concentrations is necessary to guide further dose modifications.

12.5   Environmental pH

It is well known that most drugs available are either weak acids or bases; as a consequence pH environment plays a key role in displacing acid-base equilibrium, transforming drug from its ionizable into non-ionizable form or vice versa, being the latter the one absorbable.

Latest clinical data available suggests that pH changes impact larger in compounds whose in vitro solubility varies over the pH range of 1–4. This may seem not a wide range; however it represents the normal physiological gastric acidity (pH ~1) and gastric acidity while on acid-reducing agents (pH ~4). The most relevant interaction in this sense is presented for weak bases, which have low solubility at higher pHs exposing patients to impaired absorption thus leading to reduced and variable bioavailability (Budha et al. 2012).

Stomach pH increase is common, as the use of proton pump inhibitor (PPI) and/or H2 receptor antagonists (H2RA) has rocketed. These pharmacological groups are commonly used in the clinical setting for the alleviation of gastroesophageal reflux disease. Nowadays, their overuse and misuse have raised their potential for drug interactions.

There are plenty of reports in the literature that address this topic. Oral chemo-therapeutic drugs show in general little or no interaction with acid suppression therapy. Major concerns have been raised lately with tyrosine kinase inhibitors. These anticancer drugs have warnings in their inserts to avoid PPIs because drugs require an acidic environment to be fully absorbed. Several studies have been conducted comprising erlotinib and dasatinib (Jawhari et al. 2014; Yang et al. 2017; Steinberg 2007; Keam 2008).

Another group of concern is HIV drugs. As indicated in a retrospective study, patients on a treatment of 800 mg indinavir (INV) three times daily combined with 20–40 mg omeprazole (OMPZ) presented lower plasma AUC than patients on INV alone (Burger et al. 1998). OMPZ co-administration (40 mg daily) to nelfinavir (NFV) also decreases the AUC values of the antiretroviral due to decrease in NFV solubility by gastric acid secretion suppression (Fang et al. 2008). Concomitant use of atazanavir (AZV) and OMPZ is not recommended, the use of the PPI may reduce AZV exposure by 75%, and this effect can be partially reverted by enzyme inhibition OMPZ exerts over AZV metabolism (Zhu et al. 2011).

Azole antifungals, such as KCZ, have a gastric acidity-dependent solubility. This compound has been left behind due to the high potential for drug interactions, but its analog ICZ is present in many antifungal therapies. The total exposure of ICZ, measured by AUC0-24 and C_{max}, significantly decreased (64% and 66%, respectively) by the administration of 40 mg OMPZ for 2 weeks (Johnson et al. 2003).

Altering pH may also impact on the absorption of ions such as calcium (Ca2+), magnesium (Mg2+), and iron (Fe2+/3+). In the case of PPIs by inhibiting H+/K+-ATPase, they induce hypochlorhydria which in turn decreases ingested salt solubility resulting in less ionized electrolyte in the proximal intestinal and duodenal lumen available for absorption.

However, it should also be taken into consideration that active transport of these metals could also be pH-dependent. For instance, Mg2+ homeostasis is maintained by two processes, gastrointestinal absorption and renal excretion. Gastrointestinal Mg2+ absorption occurs through both passive paracellular movement and active transport through the combined action of TRPM6/7 channels, present in the apical membrane of enterocytes. TRPM transporters and intercellular pores are pH sensitive, having the former less affinity for the ion at lower pHs.

Proton pumps have been well studied in the stomach, but they are also present in other secretory organs such as the pancreas. In this tissue they are associated not to HCl but to HCO_3 secretion. In this sense, the inhibition of the pump results in a decrease in the pH of pancreas secretion, thus decreasing pH at the duodenum. Because of this, PPI treatment is uneven, increasing the stomach pH and decreasing the intestinal one (Michalek et al. 2011); however both mechanisms impact causing lower Mg2+ absorption, the first due to Mg2+ solubility decrease and the second one owing to diminishing the affinity of TRPM6/7 channels for Mg2+ (Perazella 2013).

Given the widespread use of PPIs, it is important to know whether PPIs are a risk factor for hypomagnesemia in routine clinical practice. A study carried out in 51 hospitalized patients in which Mg2+ concentration and OMPZ treatment duration were correlated found that OMPZ given at a dose of 20 mg daily has low risk of generating hypomagnesemia even when taken for a long period of time (Maldonado et al. 2015). Nevertheless, clinicians should be aware that higher doses of OMPZ or the combination with drugs such as CYA or platinum derivatives may synergize Mg2+ loss. Apart from that, other factors such as patient characteristics (age, habits) and/or underlying conditions that impair renal uptake of this electrolyte deregulate the homeostasis resulting in low levels of Mg2+.

Regarding calcium, according to a literature review (Ito and Jensen 2010), evidence is again conflicting. Many studies reported lower calcium levels in patients treated with PPI (Yang et al. 2006; Vestergaard et al. 2006), while others showed no significant decrease of the ion (Wright et al. 2010). The most widely assumed mechanism in PPI use leads to decreased intestinal absorption of calcium resulting in negative calcium balance and osteoporosis risk. Other authors also put forward the induction of hypergastrinemia resulting in parathyroid hyperplasia/hypertrophy and increased PTH secretion with the development of secondary hyperparathyroidism (Moayyedi and Cranney 2008; Mizunashi et al. 1993; Roux et al. 2009). What most studies support is the idea that for a significant interaction between Ca2+ and PPI, two relevant factors are to be accomplished, PPI high doses and long duration of treatments (Yang et al. 2006; Corley 2009), similarly as it was mentioned for Mg2+ (Maldonado et al. 2015).

For iron, the body of evidence follows the same path as for calcium. There are numerous animal and human studies whose conclusion is that gastrointestinal pH affects Fe2+ absorption, especially when the pH changes are induced by PPIs. The mechanism of this interaction is explained with dietary sources of Fe2+, but it is applicable to medicines containing iron supplements. The absorption of all forms of iron (hem and non-hem) is markedly improved by gastric acidity. Acid pH helps to solubilize the iron salts, which allows them to be reduced to the ferrous state (Fe2+).

This facilitates the formation of complexes that facilitate Fe+2 absorption (Jensen 2006; Koop and Bachem 1992; Miret et al. 2003). In a case report, two anemic patients did not respond to oral iron treatment, and only after PPI was withdrawn, iron status improved (Sharma et al. 2004). Other studies have reported no changes on iron levels with long-term gastric acid antisecretory therapy (Stewart et al. 1998; Koop and Bachem 1992).

Case Study
Two patients, aged 51 and 83, were diagnosed with iron-deficiency anemia, which on endoscopy was found to be a result of upper GI bleeding from "erosive gastritis."

Both patients were placed on omeprazole and oral iron supplementation. After 6 months of treatment, anemia persisted in both patients, and malabsorption of iron as a result of the omeprazole was suspected. Two months after omeprazole discontinuation, hemoglobin and mean corpuscular volume were measured and had experienced significant increase.

This is based on a case report by Sharma et al. (2004).

Finally, to draw strong conclusions regarding the risk of malabsorption of minerals, there is a need for better designed studies where doses and long-term gastric acid antisecretory therapy are evaluated.

12.6 Food Effect

Pharmacokinetic food-drug interactions have been extensively reviewed in the literature (Melander 1978; Schmidt and Dalhoff 2002; Singh 1999). Given the physiological changes taking place at the gastrointestinal tract (GIT) following a meal intake, bioavailability of subsequent oral administrations can result dramatically affected with obvious implications for systemic drug exposure and pharmacodynamic outcome. Moreover, as it will be discussed, food-drug interactions can alter drug distribution and systemic elimination. Overall, food effects on drug pharmacokinetics can increase or decrease body exposure depending on drug and formulation characteristics because of a complex combination of factors affecting drug dissolution, presystemic metabolism, intestinal permeability, and/or disposition. Given its impact, this interaction has been receiving more attention in the last decades during drug development and in bioequivalence studies (Center for Drug Evaluation and Research. Food and Drug Administration 2002). Several approaches, as the Biopharmaceutical Classification System (BCS) framework, are used to predict food effects on drug bioavailability (Amidon et al. 1995). The BCS was further expanded by Wu and Benet (Wu and Benet 2005) to the Biopharmaceutics Drug Disposition Classification System (BDDCS) and applied to food-drug interaction prediction by Custodio and coworkers (Custodio et al. 2008). Pharmacodynamic interactions where food nutrients affect drug mechanism of action altering the concentration versus effect relationship have also been reported. This exceeds the scope of the present chapter and will not be analyzed here.

12.6.1 Food-Drug Interactions Based on Induced Changes in GIT Physiology

The GIT physiology is designed for food digestion, efficient nutrient absorption, and organism protection against possible threats. The magnitude of food-induced changes at this level is highly dependent on meal composition: caloric intake, protein and fat content, and physical state (liquid/solid), among other characteristics that must be considered while evaluating food effects. In addition, subject baseline characteristics are frequently not similar, and the response to a meal intake is commonly affected by a significant interindividual variability. For example, in fasting conditions men and women show significant differences in gastric pH (men<women), mean gastric residence time (men<women), and intestinal transit (men>women). In fed conditions, sex differences in GIT transit times are conserved, but gastric pH difference is minimized. Finally, drug product characteristics like dosage forms, excipients, tablet size and dissolution rate, as well as drug affinities for efflux transporters and enzymatic complex expressed at the gut wall can also influence how oral bioavailability is affected by food presence in the GIT. Therefore, although general GIT-induced changes can be summarized, the oral bioavailability outcome of the food effect should be further analyzed on a case-by-case basis.

At the gastric level, meal intake increases intraluminal volume, fluid pH, and stomach motility. A basal gastric pH of 1–2 can reach values of 3–5 after caloric or high fat content meals (Sutton et al. 2017). This effect is not similar for high-protein and high-carbohydrate meals: the former has shown to produce a smaller change both in maximum pH and time of basal pH recovery (Lennard-Jones et al. 1968). Alterations in gastric acidity can modify in vivo dissolution of pH-dependent solubility compounds, mostly BCS Class 2 drugs (low solubility/high permeability): a faster dissolution would be observed for weak acids in the fed state, while the opposite effect would be expected for weak bases. However, the most relevant food-induced change in the stomach is the delayed and extended emptying. Gastric mean residence times for high-caloric liquids and solids are 30 and 60 min in male and 54 and 92 min in female subjects, respectively (Datz et al. 1987). This increased gastric residence allows basic drugs to complete their dissolution in the stomach. The hierarchy of this factor stands in this point: a faster drug dissolution in gastric fluid can have no pharmacokinetic impact if the emptying rate is reduced, limiting the rate at which dissolved drug reaches the absorption site. This is often observed in bioequivalence studies: pharmacokinetic differences between two immediate release products showing different drug dissolution rates can be hidden in fed conditions where gastric emptying becomes the rate-limiting step for drug absorption.

At the small intestine, the transit time is usually not affected by meal intake (Kenyon et al. 1995). Relevant food effects are associated to the stimulation of pancreatic juice and bile secretion into the duodenum. Bile salts can importantly increase the solubility of lipophilic compounds by micellar solubilization. Pancreatic secretion includes lipases, amylases, and proteases. Lipases can directly affect drug release from fat−/oil-containing dosage forms and indirectly increase lipophilic compound solubility by the digestion of triglycerides present in the food. In general, poorly soluble drugs (BCS Class 2 and 4) will show enhanced intraluminal solubility when given postprandially, leading to an increased C_{max} and in some cases to a higher AUC. Meal intake also increases the cardiac output and the blood flow distribution toward the splanchnic region, enhancing drug uptake at the basolateral side of enterocytes. Drugs of high permeability (BCS Class 1 and 2) could have higher C_{max} and even AUC because of a reduction in hepatic first-pass metabolism because of such increased blood flow rate.

The pharmacokinetic outcome of food-drug interactions explained by changes in GIT physiology, as mentioned before, is wide. For drugs of high solubility and permeability (BCS Class 1), the effect is frequently limited to a reduction in absorption rate explained by a longer gastric residence time. In this case, only C_{max} became affected, having clinical significance in drugs where a rapid onset of action is pursued such as sedatives and analgesics. In some cases, drug retention at the stomach can lead to a decreased bioavailability due to compound instability at gastric conditions or intragastric metabolism. Ethanol is an example of the latter mechanism; Wu and coworkers observed in healthy male subjects a delayed gastric emptying and a decreased ethanol AUC after administration of a distilled alcohol mixed with a regular beverage, relative to the administration of the same alcohol dose mixed with "diet" beverage (Wu et al. 2006). Glucose presence in liquids has been shown to diminish gastric emptying rate, therefore increasing ethanol presystemic metabolism in stomach mediated by the gastric alcohol dehydrogenase.

Drug dosage form has also a relevant role. In a ketoprofen (KPF) bioavailability and bioequivalence study conducted in 16 healthy subjects (8 males, 8 females) between an immediate-plus-extended bilayer (BL) and a delayed-release (DR) formulations, food effect was additionally evaluated within a 4-period, 2-formulation, 2-administration design. Ingested food consisted in a standard breakfast taken 5 min before drug administration. The postprandial impact in KPF bioavailability was observed for both formulations in C_{max}, lag time (T_{lag}), and time of C_{max} (T_{max}). KPF absorbed amount assessed through the AUC resulted unaffected by food intake. As it can be observed in the mean profiles included in Fig. 12.5, the reduction in KPF C_{max} was higher for the BL formulation pharmacokinetic profile, which showed no significant changes in either T_{max} or T_{lag}. In this case, the food-induced prolonged gastric emptying reduced the rate to which the drug reached the absorption site at the gut. For the DR formulation, due to its gastro-resistant coating, drug release and dissolution in gut fluids became postponed, impacting the onset of KPF dissolution and therefore pharmacokinetic parameters T_{lag} and T_{max} in a significant way. Lag times up to 22 h were observed for some female subjects. Results of this work are partially published (Magallanes et al. 2016).

An illustrative example of the food-drug interaction complexity is LPV postprandial bioavailability increase after administration of LPV/RTV tablets (Ibarra et al.

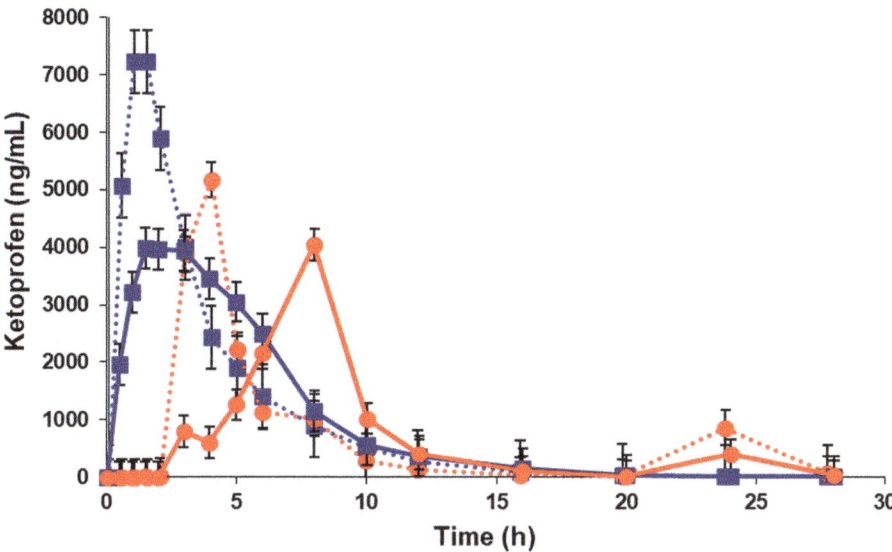

Fig. 12.5 Mean plasma levels of ketoprofen in healthy subjects (8 males, 8 females) after administration of an immediate-plus-extended bilayer formulation (blue lines) and a delayed-release formulation (red lines) in fed (continuous lines) and fasting (dotted lines) conditions

2012). In this antiretroviral combination, RTV is present as a booster of LPV bioavailability. Both are Class 2 drugs with limited oral absorption given its extensive presystemic metabolism mediated by the cytochrome CYP3A4 and the efflux transporter Pgp. RTV, which shows pH-dependent solubility (increasing at acidic conditions), has higher affinity for CYP3A4 and Pgp, being included in a subtherapeutic amount as inhibitor of LPV gut-mediated metabolism. Interestingly, Pgp gut-wall expression increases from the duodenum to the ileum, causing poorly soluble drugs with high Pgp affinity to show an "absorption window" in the proximities of the pylorus. Because of this effect, LPV co-administered with RTV increases its bioavailability when switching from oral solution to capsules and from the latter to tablets: drug absorption benefits with a slower gastric release. Observations from a LPV/RTV bioequivalence study performed under fasting conditions revealed an important inter- and intraindividual variability: 72 healthy subjects had to be included in the study to narrow down test/reference 90% confidence intervals for C_{max} and AUC and reach a bioequivalence conclusion. When the same formulations were compared under fed conditions, thanks to an increased LPV bioavailability and a dramatic reduction in pharmacokinetic variability, bioequivalence was concluded with only 16 healthy subjects. It was proposed that food intake reduced gastric pH differences among subjects (and between sexes) producing therefore a less variable RTV in vivo dissolution. In addition, the slower gastric emptying increased LPV bioavailability.

Finally, for Class 2 drugs, bioavailability is often increased in postprandial administrations thanks to an increased in vivo solubility mediated by dietary and endogenous surfactants, such as bile acids. For example, CYA bioavailability from olive oil formulation, Sandimmune® (Novartis), was shown to significantly increase when given after a fat-rich meal. However, this effect is not observed for the Neoral® formulation (Novartis) which presents the drug in a microemulsion, reducing dissolution dependence of endogenous surfactants (Mueller et al. 1994).

12.6.2 Food-Drug Interactions Based on Specific Mechanisms

Food constituents can also selectively affect oral bioavailability by different mechanisms: (i) modifying drug presystemic clearance by altering efflux transport and/or presystemic metabolism, (ii) reducing drug permeability by competition or inhibition of active uptake at the gut, and (iii) reducing drug absorption by chemical (i.e., chelation) or physical (i.e., binding) interactions at the GIT lumen. A typical example of (i) is the presystemic metabolism inhibition mediated by grapefruit juice, which causes a bioavailability increase in CYP3A4 and Pgp substrates such as CYA, felodipine, atorvastatin, and CBZ, among many others (Schmidt and Dalhoff 2002). It is important to notice that most of the interacting actions dealt in this chapter lasted more than the body residence time of the interacted drug. For instance, the inhibition of transporters and enzymes, as grapefruit juice does, may take a long period for retrieving patient's physiological basal status, and then, a delay between the ingestion of the interacting and the interacted agents does not avoid the reported pharmacokinetic changes.

When melphalan is given orally, fasting conditions are suggested due to a reduced bioavailability in postprandial conditions (from 85% to 58%) associated to absorption competition (ii) with dietary amino acids (Reece et al. 1986).

A classic example for the (iii) mechanism can be found in ciprofloxacin (CPFX) absorption, which forms chelates with multivalent cations such as aluminum, iron, magnesium, and calcium. Diminished CPFX bioavailability has been observed after co-administration with calcium-fortified orange juice, milk, and yogurt (Myers Davit and Conner 2008).

12.6.3 Food Effects on Bioequivalence and Regarding the Sex of Individuals

As mentioned above, formulation and sex-related differences in bioavailability can be less or more pronounced in the fed state. Additionally, as described elsewhere, bioequivalence evaluation in male or female subjects can sometimes led to different conclusions, a phenomenon known as sex-by-formulation interaction (Ibarra et al. 2017). This happens because men and women present a dissimilar gastrointestinal physiology, causing drug product differences to arise or result suppressed. As such, the magnitude of this interaction is commonly affected by the presence of food, which produces significant changes in gastrointestinal environment. In a EFV bioequivalence study carried out with eight female and six male subjects, a significant sex-by-formulation interaction was observed for the C_{max} test/reference mean ratio (Ibarra et al. 2016). While no differences were detected in women, men showed a point estimate of 0.705 and a CI90% of 0.54–0.92. Reference product had higher C_{max} in male subjects only. In vitro-in vivo correlations showed that this interaction was promoted by meal intake 2 h before drug administration. As a BCS Class 2 drug, EFV bioavailability is increased with the presence of surfactants. Formulation differences were observed in vitro under surfactant-free acidic media: reference product dissolved EFV more efficiently. When a surfactant was added to the dissolution media, no differences between products were detected. Due to a more rapid gastric emptying, the 2-hour period between food and drug intake was enough for men to recover a fasting, surfactant-free state, in which product differences arose. In women, the slower gastrointestinal transit conserved a fed state, and therefore the test product achieved a similar C_{max} than the brand-name drug.

12.6.4 Food Effects on Drug Disposition

Dietary components can modify plasma protein binding, drug metabolism, and excretion from the body. For example, high-protein diets have shown to accelerate oxidative hepatic metabolism of antipyrine and theophylline. Lipid intake in fat-rich diets is thought to compete for plasmatic albumin decreasing protein binding of some drugs, and although this will have no impact over drug effects, it can significantly affect drug pharmacokinetic profiles (Williams et al. 1993). Modification of urinary pH (normal value around 6.5) can alter the excretion of weak acids and bases by increasing or reducing the fraction reabsorbed by passive diffusion at the renal tubule. This has been shown measuring methylamphetamine excretion, a weak base of pKa ~ 9.8: renal excretion results decreased when urine is alkalinized (~10% of the dose excreted in 12 h post-dose) and increased when urine is acidified (~70% of the dose excreted in the same post-dosing period) (Beckett and Rowland 1965). Urine pH can be increased by chronic ingestion of sodium bicarbonate and potassium citrate and decreased by ingestion of ammonium chloride, among other substances. For subjects under specific diets, this factor should be considered upon evaluation of drug pharmacokinetics.

References

Aguado JM, Torre-Cisneros J, Fortún J et al (2009) Tuberculosis in solid-organ transplant recipients: consensus statement of the group for the study of infection in transplant recipients (GESITRA) of the Spanish Society of Infectious Diseases and Clinical Microbiology. Clin Infect Dis 48:1276–1284

Akhlaghi F, Trull AK (2002) Distribution of cyclosporin in organ transplant recipients. Clin Pharmacokinet 41:615–637

Alvariza S, Ibarra M, Vázquez M et al (2014) Different oral phenytoin administration regimens could modify its chronic exposure and its saliva/plasma concentration ratio. J Med Pharm Innov 1:35–43

Amidon G, Lennernäs H, Shah V et al (1995) A theoretical basis for a biopharmaceutic drug classification: the correlation of in vitro drug product dissolution and in vivo bioavailability. Pharm Res 12:413–420

Amundsen R, Åsberg A, Ohm IK et al (2012) Cyclosporine A- and tacrolimus-mediated inhibition of CYP3A4 and CYP3A5 in vitro. Drug Metab Dispos 40:655–661

Asberg A (2003) Interactions between cyclosporin and lipid-lowering drugs: implications for organ transplant recipients. Drugs 63:367–378

Beckett AH, Rowland M (1965) Urinary excretion kinetics of methylamphetamine in man. J Pharm Pharmacol 17:109–114

Bertz R, Hsu A, Lam W et al (2000) Pharmacokinetic interactions between Kaletra (lopinavir/ritonavir or ABT-378/r) and other non-HIV drugs. Fifth international congress on drug therapy in HIV infection. Abstract P291, Poster 438. Glasgow (UK), 22–26 Oct 2000

Boffito M, Dickinson L, Hill A et al (2004) Steady-state pharmacokinetics of saquinavir hard-gel/ritonavir/fosamprenavir in HIV-1-infected patients. J Acquir Immune Defic Syndr 37:1376–1384

Bowdle TA, Levy RH, Cutler RE (1979) Effect of carbamazepine on valproic acid clearance in normal man. Clin Pharmacol Ther 26:629–634

Budha NR, Frymoyer A, Smelick GS et al (2012) Drug absorption interactions between oral targeted anticancer agents and PPIs: is pH-dependent solubility the Achilles heel of targeted therapy? Clin Pharmacol Ther 92:203–213

Burger DM, Hugen PW, Kroon FP et al (1998) Pharmacokinetic interaction between the proton pump inhibitor omeprazole and the HIV protease inhibitor indinavir. AIDS 12:2080–2082

Canafax DM, Graves NM, Hilligoss DM et al (1991) Interaction between cyclosporine and fluconazole in renal allograft recipients. Transplantation 51:1014–1018

Carr RB, Shrewsbury K (2007) Hyperammonemia due to valproic acid in the psychiatric setting. Am J Psychiatry 164:1020–1027

Center for Drug Evaluation and Research. Food and Drug Administration (2002) Guidance for industry. Food-effect bioavailability and fed bioequivalence studies. http://www.fda.gov/cder/guidance/index.htm

Chen J, Raymond K (2006) Roles of rifampicin in drug-drug interactions: underlying molecular mechanisms involving the nuclear pregnane X receptor. Ann Clin Microbiol Antimicrob 5:3

Cohen K, Meintjes G (2010) Management of individuals requiring ART and TB treatment. Curr Opin HIV AIDS 5:61–69

Corley DA (2009) Proton pump inhibitor, H2 antagonists, and risk of hip fracture: a large population-based study [abstract]. Gastroenterology 136:A70

Custodio JM, Wu CY, Benet LZ (2008) Predicting drug disposition, absorption / elimination / transporter interplay and the role of food on drug absorption. Adv Drug Deliv Rev 60:717–733

Daneshmend TK (1982) Ketoconazole-cyclosporin interaction. Lancet 2:1342–1343

Datz F, Christian P, Moore J (1987) Gender-related differences in gastric emptying. J Nucl Med 28:1204–1207

Doig MV, Clare RA (1991) Use of thermospray liquid chromatography-mass spectrometry to aid in the identification of urinary metabolites of a novel antiepileptic drug, lamotrigine. J Chromatogr 554:181–189

Eiraldi R (1997) Interactions on digoxin. Dissertation for doctor in chemistry thesis, Universidad de la República (Uruguay)

Eiraldi R, Sánchez S, Olano I et al (2004) Study of drug interactions of Cyclosporine A in two renal transplant patients. Revista O.F.I.L. 14:13–23

European Medicines Agency (EMA) (2012) Guideline on the investigation of drug interactions. CPMP/EWP/560/95/Rev. 1 Corr. 2**

Eytan GD, Regev R, Oren G et al (1996) The role of passive transbilayer drug movement in multidrug resistance and its modulation. J Biol Chem 271:12897–12902

Fagiolino P (2004) Multiplicative dependence of the first order rate constant and its impact on clinical pharmacokinetics and bioequivalence. Eur J Drug Metab Pharmacokinet 29:43–49

Fagiolino P, Vázquez M, Eiraldi R et al (2011) Efflux transporter influence on drug metabolism: theoretical approach for bioavailability and clearance prediction. Clin Pharmacokinet 50:75–80

Fagiolino P, Vázquez M, Alvariza S et al (2014) Antiepileptic drugs: energy-consuming processes governing drug disposition. Front Biosci (Elite Ed) E6:387–396

Fan-Havard P, Liu Z, Chou M et al (2013) Pharmacokinetics of phase I nevirapine metabolites following a single dose and at steady state. Antimicrob Agents Chemother 57:2154–2160

Faught E, Morris G, Jacobson M et al (1999) Adding lamotrigine to valproate: incidence of rash and other adverse effects. Epilepsia 40:1135–1140

Fang AF, Damle BD, LaBadie RR et al (2008) Significant decrease in nelfinavir systemic exposure after omeprazole coadministration in healthy subjects. Pharmacotherapy 28:42–50

Ferguson RM, Sutherland DE, Simmons RL et al (1982) Ketoconazole, cyclosporin metabolism, and renal transplantation. Lancet 2:882–883

Ferte J (2000) Analysis of the tangled relationships between P-glycoprotein-mediated multidrug resistance and the lipid phase of the cell membrane. Eur J Biochem 267:277–294

Fitton A, Goa KL (1995) Lamotrigine. An update of its pharmacology and therapeutic use in epilepsy. Drugs 50:691–713

Food and Drug Administration (2013) FDA [7-26-2013]. Available on https://www.fda.gov/Drugs/DrugSafety/ucm362415.htm

Freeman DJ (1991) Pharmacology and pharmacokinetics of cyclosporine. Clin Biochem 24:9–14

Fricker G, Drewe J, Huwyler J et al (1996) Relevance of p-glycoprotein for the enteral absorption of cyclosporin A: in vitro–in vivo correlation. Br J Pharmacol 118:1841–1847

Funakoshi S, Murakami T, Yumoto R et al (2005) Role of organic anion transporting polypeptide 2 in pharmacokinetics of digoxin and beta-methyldigoxin in rats. J Pharm Sci 94:1196–1203

Ghodke-Puranik Y, Thorn CF, Lamba JK et al (2013) Valproic acid pathway: pharmacokinetics and pharmacodynamics. Pharmacogenet Genomics 23:236–241

Greiner B, Eichelbaum M, Fritz P et al (1999) The role of intestinal P-glycoprotein in the interaction of digoxin and rifampin. J Clin Invest 104:147–153

Gugler R, von Unruh GE (1980) Clinical pharmacokinetics of valproic acid. Clin Pharmacokinet 5:67–83

Hamer JHM, Knake S, Schomburg U et al (2000) Valproate-induced hyperammonemic encephalopathy in the presence of topiramate. Neurology 54:230–232

Hirsh J, Fuster V, Ansell J et al (2003) American Heart Association/American College of Cardiology Foundation guide to warfarin therapy. Circulation 107(12):1692–1711

Hosagrahara V, Reddy J, Ganguly S et al (2013) Effect of repeated dosing on rifampin exposure in BALB/c mice. Eur J Pharm Sci 49:33–38

Ibarra M, Fagiolino P, Vázquez M et al (2012) Impact of food administration on lopinavir-ritonavir bioequivalence studies. Eur J Pharm Sci 46:516–521

Ibarra M, Vázquez M, Fagiolino P (2014) Population pharmacokinetic model to analyze nevirapine multiple-peaks after a single oral dose. J Pharmacokinet Pharmacodyn 41:363–373

Ibarra M, Magallanes L, Lorier M et al (2016) Sex-by-formulation interaction assessed through a bioequivalence study of efavirenz tablets. Eur J Pharm Sci 85:106–111

Ibarra M, Vázquez M, Fagiolino P (2017) Sex effect on average bioequivalence. Clin Ther 39:23–33

Ito T, Jensen RT (2010) Association of long-term proton pump inhibitor therapy with bone fractures and effects on absorption of calcium, vitamin B12, iron, and magnesium. Curr Gastroenterol Rep 12:448–457

Jawhari D, Alswisi M, Ghannam M et al (2014) Bioequivalence of a new generic formulation of erlotinib hydrochloride 150 mg tablets versus tarceva in healthy volunteers under fasting conditions. J Bioequiv Availab 6:119–123

Jensen RT (2006) Consequences of long-term proton pump blockade: highlighting insights from studies of patients with gastrinomas. Basic Clin Pharmacol Toxicol 98:4–19

Johnson MD, Hamilton CD, Drew RH et al (2003) A randomized comparative study to determine the effect of omeprazole on the peak serum concentration of itraconazole oral solution. J Antimicrob Chemother 51:453–457

Keam SJ (2008) Dasatinib: in chronic myeloid leukemia and Philadelphia chromosome positive acute lymphoblastic leukemia. BioDrugs 22:59–69

Kenyon CJ, Hooper G, Tierney D et al (1995) The effect of food on the gastrointestinal transit and systemic absorption of naproxen from a novel sustained release formulation. J Control Release 34:31–36

Kerr BM, Rettie AE, Eddy AC et al (1989) Inhibition of human liver microsomal epoxide hydrolase by valproate and valpromide: in vitro/in vivo correlation. Clin Pharmacol Ther 46:82–93

Kim RB, Fromm MF, Wandel C et al (1998) The drug transporter P-glycoprotein limits oral absorption and brain entry of HIV-1 protease inhibitors. J Clin Invest 101:289–294

Kodawara T, Masuda S, Wakasugi H et al (2002) Organic anion transporter oatp2-mediated interaction between digoxin and amiodarone in the rat liver. Pharm Res 19:738–743

Koop H, Bachem MG (1992) Serum iron, ferritin, and vitamin B12 during prolonged omeprazole therapy. J Clin Gastroenterol 14:288–292

La Porte CJ, Colbers EP, Bertz R et al (2004) Pharmacokinetics of adjusted-dose lopinavir-ritonavir combined with rifampin in healthy volunteers. Antimicrob Agents Chemother 48:1553–1560

Lalic M, Cvejic J, Popovic J et al (2009) Lamotrigine and valproate pharmacokinetics interactions in epileptic patients. Eur J Drug Metab Pharmacokinet 34:93–99

Lambert C, Lamontagne D, Hottlet H et al (1989) Amiodarone-digoxin interaction in rats. A reduction in hepatic uptake. Drug Metab Dispos 17:704–708

Leather H, Boyette RM, Tian L et al (2006) Pharmacokinetic evaluation of the drug interaction between intravenous itraconazole and intravenous tacrolimus or intravenous cyclosporin A in allogeneic hematopoietic stem cell transplant recipients. Biol Blood Marrow Transplant 12:325–334

Lennard-Jones JE, Fletcher J, Shaw DG (1968) Effect of different foods on the acidity of the gastric contents in patients with duodenal ulcer. Part III: effect of altering the proportions of protein and carbohydrate. Gut 9:177–182

Lheureux PER, Penaloza A, Zahir S et al (2005) Science review: carnitine in the treatment of valproic acid-induced toxicity—what is the evidence? Crit Care 9:431–440

Loscher W (1999) Valproate: a reappraisal of its pharmacodynamic properties and mechanisms of action. Prog Neurobiol 58:31–59

Lown KS, Mayo RR, Leichtman AB et al (1997) Role of intestinal P-glycoprotein (mdr1) in interpatient variation in the oral bioavailability of cyclosporine. Clin Pharmacol Ther 62:248–260

Lu W, Uetrecht JP (2007) Possible bioactivation pathways of lamotrigine. Drug Metab Dispos 35:1050–1056

Magallanes L, Lorier M, Ibarra M et al (2016) Sex and food influence on intestinal absorption of ketoprofen gastroresistant formulation. Clin Pharmacol Drug Dev 5:196–200

Maggs JL, Naisbitt DJ, Tettey JNA et al (2000) Metabolism of lamotrigine to a reactive arene oxide intermediate. Chem Res Toxicol 13:1075–1081

Mahatthanatrakul W, Nontaput T, Ridtitid W et al (2007) Rifampin, a cytochrome P450 3A inducer, decreases plasma concentrations of antipsychotic risperidone in healthy volunteers. J Clin Pharm Ther 32:161–167

Malaty LI, Kuper JJ (1999) Drug interactions of HIV protease inhibitors. Drug Saf 20:147–169

Maldonado C, Fagiolino P, Vázquez M et al (2011) Time-dependent and concentration-dependent upregulation of carbamazepine efflux transporter. A preliminary assessment from salivary drug monitoring. Lat Am J Pharm 30:908–912

Maldonado C, de Mello N, Fagiolino P et al (2015) Safe use of a daily 20-mg dose of omeprazole in order to avoid hypomagnesemia. Int J Pharm 5:315–321

Maldonado C, Guevara N, Queijo C et al (2016) Carnitine and/or acetylcarnitine deficiency as a cause of higher levels of ammonia. Biomed Res 2016:2920108

Maldonado C, Guevara N, Silveira A et al (2017) L-Carnitine supplementation to reverse hyperammonemia in a patient undergoing chronic valproic acid treatment: a case report. J Int Med Res 45:1268–1272

Mehndiratta MM, Mehndiratta P, Phul P et al (2008) Valproate induced non hepatic hyperammonaemic encephalopathy (VNHE)-a study from tertiary care referral university hospital, North India. J Pak Med Assoc 58:627–631

Melander A (1978) Influence of food on the bioavailability of drugs. Clin Pharmacokinet 3:337–351

Michalek W, Semler JR, Kuo B (2011) Impact of acid suppression on upper gastrointestinal pH and motility. Dig Dis Sci 56:1735–1742

Miret S, Simpson RJ, McKie AT (2003) Physiology and molecular biology of dietary iron absorption. Annu Rev Nutr 23:283–301

Mizunashi K, Furukawa Y, Katano K et al (1993) Effect of omeprazole, an inhibitor of H+, K(+)-ATPase, on bone resorption in humans. Calcif Tissue Int 53:21–25

Moayyedi P, Cranney A (2008) Hip fracture and proton pump inhibitor therapy: balancing the evidence for benefit and harm. Am J Gastroenterol 103:2428–2431

Mueller EA, Kovarik JM, van Bree JB et al (1994) Influence of a fat-rich meal on the pharmacokinetics of a new oral formulation of cyclosporine in a crossover comparison with the market formulation. Pharm Res 11:151–155

Mugabo P, Els I, Smith J et al (2011) Nevirapine plasma concentrations in premature infants exposed to single-dose nevirapine for prevention of mother-to-child transmission of HIV-1. S Afr Med J 101:655–658

Myers Davit B, Conner D (2008) Food effects on drug bioavailability: implications for new and generic drug development. In: Krishna R, Yu L (eds) Biopharmaceutics applications in drug development, vol 21. Springer US, Boston, pp 317–335

Naisbitt DJ (2004) Drug hypersensitivity reactions in skin: understanding mechanisms and the development of diagnostic and predictive tests. Toxicology 194:179–196

Naylor H, Robichaud J (2013) Decreased tacrolimus levels after administration of rifampin to a patient with renal transplant. Can J Hosp Pharm 66:388–392

Nijland HM, L'Homme RF, Rongen GA et al (2008) High incidence of adverse events in healthy volunteers receiving rifampicin and adjusted doses of lopinavir/ritonavir tablets. AIDS 22:931–935

Osowski CL, Dix SP, Lin LS et al (1996) Evaluation of the drug interaction between intravenous high-dose fluconazole and cyclosporine or tacrolimus in bone marrow transplant patients. Transplantation 61:1268–1272

Palleria C, Di Paolo A, Giofrè C et al (2013) Pharmacokinetic drug-drug interaction and their implication in clinical management. J Res Med Sci 18:601–610

Perazella MA (2013) Proton pump inhibitors and hypomagnesemia: a rare but serious complication. Kidney Int 83:553–556

Ramachandran G, Hemanth Kumar AK, Rajasekaran S et al (2009) CYP2B6 G516T polymorphism but not rifampin coadministration influences steady-state pharmacokinetics of efavirenz in human immunodeficiency virus-infected patients in South India. Antimicrob Agents Chemother 53:863–868

Ramanathan S, Mathias AA, German P et al (2011) Clinical pharmacokinetic and pharmacodynamic profile of the HIV integrase inhibitor elvitegravir. Clin Pharmacokinet 50:229–244

Reece PA, Kotasek D, Morris RG et al (1986) The effect of food on oral melphalan absorption. Cancer Chemother Pharmacol 16:194–197

Riska P, Lamson M, MacGregor T et al (1999) Disposition and biotransformation of the antiretroviral drug nevirapine in humans. Drug Metab Dispos 27:895–901

Romero AJ, Le Pogamp P, Nilsson LG et al (2002) Effect of voriconazole on the pharmacokinetics of cyclosporine in renal transplant patients. Clin Pharmacol Ther 71:226–234

Roth M, Obaidat A, Hagenbuch B (2012) OATPs, OATs and OCTs: the organic anion and cation transporters of the SLCO and SLC22A gene superfamilies. Br J Pharmacol 165:1260–1287

Roux C, Briot K, Gossec L et al (2009) Increase in vertebral fracture risk in postmenopausal women using omeprazole. Calcif Tissue Int 84:13–19

Saeki T, Ueda K, Tanigawara Y et al (1993) Human P-glycoprotein transports cyclosporin A and FK506. J Biol Chem 268:6077–6080

Schachter M (2004) Chemical, pharmacokinetic and pharmacodynamic properties of statins: an update. Fundam Clin Pharmacol 19:117–125

Schlienger RG, Knowles SR, Shear NH (1998) Lamotrigine-associated anticonvulsant hypersensitivity syndrome. Neurology 51:1172–1175

Schmidt LE, Dalhoff K (2002) Food-drug interactions. Drugs 62:1481–1502

Sharma VR, Brannon MA, Carloss EA (2004) Effect of omeprazole on oral iron replacement in patients with iron deficiency anemia. South Med J 97:887–889

Sibl A, Hrudikova-Vyskocilova E, Kacirova I et al (2017) Pharmacokinetic interaction between digoxin and amiodarone. Clin Ther 39:e82

Siemes H, Nau H, Schultze K et al (1993) Valproate (VPA) metabolites in various clinical conditions of probable VPA-associated hepatotoxicity. Epilepsia 34:332–346

Singh BN (1999) Effects of food on clinical pharmacokinetics. Clin Pharmacokinet 37:213–255

Steinberg M (2007) Dasatinib: a tyrosine kinase inhibitor for the treatment of chronic myelogenous leukemia and Philadelphia chromosome-positive acute lymphoblastic. Clin Ther 29:2289–2308

Stewart CA, Termanini B, Sutliff VE et al (1998) Iron absorption in patients with Zollinger-Ellison syndrome treated with long-term gastric acid antisecretory therapy. Aliment Pharmacol Ther 12:83–98

Sutton SC, Nause R, Gandelman K (2017) The impact of gastric pH, volume, and emptying on the food effect of ziprasidone oral absorption. AAPS J 19:1084–1090

Takada T, Yamanashi Y, Konishi K et al (2015) NPC1L1 is a key regulator of intestinal vitamin K absorption and a modulator of warfarin therapy. Sci Transl Med 7:275ra23

Usach I, Melis V, Peris JE (2013) Non-nucleoside reverse transcriptase inhibitors: a review on pharmacokinetics, pharmacodynamics, safety and tolerability. J Int AIDS Soc 16:1–14

Valk-Swinkels CG, Alidjan F, Rommers MK et al (2013) Low cyclosporin levels induced by the brief use of rifampicin; immunosuppression may fail for several weeks. Ned Tijdschr Geneeskd 157:A5667

Vázquez M, Fagiolino P, Mariño EL (2013) Concentration-dependent mechanisms of adverse drug reactions in epilepsy. Curr Pharm Des 19:6802–6808

Vázquez M, Fagiolino P, Maldonado C et al (2014) Hyperammonemia associated with valproic acid concentrations. Biomed Res Int 2014:217–269

Vestergaard P, Rejnmark L, Mosekilde L (2006) Proton pump inhibitors, histamine H2 receptor antagonists, and other antacid medications and the risk of fracture. Calcif Tissue Int 79:76–83

Vourvahis M, Kashuba AD (2007) Mechanisms of pharmacokinetic and pharmacodynamic drug interactions associated with ritonavir-enhanced tipranavir. Pharmacotherapy 27:888–909

Wang E, Lew K, Casciano CN, Clement RP, Johnson WW (2002) Interaction of common azole antifungals with P glycoprotein. Antimicrob Agents Chemother 46:160–165

Warner NJ, Barnard JT, Bigger JT (1985) Tissue digoxin concentrations and digoxin effect during the quinidine-digoxin interaction. J Am Coll Cardiol 5:680–686

Wessler JD, Grip LT, Mendell J et al (2013) The P-glycoprotein transport system and cardiovascular drugs. J Am Coll Cardiol 61:2495–2502

Williams L, Davis JA, Lowenthal DT (1993) The influence of food on the absorption and metabolism of drugs. Med Clin North Am 77:815–829

World Health Organization (2012) WHO policy on collaborative TB/HIV activities – guidelines for national programmes and other stakeholders. ISBN: 9789241503006

Wright MJ, Sullivan RR, Gaffney-Stomberg E et al (2010) Inhibiting gastric acid production does not affect intestinal calcium absorption in young healthy individuals: a randomized, crossover controlled clinical trial. J Bone Miner Res 25:2205–2211

Wu CY, Benet LZ (2005) Predicting drug disposition via application of BCS: transport/absorption/ elimination interplay and development of a biopharmaceutics drug disposition classification system. Pharm Res 22:11–23

Wu KL, Chaikomin R, Doran S et al (2006) Artificially sweetened versus regular mixers increase gastric emptying and alcohol absorption. Am J Med 119:802–804

Xue L, Holford N, Ding X-L et al (2017) Theory-based pharmacokinetics and pharmacodynamics of S- and R-warfarin and effects on international normalized ratio: influence of body size, composition and genotype in cardiac surgery patients. Br J Clin Pharmacol 83:823–835

Yamashita F, Sasa Y, Yoshida S et al (2013) Modeling of rifampicin-induced CYP3A4 activation dynamics for the prediction of clinical drug-drug interactions from in vitro. PLoS One 8:e70330

Yang KM, Shin IC, Park JW et al (2017) Nanoparticulation improves bioavailability of Erlotinib. Drug Dev Ind Pharm 43:1557–1565

Yang YX, Lewis JD, Epstein S et al (2006) Long-term proton pump inhibitor therapy and risk of hip fracture. J Am Med Assoc 296:2947–2953

Yuen AWC, Land G, Weatherley BC et al (1992) Sodium valproate acutely inhibits lamotrigine metabolism. Br J Clin Pharmacol 33:511–513

Zielinski JJ, Haidukewych D (1987) Dual effects of carbamazepine-phenytoin interaction. Ther Drug Monit 9:21–23

Zielinski JJ, Haidukewych D, BJn L (1985) Carbamazepine-phenytoin interaction: elevation of plasma phenytoin concentrations due to carbamazepine comedication. Ther Drug Monit 7:51–53

Zhu L, Persson A, Mahnke L et al (2011) Effect of low-dose omeprazole (20 mg daily) on the pharmacokinetics of multiple-dose atazanavir with ritonavir in healthy subjects. J Clin Pharmacol 51: 368-377

Angela Effinger, Caitriona M. O'Driscoll, Mark McAllister, and Nikoletta Fotaki

13.1 Introduction

A study in 1997 showed that 39% of new chemical entities failed in clinical drug development due to issues related to pharmacokinetics (Kennedy 1997). This finding underlined the need for the development of tools suitable to identify compounds with a poor bioavailability at an early stage in the drug discovery process. In the last decades, a large number of in vitro and in silico tools for ADME prediction has been developed that contributed to the reduction of the drug attrition rate due to poor pharmacokinetic properties to 10% in 2000 (Kola and Landis 2004). Especially, the prediction of cytochrome P450 (CYP)-related metabolism added to this improvement. ADME prediction for drug candidates using in vitro and in silico tools helps in the selection of lead compounds before reaching clinical trials. In turn, unsuccessful drug candidates can be identified at an earlier stage resulting in saving time and costs. The knowledge of advantages and limitations of each ADME prediction tool are key to selecting the appropriate tool and to build confidence in the prediction. Considering drug absorption, solubility and dissolution studies are especially important for poorly soluble drugs, while for other compounds absorption may be limited by intestinal membrane permeability. Drug distribution can have implications on the duration of the drug effect and be associated with a risk of not reaching therapeutic

A. Effinger · N. Fotaki (✉)
Department of Pharmacy and Pharmacology, University of Bath, Bath, UK
e-mail: n.fotaki@bath.ac.uk

C. M. O'Driscoll
School of Pharmacy, University College Cork, Cork, Ireland

M. McAllister
Pfizer Drug Product Design, Sandwich, UK

© Springer Nature Switzerland AG 2018
A. Talevi, P. A. M. Quiroga (eds.), *ADME Processes in Pharmaceutical Sciences*,
https://doi.org/10.1007/978-3-319-99593-9_13

301

concentrations in vivo, especially for lipophilic drugs. The use of in vitro and in silico tools assessing plasma protein binding, partitioning into red blood cells and distribution into peripheral tissues helps to identify those issues. Drug metabolism by metabolic enzymes influences the clearance profile and is often a source of interindividual pharmacokinetic variability or drug-drug interactions. In vitro and in silico predictions of drug metabolism should consider the enzymatic reaction as well as the relevant enzymatic expression in the respective tissue. Drug exposure can also be limited by drug excretion. Therefore, in vitro and in silico tools are available for the complex processes of biliary and renal excretion. While the consideration of each of the previous ADME processes separately helps to identify issue related to one process, their mutual interaction can negate or improve the drug's pharmacokinetic profile. Physiologically based pharmacokinetic (PBPK) models take into account all ADME processes together by integration of various experimental results, and in silico predictions of unknown parameters and in vivo performance can be made.

13.2   Absorption

Drug absorption after oral administration is a very complex process influenced by drug properties, formulation-dependent factors and physiological conditions. Drug absorption includes several underlying processes, e.g. release and/or dissolution of the drug from the pharmaceutical formulation in the gastrointestinal fluids and permeation of the dissolved drug through the gastrointestinal membrane. Therefore, determination of the rate-limiting process governing the absorption of the investigated drug based on its physicochemical properties is essential.

The Biopharmaceutics Classification System (BCS), introduced in 1995 by Amidon et al. (1995), aimed to correlate in vitro drug dissolution with in vivo bioavailability. Drugs are classified based on three dimensionless parameters determining their absorption (dose number, dissolution number and absorption number) and their underlying drug properties (solubility and permeability). Four different BCS classes are defined: BCS class 1 includes compounds with high solubility and high permeability for which gastric emptying (for when drug dissolution is very rapid) or drug dissolution is the rate-limiting step to drug absorption. BCS class 2 drugs have a low solubility and a high permeability presenting solubility or dissolution rate-limited absorption. BCS class 3 contains high solubility and low permeability compounds for which the rate-limiting step to drug absorption is permeability. BCS class 4 includes low solubility and low permeability compounds which are usually challenging for oral drug delivery. Once a compound is classified according to the BCS, formulation development can be guided. For example, if the solubility of a compound is problematic, subsequent efforts in formulation development with, e.g. enabling formulations may be required.

Several 'rule of thumb' approaches have been introduced to categorise new chemical entities according to characteristics which increase their likeliness to be adequately absorbed in vivo. The most popular method to identify compounds at risk of poor absorption and permeation was developed based on an analysis of the World Drug Index and is called Lipinski's 'rule of five'. The rule implies that drugs with >5 hydrogen bond donors, a molecular weight >500 Da, an octanol-water partition coefficient (log P) >5 and >10 hydrogen bond acceptors have an increased risk for poor absorption (Lipinski et al. 2001). An exception of this rule are drug classes that are substrates for biological transporters. Another analysis by Veber et al. (2002) used oral bioavailability data of over 1000 drugs in rats and identified that drugs with $\leq$10 rotatable bonds and a polar surface area $\leq$140 $\mathring{A}^2$ (or 12 or fewer H-bond donors and acceptors) are likely to have a good oral bioavailability. The bioavailability score is another approach stating that the predominant charge at biological pH determines the properties that are important for a compound's bioavailability (Martin 2005). The important parameter for anions is the polar surface area, while for neutral, zwitterionic or cationic compounds, the previously described Lipinski's rule of five is more predictive. Additional strategies aim to identify drug-like molecules that are also expected to meet ADME profiles using simple structural rules or neural network approaches (Muegge et al. 2001).

13.2.1 In Vitro Methods

13.2.1.1 Solubility

Drug solubility can be a limiting factor in drug absorption, and its importance is highlighted by the fact that 75% drug development candidates are poorly soluble and belong to BCS class 2 or 4 (Di et al. 2009). In early development, high- throughput methods are used to determine the solubility of a large number of compounds. These methods typically include a concentrated stock solution of the investigated drug in DMSO which is either directly added to a buffer (often pH 6.5 or 7.4) or evaporated, and subsequently a buffer is added to the remaining material to reduce the effect of DMSO on solubility. The solution can be analysed by different methods including light scattering, turbidimetry, LC-UV or LC-MS. The drug concentration at which the first induced precipitate appears in a solution is called kinetic solubility. A 'semi-equilibrium' solubility refers to the solution being incubated for approximately 1 day followed by its filtration and the determination of drug concentration (Di et al. 2012a). While the latter method allows some time for equilibration between solid drug and solution, supersaturation is a frequent problem for kinetic solubility measurements. Additionally, the evaporation of DMSO can leave the drug in an energetically higher state (amorphous form) possibly resulting in a higher solubility. Consequently, high-throughput methods present the 'best case scenario' of drug solubility (Di et al. 2012a).

Equilibrium (thermodynamic) solubility refers to the concentration of the saturated solution in equilibrium with the thermodynamically stable polymorph (Bergstrom et al. 2014). Measurements of equilibrium solubility are performed with the shake flask method in later phases of drug development when crystalline drug material becomes available. The characterisation of the solid form by,

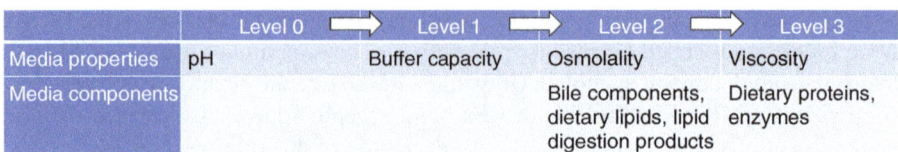

	Level 0 ⟹	Level 1 ⟹	Level 2 ⟹	Level 3
Media properties	pH	Buffer capacity	Osmolality	Viscosity
Media components			Bile components, dietary lipids, lipid digestion products	Dietary proteins, enzymes

Fig. 13.1 Levels of biorelevant media recommended for the simulation of human gastrointestinal fluids during oral formulation development. (Modified from Markopoulos et al. 2015)

e.g. polarized light microscopy and powder X-ray diffraction provides information about potential solid form changes.

At this stage, apart from simple buffers (pH 1.2, 6.5, 7.4), the solubility is also tested in biorelevant media that closely simulate the gastrointestinal fluids (Di et al. 2012a). Biorelevant media were developed since the solubility of a drug in water or simple buffers is not always reflective of the solubility in the gastrointestinal lumen (Galia et al. 1998). Especially for lipophilic drugs, biliary secretions or dietary lipids can enhance drug solubility. To consider these differences in in vitro experiments, biorelevant media can reflect the osmolality, pH and buffer capacity of gastrointestinal fluids and can include bile components, dietary lipids, lipid digestion products and enzymes. Depending on the investigated drug, not all components and properties of the medium may be necessary to reflect the in vivo solubility, and thus, the level of the biorelevant medium can be chosen accordingly (Fig. 13.1) (Markopoulos et al. 2015).

13.2.1.2 In Vitro Release and Dissolution Testing

Apart from the solubility of a drug, the dissolution rate can also be limiting for the drug absorption of poorly soluble drugs. Additionally, the release of the drug substance from a drug product can be critical for the drug product performance in vivo. For development purposes, in vitro tests should adequately represent the gastrointestinal physiology to be able to sufficiently reflect drug dissolution, degradation, supersaturation, precipitation and redissolution in vivo and to guide formulation development (Kostewicz et al. 2014b; Wang et al. 2009). Therefore, the experimental design should consider the composition, volume, flow rates and mixing patterns of the gastrointestinal fluids (Dressman et al. 1998; Fotaki and Vertzoni 2010).

Four dissolution apparatus are included in the United States Pharmacopoeia (USP) for oral drug products. In the USP apparatus 1 (basket apparatus), the investigated drug product is placed in a spinning basket in the middle of a cylindrical vessel with hemispherical bottom filled with dissolution medium (US Pharmacopoeia Convention 2005). The USP apparatus 2 (paddle apparatus) uses the same vessel, but a paddle is used as stirring element (US Pharmacopoeia Convention 2005). For both apparatus, methods usually use high volumes of dissolution medium (500–1000 mL) to generate sink conditions. Especially in the fasted state, the high volumes are unlikely to match the in vivo situation. This is particularly an issue if sink conditions are not maintained in vivo resulting in an overestimation of drug dissolution. For drugs with high solubility (BCS class 1 and 3), sink conditions are usually maintained in vivo. For drugs with low solubility, the conditions in those dissolution experiments are likely to mismatch the in vivo situation. At highest risk are BCS class 4 drugs, since the high membrane permeability of BCS class 2 drugs results in constant removal of dissolved drug from the luminal fluids (Kostewicz et al. 2014b). Furthermore, coning effects and variability in hydrodynamics depending on the investigated dosage form (size, shape, density) and its location in the vessel often result in a lack of correlation to physiological conditions (Kostewicz et al. 2014b). In the USP apparatus 3 (reciprocating cylinder), the drug product is placed in a glass tube with a mesh base that reciprocates vertically in a cylindrical, flat-bottomed glass vessel filled with dissolution medium (US Pharmacopoeia Convention 2005). Media changes can be easily performed by moving the glass tube from one vessel to another vessel but limit the usage of the apparatus to non-disintegrating dosage forms. Typical volumes of dissolution medium are 250 mL in each glass vessel. Hydrodynamics can be adjusted by changing the rate of the reciprocating movement and mesh size of the sieve. For these three USP apparatus, the temperature can be controlled with a tempered water bath surrounding the dissolution vessels. In the USP apparatus 4 (flow-through cell), the drug product is placed in a flow-through cell, through which the dissolution medium can be pumped with an adjustable flow rate (usually 4–16 mL/min) and which is immersed in a tempered water bath. The advantages of the flow-through cell are the possibility to change flow rate and media within a single experiment and to maintain sink conditions if the system is used in open mode (continuously fresh dissolution medium from the reservoir) (Fotaki 2011).

Apart from the apparatus described in the USP, various biopharmaceutical tools have been developed to simulate the dissolution process in vitro. Biphasic dissolution tests can be useful for poorly soluble compounds as membrane permeation is simulated with an organic layer constantly removing drug from the aqueous medium to maintain sink conditions. Integrated permeation systems such as the µFlux™ (Pion Inc., Woburn, MA, US) have the advantage of simultaneous measurements of dissolution and permeability. Surface dissolution imaging is used to understand surface effects during dissolution and quantify swelling, erosion and disintegration kinetics. The use of physiologically relevant bicarbonate buffers in dissolution tests was shown to be more discriminative, for e.g. enteric-coated formulations, but is laborious, and results were shown to be less reproducible (Liu et al. 2011; Boni et al. 2007). Transfer models were able to successfully predict drug precipitation of weak bases in vivo by constantly transferring medium from a gastric donor compartment to an intestinal acceptor compartment (Kostewicz et al. 2004). Complex gastrointestinal simulators such as the TNO Gastro-Intestinal Model (TNO, Zeist, Netherlands) simulate the conditions in the lumen of the gastrointestinal tract very closely by mimicking digestive fluids, constant removal of metabolites and control of pH, temperature and luminal transit.

13.2.1.3 Permeability

For the prediction of the permeability of a compound across the intestinal barrier, several methods can be used ranging from simple filter-immobilised artificial membranes, in vitro cell cultures to in situ perfusion studies (Table 13.1).

In the early stages of drug discovery, methods suitable for high-throughput screening are used such as parallel artificial membrane permeability assay (PAMPA). PAMPA consists of a microporous filter which is infused by a lipid or a lipid mixture dissolved in a nonpolar solvent and which separates two aqueous, pH-buffered solutions in a microplate sandwich (Caldwell and Yan 2014). The concentration gradient between the two compartments is the driving force for the permeability of the investigated compound. This driving force can be maintained, and the experimental time can be reduced with the use of a pH-gradient, the addition of cosolvents/solubilising agents or addition of compounds to simulate protein binding in the donor compartment (Caldwell and Yan 2014).

Other permeability assays use immortalised cell cultures with the ability to form polarised monolayers (with distinct apical and basolateral morphologies) as membranes (Alqahtani et al. 2013). The most commonly used cells are Caco-2 cells, derived from a human colon adenocarcinoma and also available for high-throughput screening. Caco-2 cells are usually cultured for at least 20 days to express high amounts of transporter enzymes, form tight junctions and obtain cell polarity (Alqahtani et al. 2013; Bohets et al. 2001). The expression of endogenous transporter systems such as P-glycoprotein (P-gp) and several drug- metabolising enzymes like aminopeptidase, esterase, sulfatase and some CYP450 isoenzymes is the main advantage of Caco-2 cells, while paracellular permeability is often underpredicted due to 'tighter' tight junctions (Alqahtani et al. 2013; Lea 2015). The use of Madin-Darby canine kidney cells (MDCK), derived from the distal tubular part of the dog kidney, has the advantage of reducing the time needed for the formation of a polarised monolayer with well-defined tight junctions. Additionally, the transepithelial electrical resistance (TEER) is lower compared to Caco-2 cells indicating increased 'leakiness' (Braun et al. 2000). The non-human origin of these cells has the disadvantage of different enzyme and transporter expression. To overcome this issue, it is possible to transfect the cells with, e.g. P-gp, MRPs (multidrug resistance-associated protein) or BCRP (breast cancer resistance protein). To only investigate passive permeability, special cell lines with low expression of endogenous canine transporters such as MDCKII-low-efflux cells can be used (Di et al. 2011). The apparent permeability observed using cultured cell lines must be normalized according to accessible intestinal surface area, paracellular permeability, pH dependence, resistance of the aqueous boundary layer and transcellular permeability to predict the effective permeability in vivo (Avdeef 2012).

Additionally, permeability can be assessed ex vivo using Ussing chambers. An excised intestinal segment (from rat, mouse, rabbit, dog, monkey or human) is mounted between two diffusion cells usually filled with Krebs-Ringer bicarbonate buffer (Alqahtani et al. 2013). Despite the supply of nutrients and carbogen gas during the experiment, tissue viability can only be maintained for 2–3 h. The method

Table 13.1 Overview of different in vitro permeability assays

Permeability assay	PAMPA	Caco-2	MDCK	Ussing chamber	In situ perfusion studies
Advantages	↓Costs ↓Experimental time Suitable for high-throughput screening	Human cell lines with tight junctions, enzyme transporters, P-gp and multidrug resistance proteins, some CYP450 isoenzymes and phase 2 enzymes Suitable for high-throughput screening	↓Time needed for cells to form a polarised monolayer with well-established tight junctions (3–5 days) ↓TEER values	Good prediction of intestinal drug absorption Influx/efflux transport Drug metabolism	Assessment of regional intestinal differences Assessment of intestinal drug transport and metabolism Assessment of dose-dependent pharmacokinetics
Disadvantages	Considers only passive transcellular permeability of compounds	Long time (ca. 21 days) to form tight junctions/express higher amount of efflux transporters Wide variation with passage number ↑ Variability between laboratories Underestimation of paracellular transport →"tighter" tight junctions compared to in vivo situation ↓ Expression of CYP3A enzymes	Canine origin (but transfection with P-gp, MRPs, BCRP is possible)	Can only be used for 2–3 h due to tissue viability	Effect of anaesthesia Not practical for high-throughput screening Overestimation of absorption for drugs with gut wall metabolism or accumulation in gut wall possible

Alqahtani et al. (2013), Caldwell and Yan (2014)

results in good predictions of intestinal drug absorption and provides information about influx/efflux transport and drug metabolism.

A labour-intensive method is the in situ perfusion model. This includes the perfusion of an isolated intestinal segment of the small bowel of rats with a solution containing the investigated drug (Alqahtani et al. 2013). The rat is unconscious during the experiment, and its body temperature is controlled with a heating pad or an overhead lamp (Stappaerts et al. 2015). Drug permeability can then be calculated by the difference between inlet and outlet flow of the investigated drug. To account for differences in drug concentration due to water absorption or secretion, non-absorbable markers (e.g. phenol red) or gravimetric methods can be used (Stappaerts et al. 2015). The method allows distinction between regional permeability differences and considers active transport mechanisms. Additionally, intestinal metabolism can also be investigated, for example, by concomitant administration of inhibitors of metabolising enzymes or by mesenteric blood sampling (Stappaerts et al. 2015). By considering only the difference in drug perfusate concentration, drug absorption may be overestimated for drugs that are accumulated in the gut wall or metabolised by intestinal enzymes. Furthermore, the use of anaesthesia can have a possible impact on drug permeability.

13.2.1.4 Active Transport

The involvement of active transport mechanisms in the membrane permeation of a drug can mediate or limit its absorption but also be responsible for drug-drug interactions. Bioavailability can be increased by drug transport from the luminal to the basolateral site via influx transporters or decreased by transport in opposite direction via efflux transporters.

The in vitro assessment of active transport mechanisms includes cell-based and subcellular assays. In both cases, cells are incubated with a drug solution followed by the monitoring of changes in drug concentration. In cell-based assays, transport proteins are overexpressed in a transfected cell line such as MDCK cells, HEK (human embryonic kidney) or LLC-PK$_1$ (Lewis lung carcinoma-pig kidney) (Caldwell and Yan 2014). Another approach is to partially or completely silence (knock down) a natively expressed transporter protein in a cell line, for example, P-gp in Caco-2 cells and compare the drug permeability to the unmodified cell line (Caldwell and Yan 2014). Other methods for active transport studies include the use of primary cells such as hepatocytes and membrane vesicles (described in detail below in Sect. 13.5.1).

13.2.1.5 Gut Wall Metabolism

The intestine with numerous metabolising enzymes is involved in the metabolism of compounds undergoing phase 1 and 2 reactions. Several in vitro methods are available to investigate intestinal drug metabolism. For drugs that are rapidly metabolised, suitable methods include the use of isolated intestinal perfusion, the everted sac method and Ussing chambers (van de Kerkhof et al. 2007). For isolated intestinal perfusions, a segment of the intestine is removed from an animal (e.g. rat) and placed in a bath filled with buffer followed by perfusion with the investigated compound from the luminal or vascular side (van de Kerkhof et al. 2007). The everted sac method includes eversion of intestine (most often from rat) and its cannulation from both sides followed by drug perfusion. Disadvantages of these two methods are their limitation to short-term incubation and the animal origin of the tissue. The Ussing chamber, as described above in Sect. 13.2.1.3, can also be used for investigations of intestinal drug metabolism.

For drugs that are slowly metabolised, more appropriate in vitro tools for gut wall metabolism are biopsies, intestinal precision-cut slices (thickness 250–400 µm) and primary cells (van de Kerkhof et al. 2007). Limitations of these methods are that it is not possible to study the direction of excretion and the very difficult isolation procedures for primary enterocytes (van de Kerkhof et al. 2007).

For mechanistic investigations of relevant metabolic enzymes, interaction studies and enzyme kinetics, in vitro assays include subcellular fractions from enterocytes and cell cultures similarly to the assays for hepatic metabolism as further discussed below in Sect. 13.4.1.

13.2.2 In Silico Methods

13.2.2.1 Solubility

In the initial stages of drug development, in silico solubility predictions are used to screen new chemical entities for drug-like characteristics. A large number of in silico tools are available for the prediction of aqueous solubility based on training sets of experimental data and either experimentally determined properties or computational 1D, 2D and 3D molecular descriptors such as hydrophobicity, molecular surface area and electron distribution (Fig. 13.2) (Dokoumetzidis et al. 2007). The lipophilicity (log P, clog P), the size of the molecule and the surface area of the nonpolar atoms have been identified as the most important predictors for aqueous solubility (Dokoumetzidis et al. 2007).

Aqueous solubility is determined by the sublimation energy and hydration energy of a drug. The extensively used modified Yalkowsky's general solubility equation (Eq. 13.1) describes the water solubility of a molecule ($S_0(M)$) using the logarithm of the octanol/water partition coefficient (log P_{Oct}) to reflect the hydration energy and the melting point (m. p.) to reflect the crystal lattice energy (Jain and Yalkowsky 2001).

$$\log S_0(M) = -\log P_{Oct} - 0.01(\text{m.p.} - 25) + 0.50 \tag{13.1}$$

For highly lipophilic drugs, micellar solubilisation can improve drug solubility, while for drugs with a high melting point, solubility can be improved by modifications of the structure resulting in a reduction of lattice energy (Sugano 2012). To consider ionisation effects, it has been proposed to use the logarithm of the distribution coefficient (log D) at pH 7.4 instead of the log P_{Oct} for the solubility prediction (Hill and Young 2010).

1-Dimensional space

$$C_{13}H_{18}O_2$$

Linear arrangement

2-Dimensional space

Planar arrangement

3-Dimensional space

Spatial arrangement

Theoretical molecular descriptors

- Atom count
- Molecular weight

- Flexibility
- Rigidity
- Electron distribution
- Hydrophilicity
- Lipophilicity

- Volume
- Polar surface area
- Non-polar surface area
- Partitioned total surface area
- Hydrophobicity/hydrophilicity balance
- Amphiphilic moment
- Critical packing parameters

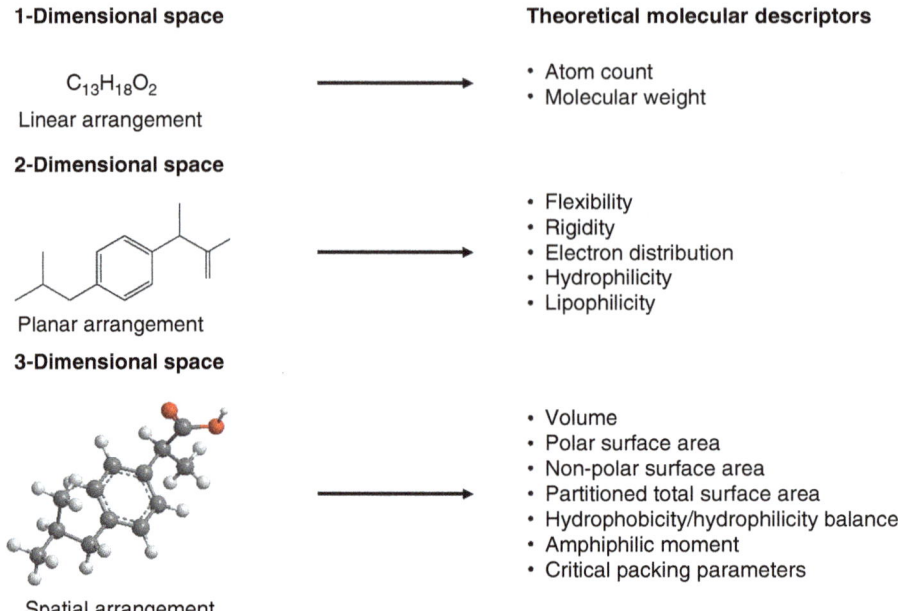

Fig. 13.2 Various types of molecular descriptors used for in silico prediction of ADME parameters. (Modified from Dokoumetzidis et al. 2007)

For weakly acidic and basic drugs, differences in drug solubility along the gastrointestinal tract can be a result of drug ionisation. pH-dependent solubility profiles can be predicted in silico using the Henderson-Hasselbalch equation (Hansen et al. 2006). Additionally, the aqueous solubility may differ from the solubility in gastrointestinal fluids, especially for lipophilic compounds, due to, e.g. luminal surfactants such as bile salts or lecithin. If reliable predictions of drug solubility in gastrointestinal fluids could be obtained using computational models, this could replace laborious biorelevant solubility studies. For several compounds, successful predictions for the increase in solubility as a function of bile salt concentrations could be made using an empirical equation introduced by Mithani et al. 1996 (Eqs. 13.2 and 13.3) (Mithani et al. 1996).

$$\log SR = 2.09 + 0.64 \log P \tag{13.2}$$

$$C_{SX} = C_{SO} + (SC_{bs})(MW)([NaTC]) \tag{13.3}$$

where SR is the solubilisation ratio, C_{SX} is the solubility [µg/mL] in the presence of taurocholate, C_{SO} is the aqueous solubility [µg/mL], MW is the molecular weight and [NaTC] is the concentration of sodium taurocholate. It should be noted that the equation only considers the effect of sodium taurocholate, but gastrointestinal fluids are more complex containing, e.g. lipids and mixed micelles as colloidal aggregates.

For solvation processes, Abraham et al. (1987) described a solvation-related property based on several parameters such as the McGowan's characteristic volume, hydrogen-bonding acidity and basicity, polarizability and an excess molar refraction descriptor. These Abraham solvation predictors have been successfully used to predict the solubility enhancement in biorelevant media (Fasted State Simulated Intestinal Fluid) compared to a simple buffer (Niederquell and Kuentz 2018). Most prominent were a positive effect of McGowan's characteristic volume and a negative effect of drug basicity on solubility enhancement.

13.2.2.2 Drug Release and Dissolution

In most oral dosage forms, the active pharmaceutical ingredient is administered in solid form. For these formulations, the drug needs to be released from the formulation and dissolve in the gastrointestinal fluids prior to its intestinal membrane permeation. If drug dissolution occurs slowly in the gastrointestinal tract, it can be the limiting step for drug absorption. Diffusion theory is widely used to describe particle dissolution assuming drug dissolution is controlled by the diffusion of the solute through a stagnant diffusion layer surrounding solid particles.

In 1897, it was shown by Noyes and Whitney (1897) that the rate of drug dissolution is proportional to the difference between the saturation solubility (C_s) and the present drug concentration at time t. This relationship was further modified to the Nernst-Brunner law:

$$\frac{dC}{dt} = \frac{DS}{Vh}(C_s - C_t) \tag{13.4}$$

where C is the concentration, t is the time, D is the diffusion coefficient, S is the surface of the solid particles, V is the volume of the dissolution medium and h is the thickness of the diffusion layer (Brunner 1904; Nernst 1904). This dissolution model is still widely used today (Dokoumetzidis et al. 2007). The diffusion coefficient can be derived from the Stokes-Einstein equation or the Hayduk-Laudie equation for non-electrolytes (Hayduk and Laudie 1974). A further improvement was made by Wang and Flanagan resulting in a generalised diffusion layer model for spherical particles (Eq. 13.5) that considers a time-dependent reduction of the particle radius, a nonlinear concentration gradient in the diffusion layer and changes in the thickness of the effective boundary layer (Wang and Flanagan 1999, 2002).

$$\frac{dC}{dt} = -4\pi r_t^2 * D * \left[\frac{1}{r_t} + \frac{1}{h}\right] * [C_s - C_t] \tag{13.5}$$

The particle radius, r_t, is time dependent and influences the thickness of the effective boundary layer h. In the case of small particles ($r < 30\ \mu m$), the particle radius is considered to be equal to h, while for larger particle radii ($r > 30\ \mu m$), h is set to 30 μm (Peters 2012).

For particles with substantially larger diameters compared to the diffusion layer thickness, the diffusion layer can be assumed as planar resulting in the cube root equation:

$$Q^{\frac{1}{3}} = Q_0^{\frac{1}{3}} - k_{\frac{1}{3}}t \tag{13.6}$$

with the cube roots of the weight of a spherical particle at time 0, $Q_0^{1/3}$, and time t, $Q^{1/3}$, and the cube root rate constant, $k_{1/3}$, as described by Hixson and Crowell (Hixson and Crowell 1931). The cube rate constant can be further described as:

$$k_{\frac{1}{3}} = \left(\frac{\pi}{6\rho^2}\right)^{\frac{1}{3}} \frac{2DC_s}{h} \tag{13.7}$$

with D as the diffusion coefficient, C_s as the equilibrium solubility, h as the thickness of the diffusion layer and ρ as the density.

For the modelling of in vitro dissolution data, fitting of the data can be obtained using empirical equations. The Weibull equation is the most commonly used equation due to its flexibility to fit almost any dissolution data:

$$W_t = W_{max}\left[1 - e^{-\left(\frac{(t-T_{lag})^b}{a}\right)}\right] \tag{13.8}$$

where W_t is the amount dissolved at time t, W_{max} is the maximum amount dissolved, t is time, T_{lag} is the lag time before the onset of dissolution, a is a scale parameter and b is a shape parameter characterising the curve (exponential curve $b = 1$, sigmoid curve $b > 1$, parabolic curve $b < 1$) (Langenbucher 1972). Other empirical approaches include gamma distribution, power laws, discrete time-step difference equations and stochastic differential equations (Dokoumetzidis et al. 2007). For in vivo predictions of drug dissolution, parameters with physical meaning derived from mechanistic models are usually used in absorption models (Dokoumetzidis et al. 2007).

In certain cases, drug absorption can also be determined by the release of a drug from the formulation. Controlled-release formulations are, for example, developed to reduce dosing frequency, to avoid toxicity for drugs with a narrow therapeutic index or to locally deliver drugs in the gastrointestinal tract. The drug release rate of these formulations is constant over a certain time and is diffusion-controlled, swelling-controlled or chemically-controlled (Siepmann and Peppas 2011). This steady release process allows the direct use of in vitro release profiles for in vivo predictions or even to use empirical equations. For the analysis of drug release data, the Higuchi model is widely used but should only be applied to the first 60% of drug release:

$$\frac{q(t)}{q_\infty} = k\sqrt{t} \qquad (13.9)$$

with $q(t)$ as the drug released at time t, q_∞ as cumulative amount of drug released at infinite time and the constant k (Higuchi 1961). Assumptions behind this model are that the carrier is of a thin planar geometry and the medium acts as a prefect sink. Adapted models for carriers with different geometries have been proposed in literature (Baker 1987). Other models used for drug release are the Peppas equation, Weibull equation, Baker and Lonsdale equation, Hixson and Crowell equation or Monte Carlo simulation methods (Carbinatto et al. 2014).

13.2.2.3 Permeability

In silico approaches to predict passive permeability of novel compounds are mostly developed based on data sets of compounds with known in vitro permeability in different cell lines (e.g. Caco-2, MDCK, PAMPA) and are used in drug discovery for the selection of novel compounds for synthesis (Broccatelli et al. 2016). Often these in silico models are based on multivariate statistical analysis (e.g. partial least-squares regression) that correlate in vitro results to 2D or 3D molecular descriptors (Zhang et al. 2006; Broccatelli et al. 2016). Other in silico approaches include mechanistic mathematical models developed for the passive transcellular drug transport which are, for example, based on simple physicochemical properties such as log P and pKa (Zhang et al. 2006).

Such mechanistic models describing passive permeability can be augmented with additional processes such as active influx and efflux transport. This more complex system description can be used to define the properties of the enterocyte as a separate compartment for absorption (Dokoumetzidis et al. 2007). Such models are, for example, implemented in PBPK models that are commercially available such as the software Simcyp® (Certara, Sheffield, UK) or GastroPlus™ (Simulations Plus, US).

13.2.2.4 Active Transport

Drug-transporter interactions can be modelled in silico either based on a set of compounds with known transporter activity (substrate-based methods) or based on the 3D structure of the transporter (transporter-based methods) (Chang and Swaan 2006). Substrate-based methods use molecular descriptors or chemical properties for pharmacophore or 3D-QSAR) modelling without the need for prior information about the structure of the transporter. Such models exist for a variety of different transporters such as P-glycoprotein, organic cation and anion transporters, bile acid transporters and nucleoside transporters (Chang and Swaan 2006). Transporter-based methods include ab initio modelling and homology modelling. Ab initio modelling generates the 3D structure of the transport protein from its primary sequence, while homology modelling uses structural information of a template protein with mutual sequence similarity (Chang and Swaan 2006).

For the modelling of active carrier-mediated transport, most often the saturable Michaelis-Menten kinetic is used. The input parameters V_{max}, the maximum reaction velocity, and k_m, the substrate concentration with 50% V_{max} (Michaelis constant) are determined in vitro. While at low concentrations, the rate of transport increases almost linear, at high concentrations, transporter saturation occurs resulting in a constant maximum transport rate.

13.2.2.5 Gut Wall Metabolism

For in silico predictions of gut wall metabolism, an allometric scale-up approach is followed if in vitro data is available. For example, the slice weight and organ weight are used for the scaling of experimental data from precision-cut intestinal slices to the in vivo situation. Alternative approaches include scaling of the information of specific enzymes determined for hepatic metabolism to the gut wall metabolism. Therefore, information about the kinetics of the specific enzymatic reaction, enzyme abundance in the in vitro assay used for the determination of the kinetics of the enzymatic reaction and enzyme scaling factors (e.g. derived from immunoquantified enzyme expression levels in the intestine and liver) are needed (Heikkinen et al. 2012). With these approaches, the intrinsic intestinal clearance is determined and can further be used to calculate the fraction of drug escaping gut wall metabolism. Using a similar approach to the well-stirred liver model, the fraction of drug escaping gut wall metabolism (F_g) can be described with the Q_{gut}-model:

$$F_g = \frac{Q_{gut}}{Q_{gut} + f_{u,g} * CLu_{int,g}} \qquad (13.10)$$

where $f_{u,g}$ is the fraction of unbound drug in the enterocytes and $CLu_{int,g}$ is the intrinsic metabolic clearance in the gut (Yang et al. 2007). Q_{gut} can further be described as:

$$Q_{gut} = \frac{Q_{villi} * CL_{perm}}{Q_{villi} + CL_{perm}} \qquad (13.11)$$

where Q_{villi} is the villous blood flow and CL_{perm} is the cellular permeability. Purely in silico approaches are used to identify the investigated compound as a substrate for specific enzymatic reactions following ligand-based or structure-based approaches as further described below in Sect. 13.4.2.

13.2.2.6 Dynamic Transit Models

Dynamic transit models are dependent on a temporal variable and include mixing tank models, the compartmental absorption transit model (CAT) and dispersion models (Yu et al. 1996). With these models, it is not only possible to predict the fraction of dose absorbed but also to predict the rate of drug absorption which can help in the simulations of plasma concentration profiles and predictions of in vivo performance.

The mixing tank model introduced by Dressman and Fleisher (1986) is based on mass balance considerations and suitable for drugs with dissolution-rate limited absorption. The model considers the gastrointestinal tract as a single well-stirred compartment with uniform drug concentration, in which transit and absorption follow first-order kinetics. The drug is administered as bolus, and the transport of solid and dissolved drug occurs at the same rate. Despite several limitations of the model such as no consideration of luminal degradation, gut metabolism or heterogeneity of the gastrointestinal tract, the model could successfully predict the determining factors limiting the absorption of digoxin (dissolution rate) and griseofulvin (solubility).

The compartmental absorption transit model (CAT) considers the gastrointestinal tract as a series of well-stirred compartments with different volumes and flow rates but equal residence time of the drug. For the small intestine, seven compartments resulted in the best fit of available literature data (Yu and Amidon 1999). Further modifications of the CAT model included addition of compartments of unreleased drug and undissolved drug, pH-dependent solubility, precipitation, gastric and colonic compartments, information of effective absorptive surface area and drug transporter processes resulting in the advanced compartmental absorption transit (ACAT™), the basis of the commercial software GastroPlus™ (Simulations Plus, US) (Kuentz 2008).

The current version of the ACAT™ model considers ionisation effects on solubility and permeability, paracellular permeability, nanoparticles effects, food effects, bile salt-enhanced solubility, precipitation and active transport. It can be used for immediate release, delayed release and controlled release formulations. Apart from human gut physiology, physiological gut models are available for a variety of species (dog, rat, mouse, rhesus monkey, cynomolgus monkey, minipig, rabbit and cat). Drug dissolution can be predicted with several dissolution models (e.g. Hintz and Johnson equation, Wang and Flanagan equation, Z-Factor Model) (Hintz and Johnson 1989; Takano et al. 2006). A similar advanced compartmental absorption model is integrated in the Simcyp® software (Certara, Sheffield, UK) under the name Advanced Dissolution, Absorption, Metabolism (ADAM) model. The ADAM model uses the Wang and Flanagan equation (described above in Sect. 13.2.2.2) as default model for drug dissolution (Wang and Flanagan 1999).

The dispersion models consider the gastrointestinal tract as a continuous single tube with constant velocity, dispersion behaviour and concentration profile across the tube diameter and spatially varying properties along the tube (Yu et al. 1996). The convection-dispersion equation is used to describe the drug absorption process:

$$\frac{\partial C}{\partial t} = \alpha \frac{\partial^2 C}{\partial x^2} - \beta \frac{\partial C}{\partial x} - \gamma C \tag{13.12}$$

where C is the concentration, x is the axial distance from the stomach, α is the dispersion coefficient, β is the linear flow velocity in the axial direction and γ is the drug absorption rate constant (Ni et al. 1980). With a modified version of this concept, drug absorption in rats and later in humans was successfully predicted

(Willmann et al. 2003; Willmann et al. 2004). The earlier versions of the absorption model of the PBPK software PK-Sim® (Open Systems Pharmacology) were evolved from this model which was later replaced by a 12 compartmental absorption model (Willmann et al. 2012).

13.3 Distribution

13.3.1 In Vitro Methods

Plasma protein binding (PPB) is an important parameter for the distribution of a drug in the body. Highly protein-bound drugs are retained in plasma and often less prone to distribute into body tissues resulting usually in a low volume of distribution. In terms of pharmacodynamics, highly protein-bound drugs may not reach therapeutic concentrations as usually only the fraction unbound is available for receptor or enzyme interaction. Plasma contains 7% proteins of which human serum albumin is the most important protein for drug binding followed by α_1-acid glycoprotein and lipoproteins (Caldwell and Yan 2014). The preferred method for the determination of plasma protein binding is equilibrium dialysis since the method is less susceptible to non-specific binding. For classical equilibrium dialysis (CED), a regenerated cellulose membrane (cut-off 12–14 kDa) separates two 1 mL paired Teflon cells filled with buffer and plasma which are tempered at 37 °C and rotated for a predetermined period (4–12 h) (Caldwell and Yan 2014). Typical methods used for drug analysis are scintigraphy or LC-MS/MS analysis. Further development of the method resulted in the rapid equilibrium dialysis (RED) with faster preparation and equilibration times and suitability for higher throughput of samples. Another method to determine PPB is ultracentrifugation where plasma is added to the device followed by centrifugation for 10–20 min at 1000–2000 × g in a fixed angle rotor (Caldwell and Yan 2014). The accuracy of this method is limited by the non-specific binding to the filtration apparatus. Additionally, high-performance affinity chromatography can be used to determine PPB using immobilised albumin or α_1-acid glycoprotein as stationary phase and correlate chromatographic retention to the percentage of drug binding to albumin or α_1-acid glycoprotein (Lambrinidis et al. 2015). Longer chromatographic retention time indicates higher percentage of protein binding.

Similar to PPB, drug partition into red blood cells (RBC), the major cellular component of blood, can influence drug distribution. For the in vitro determination of RBC partitioning, radiolabelled or unlabelled drug is mixed with whole blood followed by centrifugal separation of RBC and plasma and the determination of drug concentration in both compartments (Hinderling 1984). Measurements at several time points also permit to determine the rate of RBC partitioning.

In vitro tissue distribution can be assessed using tissue homogenates, tissue slices or isolated tissue components. After an incubation period of the tissue with the investigated drug, the tissue-to-medium distribution coefficient can be calculated using the separately measured drug concentration in tissue and medium (Ballard et al. 2003). While tissue homogenates are the most widely used method, the disruption of cellular integrity can result in an overestimation of tissue distribution for drugs mainly restricted to the extracellular space.

13.3.2 In Silico Methods

Computational models to predict plasma protein binding have been developed using ligand-based approaches with quantitative structure activity relationships and structure-based approaches focusing on the crystal structure of drug-protein complexes. Due to the predominant role of human serum albumin in PPB, most approaches only focus on albumin, and only recently advances for α_1-acid glycoprotein have been made. Due to the different binding sites of albumin, global models for a broad range of compounds are challenging, and in the beginning, in silico models focused on similar compounds using the same binding site. Based on training sets of compounds and multivariate statistical analysis, lipophilicity (log P), electronic properties, acidity, shape-modulating factors, polarity terms and fraction ionised (cationic and anionic) were identified as important predictive factors in ligand-based in silico models (Lambrinidis et al. 2015).

Drug distribution in the body has been described by different mechanistic models. The steady-state volume of distribution (V_{ss}) describes the extent of tissue distribution and can be defined as:

$$V_{ss} = \left(\sum V_t * P_{t:p}\right) + (V_e * E : P) + V_P \tag{13.13}$$

where V_t is the fractional volume of a tissue, $P_{t:p}$ is the plasma/partition coefficient, V_e is the fractional volume of erythrocytes, $E : P$ is the erythrocyte/plasma ratio and V_P is the plasma volume (Poulin and Theil 2002). The erythrocyte/plasma ratio can be described as:

$$E : P = \frac{[B : P - (1 - Ht)]}{Ht} \tag{13.14}$$

where $B : P$ is the blood/plasma ratio which can be determined in vitro (as described above in Sect. 13.3.1) and Ht is the haematocrit (volume percentage of red blood cells in blood) (Poulin and Theil 2002).

Literature data is available for the different body volumes (e.g. lung, brain, heart, liver, bone, kidney, muscle, skin, adipose) in Eq. 13.13 and for the estimation of tissue/plasma partition coefficients, the following in silico models can be used.

An in silico model developed by Poulin and Theil (2002) and modified by Berezhkovskiy (2004) accounts for plasma and tissue being composed of neutral lipids, phospholipids and water and only the unionised fraction of the drug permeating the membrane. This resulted in the following description of the tissue-partition coefficient, $P_{t:p}$, for non-adipose tissue:

$$P_{t:p} = \frac{\left[P_{o:w}\left(V_{t,nl} + 0.3V_{t,ph}\right) + 0.7V_{t,ph} + \frac{V_{t,w}}{fu_t}\right]}{\left[P_{o:w}\left(V_{p,nl} + 0.3V_{p,ph}\right) + 0.7V_{p,ph} + \left(\frac{V_{p,w}}{fu_p}\right)\right]} \qquad (13.15)$$

where $P_{o\,:\,w}$ is the n-octanol/buffer partition coefficient of the non-ionized species; V is the fractional tissue volume content of neutral lipids (nl), phospholipids (ph), and water (w) in either tissue (t) or plasma (p); and fu is the fraction unbound. The fraction unbound in tissue, fu_t, can mechanistically be estimated (Eq. 13.16) from the fraction unbound in plasma according to:

$$fu_t = \frac{1}{\left[\frac{1-fu_p}{fu_p}\right] * 0.5} \qquad (13.16)$$

as described by Poulin and Theil (2000).

Further models developed by Rodgers and Rowland differentiated between intra and extracellular space and added an acidic phospholipid fraction in tissues resulting in an improvement of the prediction for strong bases (Rodgers et al. 2005a, b; Rodgers and Rowland 2006, 2007). Additionally to passive permeability of the unionized fraction of the drug, a further modification of the model includes membrane permeability of the ionized fraction and is integrated in the Simcyp® simulator (Certara, Sheffield, UK).

13.4   Metabolism

Drug metabolism or biotransformation of orally administered drugs occurs mainly in the small intestine (as described above in Sect. 13.2.1.5) and liver. Physicochemical characteristics of drugs such as a high lipophilicity as indicated by a high $logD_{7.4}$ were shown to be associated with high metabolic clearance (van de Waterbeemd and Gifford 2003). A variety of metabolising enzymes is available to facilitate the excretion of xenobiotics by increasing their aqueous solubility. Enzymatic biotransformation can be divided in Phase 1 and Phase 2 metabolism. Phase 1 reactions are reactions of functionalisation (e.g. oxidation, hydrolysis or reduction) that introduce polar functional groups to molecules resulting in either facilitated excretion or further metabolism (Westhouse and Car 2007). In Phase 2 reactions, large polar molecules (e.g. glucuronate, acetate and sulfate) are conjugated to drug molecules further resulting in an increased aqueous solubility to facilitate excretion (Westhouse and Car 2007).

13.4.1 In Vitro Methods

Different in vitro assays are available to predict hepatic drug metabolism. Recombinant CYP enzymes expressed in different cell types can be used to identify the specific CYP enzymes involved in the metabolism of the investigated drug. Additionally, incubation of these recombinant CYP enzymes with the investigated compound provides information about the metabolic enzyme activity per mass of protein which can further be scaled up to the in vivo situation.

The homogenisation of liver and subsequent centrifugation at 1000 g and 9000 g separate the pellet with nuclei and mitochondria, respectively, from the supernatant with cytosolic and microsomal enzymes (Richardson et al. 2016). The supernatant is the hepatic S9 pool and can be used as in vitro system for investigating hepatic metabolism. An additional ultracentrifugation step at 100,000 g results in the separation of the cytosol subcellular fraction in the supernatant and the microsomal subcellular fractions in the pellet (Richardson et al. 2016). Human liver microsomes are commonly used in the pharmaceutical industry due to their richness of metabolic cytochrome P450 enzymes, low cost and ease in use (Di et al. 2012b). However, metabolic pathways in the assay can be incomplete since the present enzymes are limited to enzymes contained in endoplasmic reticulum (Phase 1 reactions) (Richardson et al. 2016).

Primary hepatocytes, taken from living tissue (e.g. biopsy material), are grown in vitro and represent more closely the in vivo situation due to the full range of metabolic enzymes (e.g. aldehyde oxidase and monoamine oxidase), cofactors and membrane transporters (Di et al. 2012b). Since it is not possible to culture primary hepatocytes indefinitely, cryopreservation of hepatocytes was introduced. This resulted in constant availability of the in vitro assay in the drug discovery setting by retaining the full activity during storage of the hepatocytes in liquid nitrogen for 1 year. When comparing assays of liver microsomes with hepatocytes, intrinsic clearance of compounds metabolised over non-CYP pathways was faster in hepatocytes (Di et al. 2012b). On the other hand, the intrinsic clearance of drugs with rate-limiting hepatic uptake was faster in microsomes (Di et al. 2012b).

13.4.2 In Silico Methods

For the prediction of hepatic metabolic clearance in vivo, the previously mentioned in vitro assays (S9 pool, liver microsomes, hepatocytes) can be scaled to the in vivo situation. For example, the intrinsic clearance of the unbound fraction of the drug in a human liver microsome assay is given in μL/min/mg protein and can be scaled up to the in vivo situation based on information about the level of microsomal proteins per gram of liver and liver weight. The hepatic clearance is also dependent on the amount of drug that comes into contact with the hepatic metabolising enzymes that can further depend on, e.g. hepatic blood flow or fraction unbound in blood. For the hepatic clearance, CL_h, most often the well-stirred liver model is used:

$$CL_h = \frac{Q_h * f_{ub} * CL_{int}}{Q_h + f_{ub} * CL_{int}} \tag{13.17}$$

where Q_h is the hepatic blood flow, f_{ub} is the fraction unbound in blood and CL_{int} is the intrinsic hepatic clearance (Pang and Rowland 1977). Assumptions behind this model are an instant equilibrium between hepatocytes and adjacent blood and a homogenous drug distribution in the liver. In contrast, the parallel tube model considers that the drug concentration decreases along the direction of the blood flow (Pang and Rowland 1977).

For the prediction of drug metabolism only with in silico methods, ligand-based approaches or structure-based approaches have been used especially for CYP enzymes. Ligand-based approaches, such as QSAR, pharmacophore-based algorithms or shape-focused models, consider the chemical structure and properties of the drugs, while structure-based approaches also model the interaction between the investigated substrate and the metabolic enzyme (Andrade et al. 2014; de Groot 2006).

13.5   Excretion

The removal of unaltered drug and its metabolites from the body is known as excretion. Apart from the rate of metabolism, drug clearance from blood is influenced by the biliary and urinary excretion rate of unchanged drug. Drug elimination occurs mainly via the highly perfused primary eliminating organs liver and kidney and is dependent on the physicochemical and structural characteristics of the drugs. Lipophilic drugs with a high molecular weight are often associated with biliary excretion (Ghibellini et al. 2006).

Drug excretion into urine via the kidney, known as renal clearance, is a complex process involving passive glomerular filtration, active tubular secretion and passive and active reabsorption (Paine et al. 2010). If the drug is only cleared by filtration, the renal clearance equals the mathematical product of fraction unbound and glomerular filtration rate ($f_u \times$ GFR). If the renal clearance exceeds this mathematical product, the drug may be a substrate for active tubular secretion by transporters. If the renal clearance is inferior to this mathematical product, it can be assumed that the drug gets reabsorbed. The importance of renal elimination is highlighted by the fact that 32% of the top 200 prescribed drugs in the United States in 2010 were at least partially excreted unchanged in urine ($\geq 25\%$) (Morrissey et al. 2013).

13.5.1 In Vitro Methods

Several in vitro methods can be used to study biliary excretion: sandwich-cultured hepatocytes, suspended hepatocytes, vectoral transport using polarised cell lines, single-cell expression systems and membrane vesicles (Ghibellini et al. 2006).

Sandwich-cultured hepatocytes, from rat or human origin, have the advantage that basolateral uptake and canalicular efflux transport can be studied and metabolic functions are retained (Ghibellini et al. 2006). In contrast to the conventionally cultured hepatocytes, the culturing of hepatocytes between two layers of gelled collagen enhances cell viability and allows the formation of functional bile canalicular networks and polarised excretory function (Swift et al. 2010). Suspended hepatocytes are relatively cheap, easy-to-handle and can be used for up to 4 h (Elaut et al. 2006). Their use is limited to the investigation of uptake mechanisms and metabolism since it is not possible to discriminate canalicular excretion from sinusoidal efflux (Swift et al. 2010). Cell lines (e.g. MDCK), transfected with transporter proteins such as multidrug resistance-associated protein 2 (MRP2) and organic anion transporting polypeptide (OATP) 1B1 and/or 1B3 and grown on a permeable membrane, are used to determine the contribution of an individual transport protein, identify driving forces and identify inhibitors (Ghibellini et al. 2006). While the extrapolation to the in vivo situation is difficult as these cell lines are less representative of hepatocytes (different expression levels of transport proteins, no complete set of transport proteins, metabolic enzymes and cofactors), the transfected systems are routinely used in drug development due to their ease in use and good availability (Ghibellini et al. 2006; Swift et al. 2010). Single-cell expression systems such as *Xenopus laevis* oocytes can transiently express membrane transporters and channels following the injection of their cRNA (Bröer 2010). These expression systems are mainly used to study the mechanism of transport and the effect of genetic diseases (Ghibellini et al. 2006). Inside-out plasma membrane vesicles from cell lines (e.g. insect or mammalian cells) transfected with specific membrane proteins were used to study polymorphisms and substrate specificity of efflux transporters (Ghibellini et al. 2006). For insect cells, the modification of the membrane composition (addition of cholesterol) results in a similar transporter function to mammalian cells (Caldwell and Yan 2014).

In terms of renal excretion, different in vitro experiments can be used for the processes of passive tubular reabsorption and active tubular secretion and reabsorption. For passive tubular permeability, similar in vitro assays as for intestinal permeability are used with cell lines such as $LLC-PK_1$, MDCK and Caco-2 (Scotcher et al. 2016a). The proximal tubule cell line, $LLC-PK_1$, derived from pig (*Sus scrofa*) kidney, is grown on permeable filter membranes and has been used for transepithelial transport studies investigating the renal disposition of drugs. Apart from the formation of polarised cell monolayers, this cell line has the benefit of expressing endogenous transport proteins (P-gp, MRP2, BCRP) (Kuteykin-Teplyakov et al. 2010; Takada et al. 2005). Recently, the bidirectional epithelial permeation of 20 compounds was studied in this cell line, and good correlations to human renal clearance of drugs were obtained after upscaling of the in vitro parameters (Kunze et al. 2014). For anionic drugs, however, clearance was underpredicted due to restricted secretion in $LLC-PK_1$ cells indicative of limited activity of organic anion transporters (Kunze et al. 2014). To closer mimic the conditions in the kidney, the apical to basolateral pH gradient should be considered in the experimental design, and the experimental results should be scaled by the corresponding tubular surface area (Scotcher et al. 2016b).

For active tubular secretion and reabsorption, similar in vitro techniques are used as for metabolism or biliary excretion. The range of in vitro assays includes membrane vesicles, transfected cells (e.g. organic anion transporter 1-expressing Chinese Hamster Ovary cells (CHO-OAT1), organic anion transporter 3-expressing human embryonic kidney cells 293 (HEK293-OAT3)), immortalised kidney cell lines, primary cultured renal tubule cells and kidney slices (Scotcher et al. 2016a). Human kidney slices can be used to investigate drug uptake at the basolateral membrane but lack information about tubular reabsorption (Watanabe et al. 2011). Their use is restricted due to the limited tissue availability but, if available, complex studies with multiple transporter substrates or inhibitors can be performed with the full set of endogenous transporters (Scotcher et al. 2016a).

13.5.2 In Silico Methods

Considering biliary excretion, several in silico models have been developed based on QSAR and compound data of in vivo rat biliary excretion. One of these models was developed using principal component regression analysis based on rat biliary excretion data from 56 compounds and 2D molecular descriptors which revealed hydrophobicity (cLogD) as most important factor for the prediction of biliary excretion (Chen et al. 2010). Another model used similar data from 50 compounds and identified a correlation of polar surface area, the presence of a carboxylic acid moiety and free energy of aqueous solvation with biliary excretion (Luo et al. 2010). A model, based on 217 compounds, was developed using a simple regression tree model with the classification and regression trees (CART) algorithm and revealed higher biliary excretion for relatively hydrophilic and large compounds, especially when anionic or cationic (Sharifi and Ghafourian 2014).

A variety of in silico approaches has been used for the prediction of renal excretion. Allometric models were developed for the prediction of renal clearance in men from animal data (Mahmood 1998; Paine et al. 2011; Lave et al. 2009). Appropriate upscaling of the in vitro result to the in vivo situation is necessary, as for in vitro assays (Kunze et al. 2014). In silico models focusing on the likelihood or extent of renal clearance were developed based on QSAR approaches using Volsurf descriptors (2D numerical molecular descriptors calculated from 3D interaction energy grid maps) or physicochemical and structural descriptors (e.g. log D, H-bond donors, ionisation potential) (Doddareddy et al. 2006; Manga et al. 2003; Dave and Morris 2015). The rate of renal clearance was predicted using in silico models developed with different statistical tools such as partial least squares (PLS) and random forests (RF) based on a human renal clearance data set of 349 drugs with active secretion and net reabsorption (Paine et al. 2010).

A mechanistic kidney model has been developed based on various physiological and anatomical parameters (e.g. nephron size and number, number of proximal tubular cells per gram of kidney, flow rates of tubular fluid and urine and pH values in tubular cells/fluid) and is incorporated in the Simcyp® simulator (Certara, Sheffield, UK) (Jamei et al. 2009). In this model, the nephron is divided into eight segments with three compartments (tubular fluid, cell mass and blood space). The processes integrated in the model include passive permeability across basal and apical membranes of each cell compartment, uptake and efflux transport across the basal and apical membranes of each proximal tubular cell compartment, metabolic clearance in proximal tubular cell compartments and bypass of a fraction of the renal blood flow (no passage through glomerulus, the Loop of Henle and subsequent segments) (Neuhoff et al. 2013). The input needed in terms of drug properties is information about drug binding and ionisation, passive permeability and transporter kinetics. The advantage of such a mechanistic model is that interindividual variability (demographics, gender, disease) can be integrated.

13.6   Physiologically Based Pharmacokinetic Models

The various previously presented in vitro and in silico ADME tools can be used separately to consider each of the ADME parameters. Linking the information from the different in vitro assays and in silico predictions offers the opportunity to predict in vivo performance, such as plasma concentration profiles and drug concentrations in specific compartments of the body, and to investigate drug-drug interactions. PBPK models were built for this purpose and consider the processes of absorption, distribution, metabolism and excretion of a drug mechanistically (Fig. 13.3).

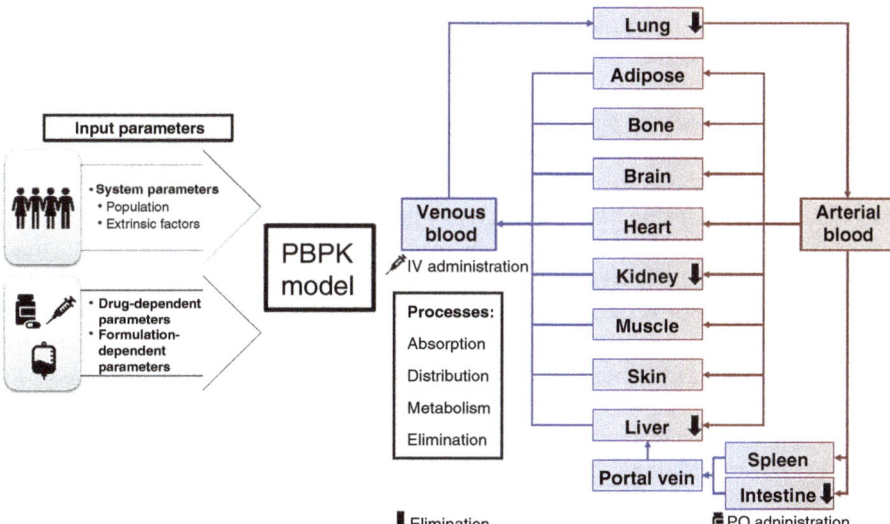

Fig. 13.3 Description of basic elements of physiologically based pharmacokinetic modelling (PO, per os; IV, intravenous)

For drug absorption, the complex compartmental absorption transit models (as described above in Sect. 13.2.2.6) are used in PBPK models together with information about various physiological parameters such as gastrointestinal transit times, luminal fluid volumes, luminal fluid pHs, regional differences in enzyme/transporter density and surface area of the gastrointestinal tract (Jamei et al. 2009). Drug release from different pharmaceutical formulation types such as controlled or modified release systems, enteric-coated granules or tablets and suspensions can also be considered.

Regarding drug distribution, PBPK models consider the whole body using predicted tissue/partition coefficients (as described above in Sect. 13.3.2) and literature data from physiological parameters such as body and organ size, blood flow rates, tissue and blood composition. Tissues include, for example, bone, brain, gut, heart, kidney, liver, lung, pancreas, muscle, skin, spleen and adipose tissue which are usually defined as perfusion-limited tissues. A modification to permeability-limited tissues and integration of active transporter processes using experimental data of transport kinetics are possible. This mechanistic approach allows tracking of the drug concentration in a specific tissue. If such a complex distribution model is not needed, simple compartmental or minimal pharmacokinetic models can be used.

Metabolism and excretion can be integrated at the enzymatic level (metabolizing enzymes and transporters) in the metabolizing and/or eliminating organs which can be considered as perfusion- or permeability-limited tissues (Kostewicz et al. 2014a). For metabolising enzymes or transporters, the input data required includes in vitro information about enzyme kinetics (e.g. V_{max} and k_m) which is scaled to the whole organ with literature information about enzymatic expression in specific organs and organ size. It is also possible to use other in vitro approaches such as hepatocytes with appropriate scaling factors as described above in Sect. 13.4.2.

Apart from physiological data based on the population, additional input data includes drug-dependent parameters, formulation-dependent parameters and information about the design of the virtual trial. For several drug-dependent parameters, it is also possible to use in silico predictions as input information instead of experimental data (e.g. log P, permeability, PPB, RBC partitioning, aqueous and bile micelle-mediated solubility) (Fotaki 2009). As more in vitro, preclinical or clinical data becomes available in the drug discovery process, these data can be used to refine the existing model. PBPK models can also be coupled with pharmacodynamic models to study the relevance of pharmacokinetic changes on therapeutic effects. In recent years, PBPK models were constantly improved by integrating more physiological processes, increasing the mechanistic background of the model and updating physiological information with newly available literature data such as gastrointestinal transit times, demographics and expression of transporters and metabolising enzymes (Rostami-Hodjegan 2012). A main advantage of PBPK models is the integration of population variability in the model to investigate drug product performance in populations that are not usually represented in clinical trials. For example, in the Simcyp® simulator (Certara, Sheffield, UK), default populations include different disease states (obesity, liver cirrhosis, renal impairment, rheumatoid arthritis), ethnicities (Chinese, Japanese, Caucasian), pregnancy and age groups (paediatric, geriatric) (Jamei et al. 2009).

13.7 Conclusion

Many interrelated processes contribute to determining the pharmacokinetic profile of a drug. A variety of useful in vitro and in silico methods to predict single ADME parameters is available to predict specific processes. The choice of the in vitro and in silico method depends on the drug discovery stage, the drug properties and the available compound data. Current approaches aim to integrate available in silico tools and experimental data from in vitro assays to predict drug plasma profiles using PBPK modelling. With the integration of physiological data from different populations in PBPK models, it is also possible to predict pharmacokinetics in special populations and to estimate interindividual variability. All these tools contribute to the reduction of drug attrition rates in later stages of drug development, to the minimisation of time and costs in drug development and to the reduction of clinical studies. Further advancements are expected when PBPK models are set up in very early stages of drug development and confidence in the model grows by further integrating in vitro, *preclinical* and *clinical* data as it becomes available.

Acknowledgements This work has received funding from Horizon 2020 Marie Sklodowska-Curie Innovative Training Networks programme under grant agreement No. 674909 (PEARRL).

References

Abraham MH, Grellier PL, McGill RA (1987) Determination of olive oil–gas and hexadecane–gas partition coefficients, and calculation of the corresponding olive oil–water and hexadecane–water partition coefficients. J Chem Soc Perkin Trans 2:797–803. https://doi.org/10.1039/p29870000797

Alqahtani S, Mohamed LA, Kaddoumi A (2013) Experimental models for predicting drug absorption and metabolism. Expert Opin Drug Metab Toxicol 9:1241–1254. https://doi.org/10.1517/17425255.2013.802772

Amidon GL, Lennernas H, Shah VP et al (1995) A theoretical basis for a biopharmaceutic drug classification: the correlation of in vitro drug product dissolution and in vivo bioavailability. Pharm Res 12:413–420

Andrade CH, Silva DC, Braga RC (2014) In silico prediction of drug metabolism by P450. Curr Drug Metab 15:514–525

Avdeef A (2012) Permeability: Caco-2/MDCK. John Wiley & Sons, Inc., Hoboken. https://doi.org/10.1002/9781118286067.ch8

Baker RW (1987) Controlled release of biologically active agents. John Wiley & Sons, New York

Ballard P, Leahy DE, Rowland M (2003) Prediction of in vivo tissue distribution from in vitro data. 2. Correlation between in vitro and in vivo tissue distribution of a homologous series of nine 5-n-alkyl-5-ethyl barbituric acids. Pharm Res 20:864–872. https://doi.org/10.1023/A:1023912318133

Berezhkovskiy LM (2004) Volume of distribution at steady state for a linear pharmacokinetic system with peripheral elimination. J Pharm Sci 93:1628–1640. https://doi.org/10.1002/jps.20073

Bergstrom CA, Holm R, Jorgensen SA et al (2014) Early pharmaceutical profiling to predict oral drug absorption: current status and unmet needs. Eur J Pharm Sci 57:173–199. https://doi.org/10.1016/j.ejps.2013.10.015

Bohets H, Annaert P, Mannens G et al (2001) Strategies for absorption screening in drug discovery and development. Curr Top Med Chem 1:367–383. https://doi.org/10.2174/1568026013394886

Boni JE, Brickl RS, Dressman J (2007) Is bicarbonate buffer suitable as a dissolution medium? J Pharm Pharmacol 59:1375–1382. https://doi.org/10.1211/jpp.59.10.0007

Braun A, Hammerle S, Suda K et al (2000) Cell cultures as tools in biopharmacy. Eur J Pharm Sci 11:S51–S60

Broccatelli F, Salphati L, Plise E et al (2016) Predicting passive permeability of drug-like molecules from chemical structure: where are we? Mol Pharm 13:4199–4208. https://doi.org/10.1021/acs.molpharmaceut.6b00836

Bröer S (2010) Xenopus laevis oocytes. In: Yan Q (ed) Membrane transporters in drug discovery and development: methods and protocols. Humana Press, Totowa, pp 295–310. https://doi.org/10.1007/978-1-60761-700-6_16

Brunner E (1904) Reaktionsgeschwindigkeit in heterogenen Systemen. Z Phys Chem 47U. https://doi.org/10.1515/zpch-1904-4705

Caldwell GW, Yan Z (2014) Optimization in drug discovery: in vitro methods, 2nd edn. Humana Press, Springer, New York

Carbinatto FM, de Castro AD, Evangelista RC et al (2014) Insights into the swelling process and drug release mechanisms from cross-linked pectin/high amylose starch matrices. Asian J Pharm Sci 9:27–34. https://doi.org/10.1016/j.ajps.2013.12.002

Chang C, Swaan PW (2006) Computational approaches to modeling drug transporters. Eur J Pharm Sci 27:411–424. https://doi.org/10.1016/j.ejps.2005.09.013

Chen Y, Cameron K, Guzman-Perez A et al (2010) Structure-pharmacokinetic relationship of in vivo rat biliary excretion. Biopharm Drug Dispos 31:82–90. https://doi.org/10.1002/bdd.692

Dave RA, Morris ME (2015) Quantitative structure-pharmacokinetic relationships for the prediction of renal clearance in humans. Drug Metab Dispos 43:73–81. https://doi.org/10.1124/dmd.114.059857

de Groot MJ (2006) Designing better drugs: predicting cytochrome P450 metabolism. Drug Discov Today 11:601–606. https://doi.org/10.1016/j.drudis.2006.05.001

Di L, Fish PV, Mano T (2012a) Bridging solubility between drug discovery and development. Drug Discov Today 17:486–495. https://doi.org/10.1016/j.drudis.2011.11.007

Di L, Keefer C, Scott DO, Strelevitz TJ et al (2012b) Mechanistic insights from comparing intrinsic clearance values between human liver microsomes and hepatocytes to guide drug design. Eur J Med Chem 57:441–448. https://doi.org/10.1016/j.ejmech.2012.06.043

Di L, Kerns EH, Carter GT (2009) Drug-like property concepts in pharmaceutical design. Curr Pharm Des 15:2184–2194

Di L, Whitney-Pickett C, Umland JP et al (2011) Development of a new permeability assay using low-efflux MDCKII cells. J Pharm Sci 100:4974–4985. https://doi.org/10.1002/jps.22674

Doddareddy MR, Cho YS, Koh HY et al (2006) In silico renal clearance model using classical Volsurf approach. J Chem Inf Model 46:1312–1320. https://doi.org/10.1021/ci0503309

Dokoumetzidis A, Kalantzi L, Fotaki N (2007) Predictive models for oral drug absorption: from in silico methods to integrated dynamical models. Expert Opin Drug Metab Toxicol 3:491–505. https://doi.org/10.1517/17425225.3.4.491

Dressman JB, Amidon GL, Reppas C et al (1998) Dissolution testing as a prognostic tool for oral drug absorption: immediate release dosage forms. Pharm Res 15:11–22. https://doi.org/10.1023/a:1011984216775

Dressman JB, Fleisher D (1986) Mixing-tank model for predicting dissolution rate control or oral absorption. J Pharm Sci 75:109–116

Elaut G, Papeleu P, Vinken M et al (2006) Hepatocytes in suspension. Methods Mol Biol 320:255–263. https://doi.org/10.1385/1-59259-998-2:255

Fotaki N (2009) Pros and cons of methods used for the prediction of oral drug absorption. Expert Rev Clin Pharmacol 2:195–208. https://doi.org/10.1586/17512433.2.2.195

Fotaki N (2011) Flow-through cell apparatus (USP apparatus 4): operation and features. Dissolut Technol 18:46–49. https://doi.org/10.14227/DT180411P46

Fotaki N, Vertzoni M (2010) Biorelevant dissolution methods and their applications in in vitro in vivo correlations for oral formulations. Open Drug Delivery J 4:2–13. https://doi.org/10.2174/1874126601004010002

Galia E, Nicolaides E, Horter D et al (1998) Evaluation of various dissolution media for predicting in vivo performance of class I and II drugs. Pharm Res 15:698–705

Ghibellini G, Leslie EM, Brouwer KL (2006) Methods to evaluate biliary excretion of drugs in humans: an updated review. Mol Pharm 3:198–211. https://doi.org/10.1021/mp060011k

Hansen NT, Kouskoumvekaki I, Jorgensen FS et al (2006) Prediction of pH-dependent aqueous solubility of druglike molecules. J Chem Inf Model 46:2601–2609. https://doi.org/10.1021/ci600292q

Hayduk W, Laudie H (1974) Prediction of diffusion-coefficients for nonelectrolytes in dilute aqueous-solutions. AIChE J 20:611–615. https://doi.org/10.1002/aic.690200329

Heikkinen AT, Baneyx G, Caruso A et al (2012) Application of PBPK modeling to predict human intestinal metabolism of CYP3A substrates - an evaluation and case study using GastroPlus. Eur J Pharm Sci 47:375–386. https://doi.org/10.1016/j.ejps.2012.06.013

Higuchi T (1961) Rate of release of medicaments from ointment bases containing drugs in suspension. J Pharm Sci 50:874–875

Hill AP, Young RJ (2010) Getting physical in drug discovery: a contemporary perspective on solubility and hydrophobicity. Drug Discov Today 15:648–655. https://doi.org/10.1016/j.drudis.2010.05.016

Hinderling PH (1984) Kinetics of partitioning and binding of digoxin and its analogues in the subcompartments of blood. J Pharm Sci 73:1042–1053

Hintz RJ, Johnson KC (1989) The effect of particle size distribution on dissolution rate and oral absorption. Int J Pharm 51:9–17. https://doi.org/10.1016/0378-5173(89)90069-0

Hixson AW, Crowell JH (1931) Dependence of reaction velocity upon surface and agitation. Ind Eng Chem 23:923–931. https://doi.org/10.1021/ie50260a018

Jain N, Yalkowsky SH (2001) Estimation of the aqueous solubility I: application to organic nonelectrolytes. J Pharm Sci 90:234–252. https://doi.org/10.1002/1520-6017(200102)90:2<234::AID-JPS14>3.0.CO;2-V

Jamei M, Marciniak S, Feng K et al (2009) The Simcyp® population-based ADME simulator. Expert Opin Drug Metab Toxicol 5:211–223. https://doi.org/10.1517/17425250802691074

Kennedy T (1997) Managing the drug discovery/development interface. Drug Discov Today 2:436–444. https://doi.org/10.1016/S1359-6446(97)01099-4

Kola I, Landis J (2004) Can the pharmaceutical industry reduce attrition rates? Nat Rev Drug Discov 3:711–715. https://doi.org/10.1038/nrd1470

Kostewicz ES, Aarons L, Bergstrand M et al (2014a) PBPK models for the prediction of in vivo performance of oral dosage forms. Eur J Pharm Sci 57:300–321. https://doi.org/10.1016/j.ejps.2013.09.008

Kostewicz ES, Abrahamsson B, Brewster M et al (2014b) In vitro models for the prediction of in vivo performance of oral dosage forms. Eur J Pharm Sci 57:342–366. https://doi.org/10.1016/j.ejps.2013.08.024

Kostewicz ES, Wunderlich M, Brauns U et al (2004) Predicting the precipitation of poorly soluble weak bases upon entry in the small intestine. J Pharm Pharmacol 56:43–51. https://doi.org/10.1211/0022357022511

Kuentz M (2008) Drug absorption modeling as a tool to define the strategy in clinical formulation development. AAPS J 10:473–479. https://doi.org/10.1208/s12248-008-9054-3

Kunze A, Huwyler J, Poller B et al (2014) In vitro-in vivo extrapolation method to predict human renal clearance of drugs. J Pharm Sci 103:994–1001. https://doi.org/10.1002/jps.23851

Kuteykin-Teplyakov K, Luna-Tortos C, Ambroziak K et al (2010) Differences in the expression of endogenous efflux transporters in MDR1-transfected versus wildtype cell lines affect P-glycoprotein mediated drug transport. Br J Pharmacol 160:1453–1463. https://doi.org/10.1111/j.1476-5381.2010.00801.x

Lambrinidis G, Vallianatou T, Tsantili-Kakoulidou A (2015) In vitro, in silico and integrated strategies for the estimation of plasma protein binding. A review. Adv Drug Deliv Rev 86:27–45. https://doi.org/10.1016/j.addr.2015.03.011

Langenbucher F (1972) Linearization of dissolution rate curves by the Weibull distribution. J Pharm Pharmacol 24:979–981

Lave T, Chapman K, Goldsmith P et al (2009) Human clearance prediction: shifting the paradigm. Expert Opin Drug Metab Toxicol 5:1039–1048. https://doi.org/10.1517/17425250903099649

Lea T (2015) Caco-2 cell line. In: Verhoeckx K, Cotter P, López-Expósito I et al (eds) The impact of food bioactives on health: in vitro and ex vivo models. Springer International Publishing, Cham, pp 103–111. https://doi.org/10.1007/978-3-319-16104-4_10

Lipinski CA, Lombardo F, Dominy BW et al (2001) Experimental and computational approaches to estimate solubility and permeability in drug discovery and development settings1PII of original article: S0169-409X(96)00423-1. The article was originally published in advanced drug delivery reviews 23 (1997) 3–25.1. Adv Drug Deliv Rev 46:3–26. https://doi.org/10.1016/S0169-409X(00)00129-0

Liu F, Merchant HA, Kulkarni RP et al (2011) Evolution of a physiological pH 6.8 bicarbonate buffer system: application to the dissolution testing of enteric coated products. Eur J Pharm Biopharm 78:151–157. https://doi.org/10.1016/j.ejpb.2011.01.001

Luo G, Johnson S, Hsueh MM et al (2010) In silico prediction of biliary excretion of drugs in rats based on physicochemical properties. Drug Metab Dispos 38:422–430. https://doi.org/10.1124/dmd.108.026260

Mahmood I (1998) Interspecies scaling of renally secreted drugs. Life Sci 63:2365–2371

Manga N, Duffy JC, Rowe PH et al (2003) A hierarchical QSAR model for urinary excretion of drugs in humans as a predictive tool for biotransformation. QSAR Comb Sci 22:263–273. https://doi.org/10.1002/qsar.200390021

Markopoulos C, Andreas CJ, Vertzoni M et al (2015) In-vitro simulation of luminal conditions for evaluation of performance of oral drug products: choosing the appropriate test media. Eur J Pharm Biopharm 93:173–182. https://doi.org/10.1016/j.ejpb.2015.03.009

Martin YC (2005) A bioavailability score. J Med Chem 48:3164–3170. https://doi.org/10.1021/jm0492002

Mithani SD, Bakatselou V, TenHoor CN et al (1996) Estimation of the increase in solubility of drugs as a function of bile salt concentration. Pharm Res 13:163–167

Morrissey KM, Stocker SL, Wittwer MB et al (2013) Renal transporters in drug development. Annu Rev Pharmacol Toxicol 53:503–529. https://doi.org/10.1146/annurev-pharmtox-011112-140317

Muegge I, Heald SL, Brittelli D (2001) Simple selection criteria for drug-like chemical matter. J Med Chem 44:1841–1846

Nernst W (1904) Theorie der reaktionsgeschwindigkeit in heterogenen systemen. Z Phys Chem 47 (1). https://doi.org/10.1515/zpch-1904-4704

Neuhoff S, Gaohua L, Burt H et al (2013) Accounting for transporters in renal clearance: towards a mechanistic kidney model (Mech KiM). In: Sugiyama Y, Steffansen B (eds) Transporters in drug development: discovery, optimization, clinical study and regulation. Springer, New York, pp 155–177. https://doi.org/10.1007/978-1-4614-8229-1_7

Ni PF, Ho NFH, Fox JL et al (1980) Theoretical model studies of intestinal drug absorption V. Non-steady-state fluid flow and absorption. Int J Pharm 5:33–47. https://doi.org/10.1016/0378-5173(80)90048-4

Niederquell A, Kuentz M (2018) Biorelevant drug solubility enhancement modeled by a linear solvation energy relationship. J Pharm Sci 107:503–506. https://doi.org/10.1016/j.xphs.2017. 08.017

Noyes AA, Whitney WR (1897) The rate of solution of solid substances in their own solutions. J Am Chem Soc 19:930–934. https://doi.org/10.1021/ja02086a003

Paine SW, Barton P, Bird J et al (2010) A rapid computational filter for predicting the rate of human renal clearance. J Mol Graph Model 29:529–537. https://doi.org/10.1016/j.jmgm.2010.10.003

Paine SW, Menochet K, Denton R et al (2011) Prediction of human renal clearance from preclinical species for a diverse set of drugs that exhibit both active secretion and net reabsorption. Drug Metab Dispos 39:1008–1013. https://doi.org/10.1124/dmd.110.037267

Pang KS, Rowland M (1977) Hepatic clearance of drugs. I. Theoretical considerations of a "well-stirred" model and a "parallel tube" model. Influence of hepatic blood flow, plasma and blood cell binding, and the hepatocellular enzymatic activity on hepatic drug clearance. J Pharmacokinet Biopharm 5:625–653. https://doi.org/10.1007/bf01059688

Peters SA (2012) Physiologically-based pharmacokinetic (PBPK) modeling and simulations. John Wiley & Sons, Inc., Hoboken

Poulin P, Theil FP (2000) A priori prediction of tissue:plasma partition coefficients of drugs to facilitate the use of physiologically-based pharmacokinetic models in drug discovery. J Pharm Sci 89:16–35. https://doi.org/10.1002/(sici)1520-6017(200001)89:1<16::aid-jps3>3.0.co;2-e

Poulin P, Theil FP (2002) Prediction of pharmacokinetics prior to in vivo studies. 1. Mechanism-based prediction of volume of distribution. J Pharm Sci 91:129–156

Richardson SJ, Bai A, Kulkarni AA et al (2016) Efficiency in drug discovery: liver S9 fraction assay as a screen for metabolic stability. Drug Metab Lett 10:83–90. https://doi.org/10.2174/1872312810666160223121836

Rodgers T, Leahy D, Rowland M (2005a) Physiologically based pharmacokinetic modeling 1: predicting the tissue distribution of moderate-to-strong bases. J Pharm Sci 94:1259–1276. https://doi.org/10.1002/jps.20322

Rodgers T, Leahy D, Rowland M (2005b) Tissue distribution of basic drugs: accounting for enantiomeric, compound and regional differences amongst beta-blocking drugs in rat. J Pharm Sci 94:1237–1248. https://doi.org/10.1002/jps.20323

Rodgers T, Rowland M (2006) Physiologically based pharmacokinetic modelling 2: predicting the tissue distribution of acids, very weak bases, neutrals and zwitterions. J Pharm Sci 95:1238–1257. https://doi.org/10.1002/jps.20502

Rodgers T, Rowland M (2007) Mechanistic approaches to volume of distribution predictions: understanding the processes. Pharm Res 24:918–933. https://doi.org/10.1007/s11095-006-9210-3

Rostami-Hodjegan A (2012) Physiologically based pharmacokinetics joined with in vitro-in vivo extrapolation of ADME: a marriage under the arch of systems pharmacology. Clin Pharmacol Ther 92:50–61. https://doi.org/10.1038/clpt.2012.65

Scotcher D, Jones C, Posada M et al (2016a) Key to opening kidney for in vitro–in vivo extrapolation entrance in health and disease: part I: in vitro systems and physiological data. AAPS J 18:1067–1081. https://doi.org/10.1208/s12248-016-9942-x

Scotcher D, Jones C, Rostami-Hodjegan A et al (2016b) Novel minimal physiologically-based model for the prediction of passive tubular reabsorption and renal excretion clearance. Eur J Pharm Sci 94:59–71. https://doi.org/10.1016/j.ejps.2016.03.018

Sharifi M, Ghafourian T (2014) Estimation of biliary excretion of foreign compounds using properties of molecular structure. AAPS J 16:65–78. https://doi.org/10.1208/s12248-013-9541-z

Siepmann J, Peppas NA (2011) Higuchi equation: derivation, applications, use and misuse. Int J Pharm 418:6–12. https://doi.org/10.1016/j.ijpharm.2011.03.051

Stappaerts J, Brouwers J, Annaert P et al (2015) In situ perfusion in rodents to explore intestinal drug absorption: challenges and opportunities. Int J Pharm 478:665–681. https://doi.org/10.1016/j.ijpharm.2014.11.035

Sugano K (2012) Biopharmaceutics modeling and simulations: theory, practice, methods, and applications. John Wiley & Sons, Inc., Hoboken. https://doi.org/10.1002/9781118354339

Swift B, Pfeifer ND, Brouwer KL (2010) Sandwich-cultured hepatocytes: an in vitro model to evaluate hepatobiliary transporter-based drug interactions and hepatotoxicity. Drug Metab Rev 42:446–471. https://doi.org/10.3109/03602530903491881

Takada T, Suzuki H, Sugiyama Y (2005) Characterization of polarized expression of point- or deletion-mutated human BCRP/ABCG2 in LLC-PK1 cells. Pharm Res 22:458–464. https://doi.org/10.1007/s11095-004-1884-9

Takano R, Sugano K, Higashida A et al (2006) Oral absorption of poorly water-soluble drugs: computer simulation of fraction absorbed in humans from a miniscale dissolution test. Pharm Res 23:1144–1156. https://doi.org/10.1007/s11095-006-0162-4

US Pharmacopoeial Convention (2005) USP 29, NF 24: the United States Pharmacopeia, the National Formulary. 711 Dissolution

van de Kerkhof EG, de Graaf IA, Groothuis GM (2007) In vitro methods to study intestinal drug metabolism. Curr Drug Metab 8:658–675

van de Waterbeemd H, Gifford E (2003) ADMET in silico modelling: towards prediction paradise? Nat Rev Drug Discov 2:192–204. https://doi.org/10.1038/nrd1032

Veber DF, Johnson SR, Cheng HY et al (2002) Molecular properties that influence the oral bioavailability of drug candidates. J Med Chem 45:2615–2623

Wang J, Flanagan DR (1999) General solution for diffusion-controlled dissolution of spherical particles. 1. Theory. J Pharm Sci 88:731–738. https://doi.org/10.1021/js980236p

Wang J, Flanagan DR (2002) General solution for diffusion-controlled dissolution of spherical particles. 2. Evaluation of experimental data. J Pharm Sci 91:534–542. https://doi.org/10.1002/jps.10039

Wang QX, Fotaki N, Mao Y (2009) Biorelevant dissolution: methodology and application in drug development. Dissolut Technol 16:6–12. https://doi.org/10.14227/Dt160309p6

Watanabe T, Kusuhara H, Watanabe T et al (2011) Prediction of the overall renal tubular secretion and hepatic clearance of anionic drugs and a renal drug-drug interaction involving organic anion transporter 3 in humans by in vitro uptake experiments. Drug Metab Dispos 39:1031–1038. https://doi.org/10.1124/dmd.110.036129

Westhouse RA, Car BD (2007) Chapter 9 - concepts in pharmacology and toxicology. In: Prendergast GC, Jaffee EM (eds) Cancer immunotherapy. Academic Press, Burlington, pp 149–166. https://doi.org/10.1016/B978-012372551-6/50073-0

Willmann S, Schmitt W, Keldenich J et al (2003) A physiologic model for simulating gastrointestinal flow and drug absorption in rats. Pharm Res 20:1766–1771

Willmann S, Schmitt W, Keldenich J et al (2004) A physiological model for the estimation of the fraction dose absorbed in humans. J Med Chem 47:4022–4031. https://doi.org/10.1021/jm030999b

Willmann S, Thelen K, Lippert J (2012) Integration of dissolution into physiologically-based pharmacokinetic models III: PK-Sim®. J Pharm Pharmacol 64:997–1007. https://doi.org/10.1111/j.2042-7158.2012.01534.x

Yang J, Jamei M, Yeo K et al (2007) Prediction of intestinal first-pass drug metabolism. Curr Drug Metab 8:676–684. https://doi.org/10.2174/138920007782109733

Yu LX, Amidon GL (1999) A compartmental absorption and transit model for estimating oral drug absorption. Int J Pharm 186:119–125. https://doi.org/10.1016/S0378-5173(99)00147-7

Yu LX, Lipka E, Crison JR et al (1996) Transport approaches to the biopharmaceutical design of oral drug delivery systems: prediction of intestinal absorption. Adv Drug Deliv Rev 19:359–376

Zhang X, Shedden K, Rosania GR (2006) A cell-based molecular transport simulator for pharmacokinetic prediction and cheminformatic exploration. Mol Pharm 3:704–716. https://doi.org/10.1021/mp060046k

Drug Transporters

14

Alan Talevi, Carolina Leticia Bellera, and Guido Pesce

14.1 Introduction

Membrane transporters perform a central function in many essential body functions, including traffic and compartmentalization of physiologic compounds, absorption of physiologic compounds (and, occasionally, structurally related xenobiotics, including drugs), and protection from potentially toxic exogenous compounds, which includes limiting their absorption and/or distribution and promoting their elimination (with no chemical modifications or as metabolites). The impact of drug biotransformation enzymes on pharmacokinetics and clinical outcome is much better understood than that of drug transporters, possibly due to the earlier discovery of metabolizing enzymes. The science of drug transporters can still be described as an emerging field, with knowledge on their structure, function, and regulation being continuously expanded.

A. Talevi (✉)
Laboratory of Bioactive Research and Development (LIDeB), Department of Biological Sciences, Faculty of Exact Sciences, University of La Plata (UNLP), La Plata, Buenos Aires, Argentina

Consejo Nacional de Investigaciones Científicas y Técnicas (CONICET), La Plata, Buenos Aires, Argentina
e-mail: atalevi@biol.unlp.edu.ar

C. L. Bellera
Medicinal Chemistry/Laboratory of Bioactive Research and Development (LIDeB),
Faculty of Exact Sciences, Universidad Nacional de La Plata (UNLP),
Consejo Nacional de Investigaciones Científicas y Técnicas (CONICET), Buenos Aires, Argentina

G. Pesce
Department of Pharmacology, Argentinean National Institute of Medications (INAME),
Administración Nacional de Alimentos, Medicamentos y Tecnología Médica (ANMAT),
Buenos Aires, Argentina
e-mail: guido@anmat.gov.ar

© Springer Nature Switzerland AG 2018
A. Talevi, P. A. M. Quiroga (eds.), *ADME Processes in Pharmaceutical Sciences*,
https://doi.org/10.1007/978-3-319-99593-9_14

▶ **Important** Although ubiquitously expressed throughout the body, the highest levels of drug transporters are found in those organs and tissues with barrier and/or elimination functions, among them the gut, liver, kidney, blood-brain barrier, and placenta. Their regulation also plays a significant role in drug resistance issues.

Transporters may have different functions: (a) helping their substrates to move across diffusional barriers which they may not be able to (efficiently) cross otherwise (typically, cell or organelle membranes). This is usual for polar (including charged) substrates; (b) moving their substrates against an electrochemical gradient, which might be useful to build up reserves of a given chemical substance in a certain cell or cell compartment and; (c) acting in cooperation with metabolizing enzymes, regulating the level of substrate to which they are exposed, and thus getting the most out of metabolizing systems. As we will see later, more than one of the precedent general functions might be facilitated by the same transporter, depending on the substrate under consideration and the location in the body.

The term transporter refers to a diversity of membrane proteins with diverse functions, structures, and subcellular locations. As seen in Chap. 2, they can mediate either facilitated or active transport. Facilitated transport implies movement of drug across a membrane down an electrochemical gradient. In contrast, active transport relies on energy coupling that either directly or indirectly implies ATP consumption.

The most commonly invoked model to explain the substrate transport by transporters is the *alternating access model*. It proposes a substrate binding site that can alternatively access either the extracellular (or extra-organellar) side or the intracellular (or intra-organellar) side of the membrane, corresponding to the outward and inward facing conformations of the transporter (Forrest and Rudnick 2009; Rees et al. 2009). After substrate/s binding (coupled with energy consumption mechanisms in the case of active transport) conformational changes are induced that result in the alternating exposure of this binding site to the other side of the membrane, producing substrate translocation. The relative binding affinities for substrate of the two conformations largely determine the net direction of transport (e.g., the outward facing conformation of an importer is expected to have a higher affinity for substrate than the inward facing conformation, whereas the opposite holds for exporters).

Transporters can be classified or grouped following different criteria. For instance, they may be grouped on the basis of the direction of the flow of their substrates, their substrate specificity, or their sequence homology/evolutionary relationship.

▶ **Definition** Taking the flow direction of their substrates as classification criterion, transporters may be classified as *efflux transporters* (which export their substrates out of the cells or, in some cases, an organelle), *influx* or *uptake transporters* (which import their substrates to the cell or organelle), and *bidirectional transporters* (which, as the name suggests, display bidirectional fluxes, e.g., hepatic glucose transporter GLUT2 facilitates bidirectional glucose transport across the hepatocyte plasma membrane under insulin regulation).

The two most important transporter superfamilies in relation to drug transport are the ATP-binding cassette (ABC) transporters and the solute carrier (SLC) transporters. In the case of the SLCs, the term "superfamily" may be misleading, because it generally implies a common ancestor as well as detectably similar sequences and structures. However, phylogenetic analysis of human SLC proposed that they consist of 15 related families and 32 additional unlinked families (Schlessinger et al. 2010). Figure 14.1 shows how the alternating access model may be adapted to ABC transporters or to SLC transporters that work facilitating diffusion, as symporters or antiporters.

As always when saturable systems are involved in drug kinetics, the existence of drug transporters gives rise to the possibility of drug-drug and food-drug interactions involving transport inhibition and transporter induction, the possibility of non-linear kinetics and the existence of genetic polymorphism that results in interindividual variability in drug sensitivity. We will also see that some drug transporters are responsible for multidrug resistance issues. The potential involvement of some important drug transporters in drug-drug interactions is clearly reflected in different FDA guidances that describe the Agency's current thinking on the matter (US Food and Drug Administration 2017a, b).

Besides discussing the general features of the members of the ABC and SLC superfamilies more directly involved in drug transport, this chapter will discuss the coordination of drug transporters and metabolizing enzymes. Although the scope of the chapter are those transporters involved in drug kinetics, we will also present an overview of other transporters that are valuable as drug targets. The reader might however resort to pharmacology textbooks for a deeper insight into this latter topic.

14.2   ATP-Binding Cassette (ABC) Transporters

All the transporters from the ABC superfamily have a characteristic (minimal) modular architecture, which consists of four domains: two transmembrane domains (TMDs) which are embedded in the membrane bilayer and two ABCs (or nucleotide-binding domains, NBDs) located in the cytoplasm (Rees et al. 2009; ter Beek et al. 2014).

At the sequence level, ABC transporters superfamily is identified by the highly conserved motifs present in the ABCs (that give name to the superfamily); conversely, the sequences and architectures of the TMDs, where the substrate binding site/s is/are located, are variable, which reflects the chemical diversity of the translocated substrates. NBDs do not always associate with TMDs and are logically involved in various functions that do not occur at the membrane (ter Beek et al. 2014). Nevertheless, the name "ABC transporter" is only used when the NBDs are associated with TMDs.

ATP binding to the ABCs and its subsequent hydrolysis is coupled with to the conformational states of the TMDs; accordingly, ABC transporters produce primary active transport.

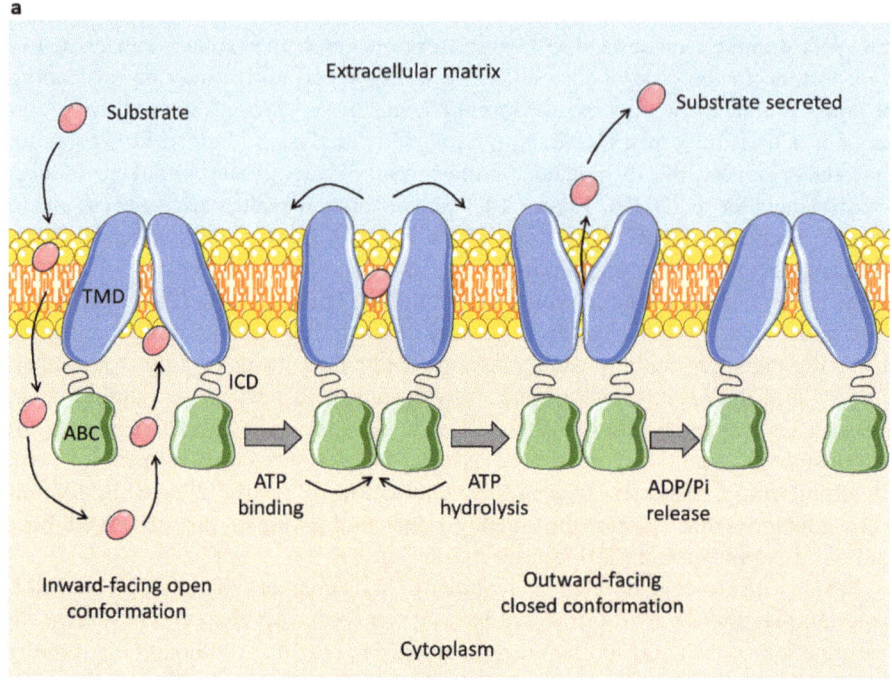

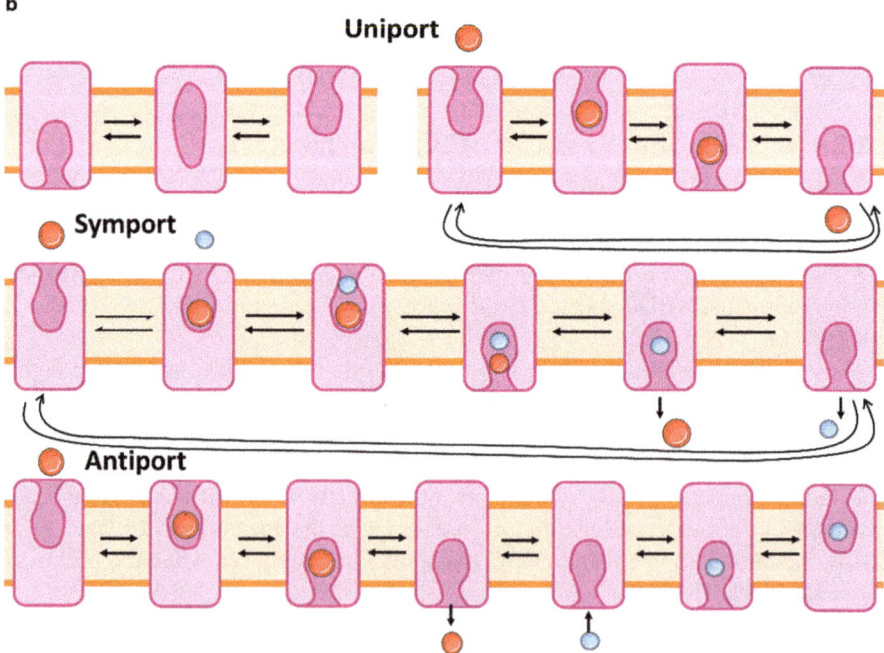

Fig. 14.1 Diagram showing how the alternating access model adapts to different transportation mechanisms. (**a**) ABC transporters. (**b**) SLC transporters

▶   **Important** To our current knowledge, ABC transporters in mammals (and, more generally, in eukaryotes) act exclusively as exporters, that is, they remove their substrates from the cell (or organelle) to the outside of the cell (or organelle). In prokaryotes, they can act as exporters or importers.

The subcellular location of the transporter (e.g., apical or basolateral membranes of polarized cells) will determine the direction of the substrate flux. For instance, an ABC transporter expressed in the apical membrane of an enterocyte will export its substrates to the gut lumen, whereas another ABC transporter located at the basolateral membrane will export its substrates in the opposite direction, facilitating access to the capillary blood. In the first case, absorption would be limited by the transporter. In the second, absorption will be favored.

Around 50 ABC transporters have been described in humans, which participate in cholesterol and lipid transport, antigen presentation, mitochondrial iron homeostasis, and other fundamental processes. Accordingly, mutations or dysfunction of these proteins have been associated with a diversity of disorders including cystic fibrosis, diabetes, hypercholesterolemia, and Alzheimer's disease, among many others (Rees et al. 2009; Pereira et al. 2018).

In relation to drug pharmacokinetics, the most relevant transporters are *p-glycoprotein* (Pgp) (the product of the ABCB1 gene, also known as *multidrug resistance protein 1*, MDR1), nine members of the *multidrug resistance-associated protein* (MRP) family (also known as MRPS or ABCCs), including MRP1 to MRP9, and the *breast cancer resistance protein* (BCRP) (also known as ABCG2).

▶   **Important** The nomenclature of the ABC transporters that have an impact on drug pharmacokinetics (MDR1, MRPs, BCRP) reflects the fact that, initially, high expression levels of these transporters were identified in cancerous cells and were associated to cross-resistance (where a patient is being treated with one drug and the tumor acquires a type of drug resistance that enables it to avoid being killed by a variety of drugs). Later though, it was observed that ABC transporters are ubiquitously expressed and have a privilege role as detoxifying systems which, as we will see, act in coordination with metabolizing enzymes.

Their involvement in multidrug resistance phenomena and their ability to protect the body from a wide diversity of structurally and functionally unrelated xenobiotics (drugs included) emerge from their individual and collective polyspecificity (or *wide substrate specificity*). For example, Pgp alone can transport diverse substrates as aldosterone, amprenavir, bilirubin, cyclosporine, digoxin, doxorubicin, erythromycin, etoposide, itraconazole, paclitaxel, or verapamil, among many, many others (Kim 2002). The ability to recognize such variety of substrates arises from the presence of multiple binding sites (Safa 2004; Ferreira et al. 2013).

Remarkably, the sets of substrates of Pgp and CYP3A4 (the most promiscuous CYP450 member) are partially overlapping (Wacher et al. 1995; Katoh et al. 2001; Zhou 2008), and also overlapping is their tissue distribution (Wacher et al. 1995). They also share some common inducers, being possibly the best known example of coordinated enzyme and transporter regulation (Salphaty and Benet 1998; Matheny et al. 2004; Huwyler et al. 2006; Annaert et al. 2007).

Despite their wide substrate specificity, each transporter prefers certain types of substrates. The general features of the substrates for each transporter are shown in Table 14.1.

Besides their well-known role as major determinants of multidrug resistance in cancer (Pérez-Tomás 2006; Sun et al. 2012), they have also been linked to refractoriness in other disorders, such as epilepsies (Feldmann and Koepp 2016). Several possible therapeutic strategies have been conceived to prevent or cope with constitutive or acquired cross-resistance linked to ABC transporters (Potschka and Luna-Munguia, 2014; Couyoupetrou et al. 2017). The first explored approach was the coadministration of chemosensitizers capable of reversing the transporter effect. However, there are currently no approved drugs available for clinical use to reverse multidrug resistance by inhibiting ABC transporters (Nanayakkara et al. 2018). Although some Pgp inhibitors reached clinical trials, discouraging results were obtained, mainly due to toxicity issues and limited efficacy (Lhommé et al. 2008; Tiwari et al. 2011; Choi and Yu 2014). The reader should have in mind the physiologic role of ABC transporters as a general detoxification mechanism and their involvement in the traffic of physiologic compounds, which possibly discourages the use of add-on inhibitors in the context of long-term therapies. The

Table 14.1 Some (very) general features of the known substrates of pharmacokinetically relevant ABC transporters. Note that, in some way, the universe of substrates of Pgp and MRPs seem to be complementary. As will be discussed later in the chapter, the specific substrate preferences of each transporter coupled with their cellular location in different organs will contribute to their coordinated function as well as their articulation with metabolizing enzymes

Transporter/s	Substrates characteristics
Pgp	Relatively hydrophobic drugs. Cationic molecules. Many substrates possess planar aromatic rings and positively charged tertiary N atoms (Sarkadi et al. 2006; Subramanian et al. 2016). Strong overlap with BCRP
MRPs	Anionic organic anions and drug conjugated to glutathione, glucuronate, and sulfate. MRP1 to MRP3 can transport unconjugated neutral organic drugs by (co)transporting them with free glutathione. MRP1 can even confer resistance to arsenite and MRP2 to cisplatin. MRP4 overexpression is associated with high-level resistance to nucleoside analogs (Borst et al. 2000; Homolya et al. 2003). Comparatively low overlap with Pgp
BCRP	Large, hydrophobic, both cations and anions. Sulfate and glucuronic conjugates, such as estrone-3-sulfate and 17β-estradiol 17-(β-d-glucuronide). In general, sulfated conjugates seem to be better BCRP substrates than glutathione and glucuronide conjugates. In addition, phosphorylated nucleosides and nucleotides. More than 200 verified substrates (Sarkadi et al. 2004; Mao and Unadkat 2015)

study of next-generation, highly selective inhibitors of ABC transporters continues being an active area of research, with many ongoing clinical trials. Research focus has lately been directed to therapeutic agents targeting to the signaling cascade that regulates efflux transporters expression. Other possibly safer approaches include the use of a Trojan Horse stratagem to deliver therapeutic levels of ABC transporters substrates to the targeted tissue. By encapsulating ABC transporter substrates in particulate delivery systems (primarily, pharmaceutical nanocarriers), the recognition and translocation by the efflux pumps can be avoided or minimized (Blanco et al. 2015). Alternatively, the design of novel therapeutics which are not recognized by the transporters (disregarding their substrates early at the drug discovery and development process) may be possible, thus considering ABC transporters as *anti-targets*.

Example

A 11-year and, so far, healthy patient was admitted to the emergency room with simple partial motor seizures accompanied by fever and headache. The seizures were immediately terminated with intravenous diazepam. The next day, the child had again a simple partial motor seizure, which was unresponsive to diazepam, lorazepam, and phenytoin. The episode evolved to generalized convulsive status epilepticus (a medical emergency associated with high mortality and morbidity, characterized by continuous seizure lasting more than 30 min, or two or more seizures without full recovery of consciousness between any of them), which the physicians were not able to control despite sequential administration of lorazepam, fentanyl, and phenobarbital. The boy became comatose. After trying other unsuccessful therapeutic options, verapamil was administered on day 37, based on the evidence that it has anticonvulsant activity in animal models and the fact that the boy had a supraventricular tachycardia. The patient regained consciousness and was able to breathe spontaneously, and the status epilepticus promptly disappeared. The dramatic response to verapamil may be explained by its direct anticonvulsant action and the fact that this drug is a known Pgp inhibitor which may have acted in the cerebrovascular endothelium and in the epilepticus focus by facilitating the brain penetration of the antiepileptic drugs that the patient was simultaneously receiving (at that time, valproic acid and phenobarbital).

Note that phenobarbital has been shown to act as a (weak) Pgp substrate (Luna-Tortós et al. 2008). Example is based on a case report by Iannetti et al. (2005).

14.3 Solute Carriers (SLC)

A second important family of drug transport proteins are the SLC transporters. Today, the Gene Nomenclature Committee of the Human Genome Organization classifies about 400 human SLC transporters into over 60 families, based on their sequences, number of transmembrane α-helices and biological functions (Schlessinger et al. 2010, 2013) (note however, that their classification and organization can sometimes be highly dynamic). SLC function as facilitative transporters (allowing solutes to flow downhill with their electrochemical gradients) or as secondary active transporters (allowing their substrates to flow uphill against their electrochemical gradient by coupling to transport of a second solute, either as cotransporters and exchangers). Despite weak sequence similarities, solute carriers can share similar structural features (Taft 2009). Most members of this superfamily locate in the cell membrane, though some of them are embedded in organelle membranes. They are typically composed of one large domain consisting of 10–14 transmembrane α-helices. The nomenclature for these transporters is in general similar to that already discussed for CYP450 members in Chap. 4. A protein from this superfamily is generically denoted as SLCnXm, where SLC is the root name that specifies membership to the superfamily, n is an integer representing a family, X is a letter denoting a subfamily, and m is another integer that specifies an individual family member.

Members of the SLC1 (glutamate and neutral amino acid transporter), SLC6 (Na^+ - and Cl^--dependent neurotransmitter transporters), SLC17 (vesicular glutamate transporters), SLC18 (vesicular amine transporters), and SLC32 (vesicular inhibitory amino acid transporters) regulate the concentration of neurotransmitters, thereby regulating neurosignaling pathways (Gether et al. 2006); therefore, several members of these families are exploited as drug targets (see Sect. 14.3.1 later in this chapter).

Members of several SLC families, such as SLCO (organic anion transporting polypeptide superfamily, formerly SLC21), SLC15 (oligopeptide transporters), SLC22 (organic cation/anion/zwitterion transporters), and SLC47 (multidrug and toxin extrusion (MATE) transporters), abound in the liver, kidney, and blood-brain barrier and thus impact on drug distribution, metabolism, and excretion (Schlessinger et al. 2013; Hagenbuch and Stieger 2013; Koepsell 2013).

Some members of the SLC are particularly relevant in terms of drug absorption and/or disposition. SLC22 members are characterized by a wide substrate specificity and wide tissue distribution. OAT1 is a dicarboxylate exchange (sodium dependent) protein responsible for basolateral uptake of organic anions into the tubular epithelial cells in the kidney (Taft 2009). It mediates the urinary excretion of a wide array of substrates, including antibiotics, antiviral nucleoside analogs, and nonsteroidal anti-inflammatory medications. More than 100 drugs and toxins have been found to interact with OAT1. OAT3 is a basolateral transporter similarly to OAT1. There is some degree of overlap of substrate specificity between both, OAT3 transports sulfate- and glucuronide steroid conjugates. OAT1 and OAT3 are inhibited by probenecid. OATs function as antiporters. Other members of SLC22 specialize in the translocation of cationic drugs, mainly the organic cation transporters (OCT1/2) that function as uniporters and the organic cation and carnitine transporters (OCTN1/ 2) (Russel 2010). OCT1 is most strongly expressed in the liver. OCT2 abounds in the kidney. Both mediate cellular uptake of rather small monovalent organic cations, by facilitated diffusion down the electrochemical gradient. Other cation transporters belonging to the SLC47 family have been identified more recently. They are proton-cation antiporters from the multidrug and toxic compound extrusion family (MATE). Among them, MATE1 and MATE2-K are efflux transporters fueled by an inwardly directed proton gradient and (partially) responsible of the final step in the excretion of metabolic waste and xenobiotic organic cations in the kidney and liver.

At last, the SLCO family consists of organic anion transporting polypeptides (OATPs) that are involved in the cellular uptake of bulky (MW > 450 Da) and relatively hydrophobic organic anions. In addition, the substrate specificity of the OATPs covers a wide range of amphipathic compounds, including bile salts, steroid conjugates, thyroid hormones, and several drugs as enalapril and digoxin. OATPs are widely expressed in different physiological barriers, including the gut, renal tubules, blood-brain barrier, and placenta.

14.3.1 Examples of Some Pharmacodynamically Relevant SLC Transporters

As already mentioned, some other members of the SLC superfamily are involved in the transport of neurotransmitters and are therefore very relevant as drug targets. As examples, we will briefly discuss the serotonin, dopamine, and monoamine vesicular transporters. Further information on this and other pharmacodynamically relevant SLCs may be found in pharmacology textbooks.

14.3.1.1 Serotonin Transporter

5-hydroxytryptamine (serotonin, 5HT) is a neurotransmitter located in several structures of the central nervous system and peripheral organs. It is found in high concentrations in the chromaffin cells of the gastrointestinal tract and platelets and is involved in numerous biological processes that include cell division, differentiation and migration, and neuronal synaptogenesis (Daws and Gould 2011). Serotonin levels are controlled by a set of proteins, including a multiplicity of pre- and post-synaptic serotonergic receptors, heteroreceptors, enzymes, and, particularly, specific transporters located in the cell membrane that capture the released 5HT.

The serotonin transporter (TDS, SLC6A4) is distributed in the CNS along the fibers that run through the anterior brain and raphe nuclei (Qian et al. 1995; Carneiro and Blakely 2006), while peripherally it is located in platelets, placenta, gastrointestinal wall, and lungs. Studies of cellular fractionation and immunohistochemistry have shown that the 5HT transporter can locate in different cellular compartments depending on the regulatory phase and the cell type in which it is found.

TDS is the most efficient regulator of the concentration of extracellular 5HT, exerting a rigorous control over the intensity and duration of serotonergic neurotransmission. Dynamic changes in the number and function of TDS regulate the rate by which the 5HT released by exocytosis is removed from the synapse. The TDS-mediated uptake of 5HT is critical for the recycling of the neurotransmitter. Once in the cytoplasm, 5HT is sequestered by the vesicular monoamine transporters (TMV2) and stored inside the synaptic vesicles (Kristensen et al. 2011).

The dysregulation of the transporter activity has been related to some psychiatric disorders (Daws and Gould 2011). Clinically, the importance of TDS is reflected by the action of 5HT reuptake inhibitors, widely used in the treatment of depressive states. On the other hand, psychostimulant drugs such as cocaine, amphetamines, methamphetamine, and MDMA exert their actions at TDS (Spiller et al. 2013).

The transport mechanism that best describes the functioning of TDS is the alternating access model. It involves two transport processes dependent on the direction in which they move the substrate (cotransporter or symport) and counter-transport (or antiport).

TDS uses the Na^+ transmembrane concentration difference to favor the accumulation of 5HT within the cells. The energetically favorable influx of a Na^+ is coupled to the transport of a molecule of 5HT and a Cl^- ion through the plasma membrane. This process of cotransport is followed by a counter-transporting process in which a K^+ ion is released into the extracellular medium, an exchange that contributes with additional energy necessary for the accumulation of 5HT in the cytosol.

The serotonin transporter catalyzes a complex reaction by simultaneously incorporating the processes of cotransport and counter-transport. In the process of cotransport, Na^+, Cl^-, and 5HT are associated in a 1:1:1 stoichiometry, generating occlusion of the transporter in the external medium and a conformational change that exposes it to the cytosol. The dissociation of Na^+, Cl^-, and 5HT to the cytosol allows the transporter to return to its original conformation but only after binding a cytoplasmic K^+ ion and releasing it to the extracellular medium (Rudnick 1998). The total stoichiometry of the reaction is consequently 1:1:1:1 resulting in an electrically neutral exchange.

14.3.1.2 Dopamine Transporter

Dopamine (DA) is a catecholaminergic neurotransmitter located in several central nervous structures including the nigrostriatal and mesolimbic systems. It is an essential mediator in many cognitive and motor functions, including functions of movement, learning, humor, reward, and learning. The imbalance in the levels of DA leads to the appearance of different disorders such as Parkinson's disease, addictions, or bipolar disorder.

Common to other neurotransmitters such as noradrenaline (NE) and adrenaline (A), DA follows a synthesis pathway from the essential amino acid tyrosine that is converted to L-DOPA by the action of tyrosine hydroxylase, an enzyme that limits the synthesis of the catecholaminergic pathway, and then in the cytoplasm the L-DOPA is converted into DA by the enzyme decarboxylase of the aromatic amino acids. The cytoplasmic DA is captured by the vesicular monoamine transporter 2 (TVM2) to the synaptic vesicles of the presynaptic nerve terminals. After releasing the DA to the extra-synaptic space and exerting its action on the pre- and post-dopaminergic synaptic receptors, it is recaptured by the nerve terminals through the DA transporter (ADT, SLC6A3) and once in the cytoplasm is degraded or again stored in the synaptic vesicles by the TVM2.

As a consequence of the coordinated action of the tyrosine hydroxylase (TH), the TVM2 and the TDA, the quantal size of DA in the nerve terminal established as well as the extension and duration of the dopaminergic signal. ADT, in addition to modulating the dynamics of the dopaminergic signal, is responsible of recycling the DA molecules that it transports into the cytosol, directly influencing the cytoplasmic concentration of DA and indirectly the recharging of the synaptic vesicles.

TAD belongs to the family of secondary active transporters that cotransport (or symport) a DA molecule together with two Na^+ ions and one Cl^- ion (Kristensen et al. 2011) against a concentration gradient of DA and in favor of an electrochemical gradient created by the membrane Na^+/K^+-ATPase pump, the latter constituting the driving force for the uptake of DA. DA translocation from the extra-synaptic space to the cytosol occurs through an alternating access mechanism in which the protein cyclizes between a conformation with an open orientation toward the synaptic cleft and a conformation with an open orientation toward the cytoplasm (Singh et al. 2008). This process is initiated by the binding of the DA to the pocket formed by the transmembrane regions 1,3,6, and 8 in the conformation of the TAD with open orientation toward the synaptic cleft and binding of two Na^+ ions and a Cl^- ion to the same pocket. The binding of the substrates generates the subsequent occlusion of the transporter (closed state) avoiding changes in its conformation and allowing the substrates to reach the intracellular space; the process continues with the change of conformation of the TAD to the open orientation toward the cytosol releasing DA, Na$^+$, and Cl^-; once the substrates are released, the TAD returns to the conformation open toward the synaptic cleft to start a new cycle (Reith et al. 2015).

A variant to the alternating access model has been described, which is the exchange mode, prototype model mediated by amphetamine, where TAD undergoes the same conformational changes described in the alternating model, but the cycle is interrupted when it acquires a conformation with internal orientation. The amphetamine transported by the TAD is released to the cytosol, and the transporter in its open orientation to the cytoplasm binds the DA and cytosol ions and cyclizes to the conformation with external orientation releasing its substrate load to the synaptic cleft. In this exchange mode, TAD functions as a pendulum between cycle halves and, unlike the alternating access model, TAD is always occupied by a substrate. Consequently, in the exchange model, the DA is released into the synaptic space (Sitte and Freissmuth 2015).

14.3.1.3 Vesicular Monoamine Transporter 2

The vesicular monoamine transporter 2 (SLC18A2) is responsible for the storage of the cytosolic DA, 5HT, NA within the synaptic vesicles (dense core) of brain monoaminergic neurons (Erickson and Eiden 1993; Eiden and Weihe 2011). TVM2 is a member of the SLC18 family, to which also belong TMV1 and VAChT (vesicular acetylcholine carrier) (Lawal and Krantz 2013). It is preferentially expressed in the monoaminergic neurons of the central nervous system, sympathetic nervous system, mast cells, and histaminergic cells of the intestine.

The synaptic vesicles are small bilaminar lipid spherical formations formed in the Golgi apparatus and transported to the nerve terminal (Südhof 2004) where they are responsible for the storage and concentration of the neurotransmitter for subsequent release to the synaptic space.

TVM2 functions as a stoichiometric counter-transport, transporting an H^+ ion out of the vesicle to transport a monoamine molecule inside the vesicle. The activity of the H^+-ATPase establishes the high concentration of intravesicular H^+ ions on which the transport of TVM2 depends (Sulzer et al. 2005). TMV2 plays an important role in the storage capacity of the vesicular neurotransmitter and in the regulation of vesicular volume, both mechanisms being compensatory to the different changes in the dopaminergic system (Lohr et al. 2017). The movement of H^+ ions creates an electrochemical gradient and establishes an acidic environment inside the vesicle (pH 5.5) (Rudnick et al. 1990). Amphetamine and its analogs modify the intravesicular pH which contributes to the dysfunction of TVM2 (Sulzer et al. 1993; Omiatek et al. 2013).

The presynaptic localization of TVM2 plays a critical role in regulating the capacity of the transporter in the uptake of monoamines. This has shown that psychostimulants redistribute the location of TVM2 causing an alteration in the uptake and vesicular content (Ferman et al. 2015).

14.4   Cooperation Is Key

ABC and SLC transporters associated to drug absorption and disposition and biotransformation enzymes usually operate in a highly concerted manner.

▶   **Important** We may mention three general ways of interplay between transporters and metabolic enzymes.

In the first case, an efflux transporter may (down) regulate the rate at which a given xenobiotic is uptaken by a cell, so that the enzymatic machinery is not saturated by the transient exposure to high xenobiotic levels (Fagiolino 2017). This type of transporter/enzyme cooperation may have a high impact on drug absorption in the gut. Imagine, for instance, the cooperation of Pgp and CYP3A4 (we have already underlined that these two systems have a strongly overlapping substrate specificity). Pgp locates in the apical membrane of the enterocyte, while CYP3A4 is expressed at the endoplasmic reticulum (Fig. 14.2). Have in mind that (a) Pgp removes lipophilic drugs from the cell and (b) the drug concentration gradient across the intestinal epithelia can be quite high, especially when high doses of a drug are orally administered. Lipophilic drugs will tend to be highly permeable; when administered in high doses, it is highly probable that the enterocyte metabolizing enzymes are overwhelmed by the high drug influx rate (the number of drug molecules entering the enterocyte per unit of time exceeds the number of available enzyme copies; accordingly, a fraction of the drug molecules will evade the intestinal first pass effect). The situation is illustrated in Fig. 14.3. The efflux transporter in the apical membrane will limit the number of xenobiotic molecules entering the cell per unit of time, so that

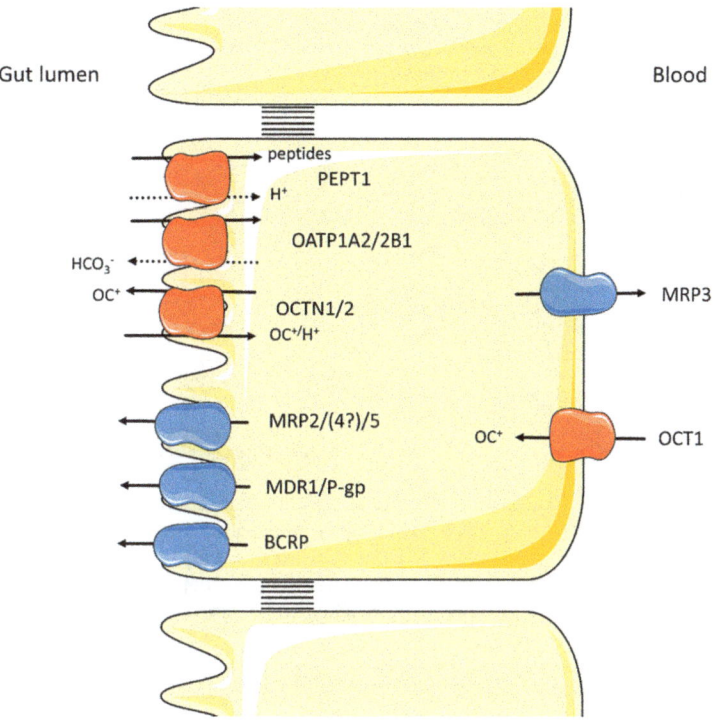

Fig. 14.2 Diagram showing the expression of different transporter systems in the gut wall

the system does not get saturated (the same principle impacts in the number of molecules that reach the liver per time unit, with the consequent optimization of the hepatic first pass effect).

A second way of cooperation between transporters and enzymes can be observed in eliminating organs, especially for drugs with high extraction ratios. Passive permeability can be rate limiting for (hepatic) clearance if the intrinsic metabolism is moderately fast. Therefore, uptake transporters in the sinusoidal membrane operate by improving the drug influx into the cell so that the high levels of the enzymatic systems within the liver can be better exploited (Annaert et al. 2007; Di et al. 2012) (see Fig. 14.4).

Finally, we must consider the contribution of efflux transporters to get rid of (potentially toxic) waste products (Annaert et al. 2007). In particular, note that phase II metabolites, due to their high polarity, are not able to readily diffuse through cell membranes. If not with the help of transporters, and even considering their favorable concentration gradient, they would find difficulties to leave the hepatocyte for subsequent urinary or biliary excretion (Fig. 14.4), to reach the cytosol of the tubular cells (Fig. 14.5), and to move from the tubular cell to the tubular lumen.

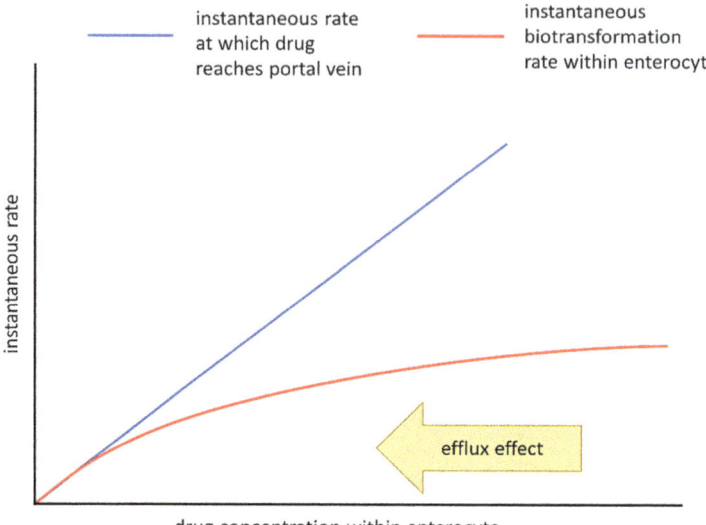

Fig. 14.3 Diagram illustrating how the efflux transporters cooperate with biotransformation enzymes by limiting the flux of the drug to the metabolizing organs. Adapted from Fagiolino (2017)

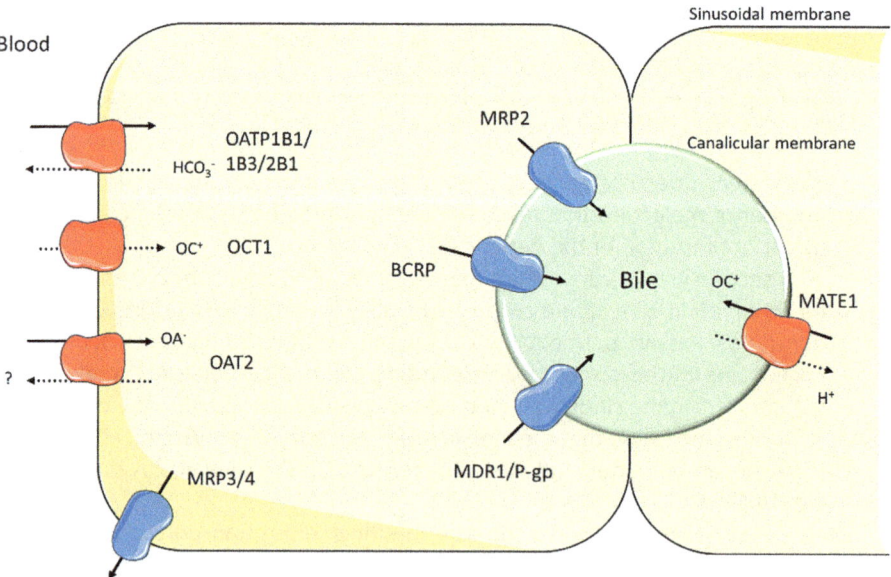

Fig. 14.4 Diagram showing the expression of different transporter systems in hepatocyte sinusoidal and canalicular membranes

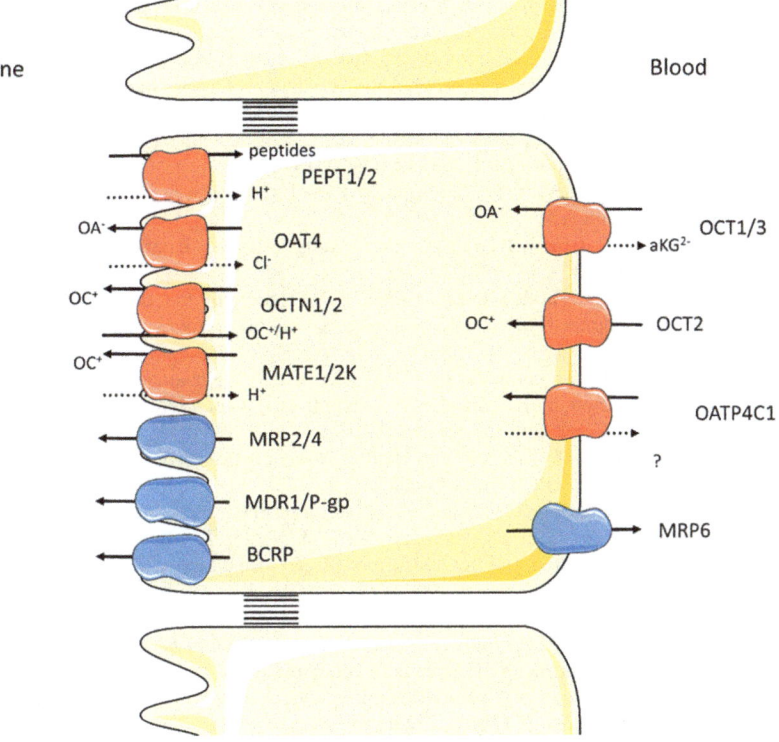

Fig. 14.5 Diagram showing the expression of different transporter systems in the tubular epithelium

References

Annaert P, Swift D, Lee JK et al (2007) Drug transport in the liver. In You G, Morris ME (eds). John Wiley & Sons, Inc., Hoboken

Blanco E, Shen H, Ferrari M (2015) Principles of nanoparticle design for overcoming biological barriers to drug delivery. Nat Biotechnol 33:941–951

Borst P, Evers R, Kool M et al (2000) A family of drug transporters: the multidrug resistance-associated proteins. J Natl Cancer Inst 92:1295–1302

Carneiro A, Blakely R (2006) Serotonin-, protein kinase C-, and Hic-5-associated redistribution of the platelet serotonin transporter. J Biol Chem 281:24769–24780

Choi YH, Yu AM (2014) ABC transporters in multidrug resistance and pharmacokinetics, and strategies for drug development. Curr Pharm Des 20:793–807

Couyoupetrou M, Gantner ME, Di Ianni ME et al (2017) Computer-aided recognition of ABC transporters substrates and its application to the development of new drugs for refractory epilepsy. Mini Rev Med Chem 17:205–215

Daws L, Gould G (2011) Ontogeny and regulation of the serotonin transporter: Providing insights into human disorders. Pharmacol Ther 131:61–79

Di L, Keefer C, Scott DO et al (2012) Mechanistic insights from comparing intrinsic clearance values between human liver microsomes and hepatocytes to guide drug design. Eur J Med Chem 57:441–448

Eiden LE, Weihe E (2011) VMAT2: a dynamic regulator of brain monoaminergic neuronal function interacting with drugs of abuse. Ann N Y Acad Sci 1216:86–98

Erickson JD, Eiden LE (1993) Functional identification and molecular cloning of a human brain vesicle monoamine transporter. J Neurochem 61:2314–2317

Fagiolino P (2017) Farmacocinética y biofarmacia. Parte I: principios fundamentales. UdelaR-FQ; FUNDAQUIM, Montevideo

Feldmann M, Koepp M (2016) ABC transporters and drug resistance in patients with epilepsy. Curr Pharm Des 22:5793–5807

Ferman C, Baladi M, McFadden L et al (2015) Regulation of the dopamine and vesicular monoamine transporters: Pharmacological targets and implications for disease. Pharmacol Rev 67:1005–1024

Ferreira RJ, Ferreira MJ, dos Santos DJ (2013) Molecular docking characterizes substrate-binding sites and efflux modulation mechanisms within P-glycoprotein. J Chem Inf Model 53:1747–1760

Forrest LR, Rudnick G (2009) The rocking bundle: A mechanism for ion-coupled solute flux by symmetrical transporters. Physiology (Bethesda) 24:377–386

Gether U, Andersen PH, Larsson OM et al (2006) Neurotransmitter transporters: molecular function of important drug targets. Trends Pharmacol Sci 27:375–383

Hagenbuch B, Stieger B (2013) The SLCO (former SLC21) superfamily of transporters. Mol Asp Med 34:396–412

Homolya L, Váradi A, Sarkadi B (2003) Multidrug resistance-associated proteins: Export pumps for conjugates with glutathione, glucuronate or sulfate. Biofactors 17:103–114

Huwyler J, Wright MB, Gutmann H, Drewe J (2006) Induction of cytochrome P450 3A4 and P-glycoprotein by the isoxazolyl-penicillin antibiotic flucloxacillin. Curr Drg Metab 7:119–126

Iannetti P, Spalice A, Parisi P (2005) Calcium-channel blocker verapamil administration in prolonged and refractory status epilepticus. Epilepsia 46:967–969

Katoh M, Nakajima M, Yamazaki H et al (2001) Inhibitory effects of CYP3A4 substrates and their metabolites on P-glycoprotein-mediated transport. Eur J Pharm Sci 12:505–513

Kim RB (2002) Drugs as p-glycoprotein substrates, inhibitors, and inducers. Drug Metab Rev 34:47–54

Koepsell H (2013) The SLC22 family with transporters of organic cations, anions and zwitterions. Mol Asp Med 34:413–435

Kristensen A, Andersen J, Jørgensen T et al (2011) SLC6 neurotransmitter transporters: Structure, function, and regulation. Pharmacol Rev 63:585–640

Lawal HO, Krantz DE (2013) SLC18: vesicular neurotransmitter transporters for monoamines and acetylcholine. Mol Asp Med 34:360–372

Lhommé C, Joly F, Walker JL et al (2008) (PSC 833) combined with paclitaxel and carboplatin compared with paclitaxel and carboplatin alone in patients with stage IV or suboptimally debulked stage III epithelial ovarian cancer or primary peritoneal cancer. J Clin Oncol 26:2674–2682

Lohr K, Masoud S, Salahpor A et al (2017) Membrane transport as mediators of synaptic dopamine dynamic: implications for disease. Eur J Neurosci 11:3499–3511

Luna-Tortós C, Fedrowitz M, Löscher W (2008) Several major antiepileptic drugs are substrates for human P-glycoprotein. Neuropharmcology 55:1364–1375

Mao Q, Unadkat JD (2015) Role of the breast cancer resistance protein (BCRP/ABCG2) in drug transport--an update. AAPS J 17:65–82

Matheny CJ, Ali RY, Yang X et al (2004) Effect of prototypical inducing agents on P-glycoprotein and CYP3A expression in mouse tissues. Drug Metab Dispos 32:1008–1014

Nanayakkara AK, Follit CA, Chen G, Williams NS, Vogel PD, Wise JG (2018) Targeted inhibitors of P-glycoprotein increase chemotherapeutic-induced mortality of multidrug resistant tumor cells. Scientific Reports 8 (1)

Omiatek DM, Bressler AJ, Cans AS et al (2013) The real catecholamine content of secretory vesicles in the CNS revealed by electrochemical cytometry. Sci Rep 3:1447

Pereira CD, Martins F, Wiltfang J et al (2018) ABC transporters are key players in Alzheimer's disease. J Alzheimer Dis 61:463–485

Pérez-Tomás R (2006) Multidrug resistance: retrospect and prospects in anti-cancer drug treatment. Curr Med Chem 13:1859–1876

Potschka H, Luna-Munguia H (2014) CNS transporters and drug delivery in epilepsy. Curr Pharm Des 20:1534–1542

Qian Y, Melikian HE, Rye DB et al (1995) Identification and characterization of antidepressant-sensitive serotonin transporter proteins using site-specific antibodies. J Neurosci 15:1261–1274

Rees DC, Johnson E, Lewinson O (2009) ABC transporters: The power to change. Nat Rev Mol Cell Biol 10:218–227

Reith ME, Blough BE, Hong WC et al (2015) Behavioral, biological, and chemical perspectives on atypical agents targeting the dopamine transporter. Drug Alcohol Depend 147:1–19

Rudnick G, Steiner-Mordoch SS, Fishkes H et al (1990) Energetics of reserpine binding and occlusion by the chromaffin granule biogenic amine transporter. Biochemistry 29:603–608

Rudnick G (1998) Bioenergetics of neurotransmitter transport. J Bioenerg Biomembr 30:173–185

Russel FGM (2010) In: Pang KS, Rodrigues AD, Peter RM (eds) Transporters: importance in drug absorption, distribution, and removal. Springer, New York

Safa A (2004) Identification and characterization of the binding sites of P-Glycoprotein for multidrug resistance-related drugs and modulators. Curr Med Chem Anticancer Agents 4 (1):1–17

Salphaty L, Benet LZ (1998) Modulation of P-glycoprotein expression by cytochrome P450 3A inducers in male and female rat livers. Biochem Pharmacol 55:387–395

Sarkadi B, Ozvegy-Laczka C, Német K et al (2004) BCG2 -- a transporter for all seasons. FEBS Lett 567:116–120

Sarkadi B, Homolya L, Szakács G et al (2006) Human multidrug resistance ABCB and ABCG transporters: participation in a chemoimmunity defense system. Physiol Rev 86:1179–1236

Singh SK, Piscitelli CL, Yamashita A et al (2008) A competitive inhibitor traps LeuT in an open-to-out conformation. Science 322:1655–1661

Sitte H, Freissmuth M (2015) Amphetamines, new psychoactive drugs and the monoamine transporter cycle. Trends Pharmacol Sci 36:41–50

Spiller HA, Hays HL, Aleguas A Jr (2013) Overdose of drugs for attention-deficit hyperactivity disorder: Clinical presentation, mechanisms of toxicity, and management. CNS Drugs 27:531–543

Schlessinger A, Matsson P, Shima JE et al (2010) Comparison of human solute carriers. Protein Sci 19:412–428

Schlessinger A, Khuri N, Giacomini KM et al (2013) Molecular modeling and ligand docking for solute carrier (SLC) transporters. Curr Top Med Chem 13:843–856

Subramanian N, Schumann-Gillett A, Mark AE, O'Mara ML (2016) Understanding the accumulation of P-glycoprotein substrates within cells: The effect of cholesterol on membrane partitioning. Biochimica et Biophysica Acta (BBA) - Biomembranes 1858 (4):776–782

Sun YL, Patel A, Kumar P et al (2012) Role of ABC transporters in cancer chemotherapy. Chin J Cancer 31:51–57

Südhof TC (2004) The synaptic vesicle cycle. Annu Rev Neurosci 27:509–547

Sulzer D, Maidment NT, Rayport S (1993) Amphetamine and other weak bases act to promote reverse transport of dopamine in ventral midbrain neurons. J Neurochem 60:527–535

Sulzer D, Sonders MS, Poulsen NW et al (2005) Mechanisms of neurotransmitter release by amphetamines: a review. Prog Neurobiol 75:406–433

Taft DR (2009) In: Hacker M, Bachmann K, Messer W (eds) Drug excretion. Academic Press, Burlington

Ter Beek J, Guskov A, Slotboom DJ (2014) Structural diversity of ABC transporters. J Gen Physiol 143:419–435

Tiwari AK, Sodani K, Dai CL et al (2011) Revisiting the ABCs of multidrug resistance in cancer chemotherapy. Curr Pharm Biotechnol 12:570–594

US Food and Drug Administration (2017a) In vitro metabolism and transporter-mediated drug-drug interaction studies. Guidance for industry

US Food and Drug Administration (2017b) Clinical drug interaction studies - Study design, data analysis, and clinical implications. Guidance for industry

Wacher VJ, Wu C-Y, Benet LZ, (1995) Overlapping substrate specificities and tissue distribution of cytochrome P450 3A and P-glycoprotein: Implications for drug delivery and activity in cancer chemotherapy. Molecular Carcinogenesis 13 (3):129–134

Zhou SF (2008) Drugs behave as substrates, inhibitors and inducers of human cytochrome P450 3A4. Curr Drug Metab 9:310–322

Further Reading

Transporter science is an expanding field, and this chapter is only intended as an introduction to the topic. The reader may find deeper insight into some excellent volumes that specifically deal with the subject, such as the ones edited by You and Morris (Drug Transporters: *Molecular Characterization and Role in Drug Disposition*, Second Edition, Wiley, 2014); Pang, Rodrigues, and Peter (*Enzyme- and Transporter-based Drug Drug Interactions. Progress and Future Challenges*, Springer, 2010); or Ecker and Chiba (*Transporters as Drug Carriers. Structure, Function, Substrates*, Wiley-VCH, 2009)

Index

© Springer Nature Switzerland AG 2018
A. Talevi, P. A. M. Quiroga (eds.), *ADME Processes in Pharmaceutical Sciences*,
https://doi.org/10.1007/978-3-319-99593-9

CPSIA information can be obtained
at www.ICGtesting.com
Printed in the USA
BVHW011913260122
627279BV00002B/14

9 783319 995922